AF327029

THERMOLUMINESCENCE IN SOLIDS AND ITS APPLICATIONS

K. Mahesh
Professor of Physics
Kurukshetra University
Kurukshetra-132119, India.

P. S. Weng
Professor of Physics
Director, Health Physics Division
Institute of Nuclear Science
National Tsing Hua University
Hsinchu, Taiwan-300
Republic of China.

C. Furetta
Research Physicist
Department of Physics
University of Rome 'La Sapienza'
00185 Rome, Italy.

Nuclear Technology Publishing
Ashford, Kent TN25 4NW
England, (U.K.)

THERMOLUMINESCENCE IN SOLIDS
AND ITS APPLICATIONS

COPYRIGHT © 1989
Nuclear Technology Publishing

ISBN 1 870965 00 0 (hardback)
ISBN 1 870965 01 9 (paperback)

THERMOLUMINESCENCE IN SOLIDS AND ITS APPLICATIONS

ACKNOWLEDGEMENTS

The authors are grateful to Mr E. P. Goldfinch, Executive Editor, *Radiation Protection Dosimetry*, for his constant efforts and sincere help towards improvement of the book manuscript. One of the authors (K. Mahesh) acknowledges the opportunity provided by Dr A. K. Singhvi (Physical Research Laboratory, Ahmedabad, India) to attend the National TL Symposium at Ahmedabad during February 1984, enabling the author to use the library facilities of PRL. He also profited immensely from the symposium lectures of Dr A. K. Singhvi, Prof. M. J. Aitken (Oxford), Dr S. A. Durrani (Birmingham), Dr V. Mejdahl (Denmark), Dr K. S. V. Nambi (BARC, Bombay) and Dr C. M. Sunta (BARC, Bombay). He is further obliged to Mr E. P. Goldfinch for revising Appendix II and for writing its Bibliography. Dr P. Braunlich (Washington State University, USA) kindly made available reprints of his work on laser heating of TL phosphors.

THE AUTHORS

K. Mahesh

Born June 1940, Dr Mahesh obtained his M.Sc. degree from the Muslim University, Aligarh, and his B.Sc. and Ph.D. from the University of Agra, India. He holds the position of Professor of Physics in the University of Kurukshetra, India. He has held visiting appointments at the United Kingdom Atomic Energy Research Establishment, Harwell, the University of Manchester, England, and The University of Mosul, Iraq. He has done research work in the fields of nuclear detectors, Mossbauer spectroscopy, thermoluminescence physics and electrets and has published about 45 research papers in international journals besides supervising Ph. D. and M. Phil. dissertations in these areas. He has authored several books and has contributed to and co-edited *Techniques of Radiation Dosimetry*, published by Wiley (New Dehli and New York). Dr Mahesh has, during the past 25 years, taught the subjects of atomic physics, statistical mechanics, radiation physics and dosimetry, thermoluminescence in solids and solid state physics in the universities of Kurukshetra and Mosul. His current research activities are in thermoluminescence physics, radiation dosimetry and Mossbauer studies of the hyperfine fields.

P. S. Weng

Born September 1931, Professor Weng obtained the degree of M.S. from the National Tsing Hua University, Hsinchu, Taiwan, and a Ph. D in nuclear engineering from Texas A&M University, USA. Prof. Weng was Director of the Health Physics Division, Chairman of the Nuclear Engineering Department and the Director of the Institute of Nuclear Science of the National Tsing Hua University. He has held visiting appointments at the Karlsruhe Nuclear Science Centre, Federal Republic of Germany, King Saud University, Saudi Arabia, and was until recently a visiting Scientist at the Oak Ridge National Laboratory, USA. He has carried out research in the field of thermoluminescence dosimetry and environmental monitoring and has published about 100 research papers and some books. Prof. Weng has taught various subjects such as nuclear reactor engineering, health physics, and radiology. He is Editor-in-Chief of the *Nuclear Science Journal* which is an official journal of the Nuclear Energy Society of the Republic of China.

C. Furetta

Born October 1937, Dr Furetta obtained the degree of Doctor of Physics from the University of Rome and has been a research Physicist at this University since 1971. He has held research assignments in Belgium and at CERN, Geneva, Switzerland. He was a visiting professor at the National Tsing Hua University, Hsinchu, Taiwan, in 1982-83. Dr Furetta has done research in the field of nuclear detectors, radiation dosimetry and thermoluminescence and has published about 75 research papers in international journals. In addition to this he has contributed to several books and has authored *Thermoluminescence Processes: Theory and Methods*, published by National Tsing Hua University, Taiwan, 1983, and *An Elementary Course of Physics for Professional Nurses* (in Italian), 1985. His current research interest is in thermoluminescence dosimetry and radiation protection. He has taught TL dosimetry and medical physics at the University of Rome and at National Tsing Hua University.

Contents

Contents

Preface

Thermoluminescence in solids (insulators) is a physical phenomenon involving the emission mainly of visible light when these solids are heated after having been previously exposed to some source of ionising radiation. The underlying mechanism involves the role of (i) crystal defects which allow the storing of energy derived from exposure to radiation through the trapping of carriers at these defect-sites, and (ii) subsequent release of the stored energy as visible light when these trapped carriers, after having been freed by thermal stimulation (heating), recombine at the luminescent centres provided by the impurity atoms that have been added to (preferably) or are present naturally in the solid. Thermoluminescence (TL), therefore, provides a simple and relatively inexpensive experimental technique to study various aspects of the role of defects and impurities in solids with considerable reliability.

There has been a rapid development in the research and methodology during the past 25 years on account of the fact that the light output from an irradiated solid is proportional to the radiation dose (dose rate $\times$ time of exposure). This feature allows two very important applications: (i) measurement of radiation dose required in the application of radiation in medicine, industry, biology and agriculture, and (ii) the measurement of radiation exposure time (if the dose rate is known) which allows the dating of ancient pottery sherds and certain rocks. It is, therefore, only appropriate that the subject of TL constitutes an integral feature of the study of solid state physics and spectroscopy and of experimental techniques in nuclear science, radiation dosimetry and dating of archaeological, geological and meteorite samples. In view of these facts, teaching courses as well as research facilities in TL and its applications are increasingly being offered in various universities and nuclear establishments in different countries.

The present book, highly suitable for private study, aims to clarify concepts and has been written in principles by highlighting the established features of TL and offering, therefore, a unique advantage to students and teachers in solid state physics and chemistry, radiation dosimetry and nuclear science and techniques. The book, at the same time, retains all the features of a reference work which will appeal equally to researchers in TL and users of this technique in a wide spectrum of applications.

K. Mahesh, P. S. Weng and C. Furetta

CHAPTER 1

Historical Background of Luminescence and Current Trends in Thermoluminescence

K. Mahesh

1.1.   Introduction

Materials and substances that are capable of emitting light, particularly in the visible range, are termed 'luminescent'. There is a wide variety of such substances: minerals, crystals, chemical substances, and biomaterials of plant and animal origin, and they can be natural or manufactured in laboratories. The underlying phenomenon that causes these materials to luminesce is called luminescence.

Luminescence, unlike natural radioactivity, does not occur spontaneously but, generally, requires energy to excite it. The excitation can be caused by ultraviolet light, X rays, gamma rays, electrons, or alpha particles, all of which can deposit energy into the material that is subsequently re-emitted in the form of visible light. Luminescence phenomena essentially involve electronic transition of the atom or a molecule in a solid, so that the absorption and release of energy is a quantum process. Application of an electric field or even some mechanical stress can also energise electrons or ions. In some chemical reactions, the liberated energy can also serve to excite luminescence in the appropriate participating chemical substances.

There are numerous luminescent materials but not all are efficient enough to be put to practical use. The more efficient materials are those which are prepared in laboratories and have a specific chemical composition. These materials exhibit a degree of repeatability and reliability in their performance which qualifies them for practical application.

Other accompanying effects that are luminescent in origin should be distinguished. These are 'incandescence', which is classical in nature and is exhibited when the material is very hot. Heating agitates the atoms strongly and, by collision with the neighbouring atoms, results in the excitation and subsequent emission of light. To begin with, this radiation is in the invisible far infrared but, at higher temperatures, shifts to the visible region. The other effect is the so-called 'pigment action' in which a given dye reflects a particular colour of the incident white light while absorbing the remainder. The energy of the absorbed light is simply converted into heat which elevates the temperature and increases the thermal radiation of the dye by an insignificant amount. Some materials can be luminescent and exhibit the dye action simultaneously. It is possible for a dye to absorb a particular colour of the incident light and excite the same radiation by both the luminescence process and by reflection, thus adding to the intensity of the reflected light. Such luminescent dyes are particularly attractive for outdoor advertising displays. The other distinguishing feature of the luminescence process is the fact that the colours from a mixture of luminescent materials are additive, thus producing a net effect of the sum of the emitted colours, whereas the colours of the simple dyes and pigments are subtractive, giving as a net effect the difference of the reflected colours when viewed under white light.

1.2.   **Historical development**

The phenomenon of luminescence has been known since early times[1]. The existence of luminous organisms such as bacteria in the sea and in decaying organic matter, glow-worms and fireflies, have mystified and thrilled man since time immemorial. The first attempt to enquire into the fascinating light emitted during the night concerns the legendry Bolognian stone (the mineral barite, now identified as barium sulphate) found around 1603 by an Italian alchemist, V. Cascariolo of Bologna, Italy. G. C. Lagalla of Italy wrote about this stone and the light emitted from it in his book *De Phaenomenis in Orbe Lunae* (Venice, 1612). Several names were given to the Bolognian stone until the modern name 'phosphor' was used to designate the stone and, subsequently, all other solid or liquid luminescent materials.

A systematic scientific study of the subject of luminescence is, however, of recent origin, dating approximately from the middle of the nineteenth century. The application of luminescence in general and of thermoluminescence in particular, is of still more recent origin. The bulk of investigations leading to various applications have been carried out within the past 25 years or so.

In 1852 the English physicist G. G. Stokes identified this phenomenon and formulated his law of luminescence (now known as Stokes' law) which states that the wavelength of the emitted light is greater than that of the exciting radiation. It was the German physicist E. Weidemann who, in 1888, introduced the term 'luminescence' for the observation of excess emission over and above the thermal background. Experimental techniques for the study of emission spectra, efficiency of excitation, and duration of luminescence afterglow, were initiated by A. E. Becquerel.

1.3.   Luminescence classification

1.3.1.   *Phosphorescence*

The early experiments with phosphors depended on sunlight as the source of exciting radiation and the 'phosphor quality' of the material was judged by the ability of the material to display considerable luminescence after a sample was withdrawn from sunlight into a dark place. This persisting ability of the material was termed 'phosphorescence'. The early belief, that the phosphors absorbed light/sunlight and then re-emitted it in much the same way as a sponge absorbs water and releases it when squeezed, was proved false by N. Zucchi when in 1652 he showed that the colour of luminescence emission was the same whether the sample was excited by white light or light of any other colour (obtained by passing the white light through a filter).

1.3.2.   *Fluorescence and luminescence*

In the course of studies by the French mineralogist R. J. Hauy, the English astronomer J. Herschel, the Scottish physicist D. Brewster and the English mathematician-physicist G. G. Stokes, the usage and definition of the terms fluorescence and luminescence were adopted. It is now accepted that the term fluorescence be used to denote short duration luminescence, typically the one in which light emission persists to around 10^{-8} s after the source of excitation is switched off. The emission that persists for a longer period, extending even to hours, is phosphorescence. The luminescence phenomenon covers ultraviolet and infrared radiation in addition to visible light.

1.3.3.   *Bioluminescence*

The earliest notice of luminescence was the observation of light emission by glow-worms and fireflies. Recorded evidence from a later period (1744 and 1747) is attributed to J. B. Beccari of Bologna, Italy, who observed light emission in glow-worms, certain marine life, rotten wood, and the putrefying flesh of some birds and animals. The luminescence of natural organic matter, both animate and inanimate, falls into the category of chemiluminescence. In this process the chemical reaction of the matter with oxygen or water in the air provides the excitation energy. When chemiluminescence occurs in living organisms we call it 'bioluminescence'. In this process the oxygen of the air reacts with certain specific organic fluids secreted by the animal. For instance, when crustaceans release luciferin granules, the bioluminescence arises from the oxidation of luciferin in sea water in the presence of an enzyme known as *luciferase* which catalyses the oxidation. While the role of luminescence in the life of crustaceans is not understood, glow worms and firefly beetles regulate their luminescence mechanism to exhibit periodic light emission chiefly for the purpose of attraction during courtship (sometimes also for attracting a meal). More sophisticated glow-worms also exist that emit red light from a spot on the head and greenish yellow light from spots located on other parts of the body.

The emission spectrum of firefly beetles has also been studied and is found to lie in the range of 520-590 nm with a peak at 570 nm. The spectral distribution of the emission of the female firefly, though covering the same range, differs significantly from that of the male. In the latter case the peak is more pronounced.

1.3.4. *Triboluminescence*

The prefix 'tribo' is a Greek word meaning 'to rub'. Francis Bacon, in his book *Advancement of Learning* (1605), mentioned the luminescence from large crystals of sugar upon being scratched or broken. The surfaces that are separated by breaking, scratching or tearing, produce an electric discharge in the thin layer of air between the separating surfaces, owing to the generation of opposite charges on the two surfaces. Some adhesive tapes, when unrolled in the dark, emit a blue glow due to an electric discharge in atmospheric nitrogen.

In the process of grinding of materials, the mechanical energy expended provides the excitation energy and can lead to an observable glow in the dark. Triboluminescence caused by this process is an important factor when powder samples are prepared by grinding and must be accounted for in the more reliable determination of various physical parameters.

1.3.5. *Photoluminescence*

The process in which the light radiation causes a phosphor to luminesce is known as 'photoluminescence'. Around 1800, J. W. von Goeth and T. J. Seebeck observed a strong emission when the Bolognian phosphor was excited by the invisible radiation lying just beyond the visible violet through the prism. This experiment thus heralded the existence of invisible ultraviolet and, similarly, of the infrared radiation and their photo-exciting (by ultraviolet) and photo-quenching (by infrared) influence on phosphor luminescence. Photo-excitation by ultraviolet radiation is common for the study of luminescence and associated characteristics in luminescent materials.

1.3.6. *Electroluminescence*

Benjamin Franklin, in 1752, identified the cause of lightning in the atmosphere as an electric discharge. The excitation of air molecules is caused by the energetic electrons in the electric discharge. Lightning is nothing but luminescence in the air. Thus, electroluminescence is the luminescence phenomenon caused by electric fields. Electric fields across gases, liquids or solids can accelerate electrons to an energy sufficient to excite the luminescent materials. The source of electrons can be inside or outside the material. The modern fluorescent tube owes its development to the phenomenon of light emission in phosphors by exposure to electrical discharge, as demonstrated by J. Canton in 1768. The fluorescent tube utilises the combination of electroluminescence and photoluminescence. The electrons excite the mercury vapour atoms which emit ultraviolet radiation which, in turn, excites a phosphor coating, giving the visible light. Light emitting diodes are also an example of electroluminescence.

1.3.7. *Cathodoluminescence*

This is a phenomenon similar to electroluminescence except that the electrons in cathodoluminescence act in an isolated manner and not in association with an electric discharge or spark.

The effect of electrons in producing a spot of light was first observed on the glass wall of a cathode ray tube. This served as the visible indication of the invisible cathode rays discovered by Goldstein in 1876 and named electrons by G. J. Stoney in 1881. The cathode ray oscilloscope tube (invented by K. F. Braun in 1897) and

TV sets are a practical example of cathodoluminescence. The viewing screen of an electron microscope uses the luminescence caused by electrons on the phosphor coating, the invisible electron pattern thus turns into visible images showing minute details.

1.3.8.　*Radioluminescence and roentgenoluminescence*

In 1896 A. H. Becquerel experimented to find out whether phosphors irradiated by visible light also emitted penetrating radiation. His negative results, however, led to the discovery of radioactivity. Later, the luminescence effect caused by alpha particles in certain phosphors (ZnS) became the standard detecting technique for alpha particles, and was used extensively by Rutherford in his alpha ray scattering experiments. The luminous markings, made of phosphor paint containing an alpha source, on clock and watch dials are scintillations caused by individual alpha particles. Similarly, Roentgen rays (X rays) can cause luminescence in the phosphors used as coatings on X ray screens for visualising details of bone structure.

1.4　**Thermoluminescence**

The description of thermoluminescence (TL) is the theme of this book, and details of the processes underlying TL phenomena will be treated. It is, however, pertinent to place the topic in its historical context. Several useful applications of TL are of relatively recent origin. The main TL dating work started some 25 years ago when the observation of TL in ancient pottery was reported (in 1960) by Grogler, Houtermans and Stanffer from Berne, Switzerland, and by Kennedy and Quopff from Los Angeles, USA. TL of lunar materials was reported in the early seventies. TL dosimetry has made extensive progress only very recently.

The release, in the form of light, of energy already stored in a phosphor when it is heated is the phenomenon of TL. In other words, a TL phosphor has been pre-excited by some suitable radiation (ultraviolet, nuclear, cosmic rays) whose energy is retained within the sample; when the phosphor is heated to an appropriate temperature, this energy is released as visible light. The emitted light can easily be detected by a photomultiplier and associated electronic equipment. Robert Boyle seems to have observed such an emission in diamond as early as 1663. A proper understanding of the TL phenomenon developed only after more than two centuries of experimental investigation. In TL methods, the phosphorescent solids, after excitation, are heated and the corresponding intensity of the emitted light is recorded as the temperature increases at a linear rate. It is as if the excitation energy remains 'frozen' in a thermoluminescent material until heating stimulates its release in the form of light. The basic process and the models used to explain TL will be described later.

1.5 Lyoluminescence

The phenomenon of light emission on the dissolution of irradiated alkali chlorides in water was first observed in 1895 by E. Wiedman and G. C. Schmidt and was termed 'lyoluminescence'. The same effect was observed by T. Westermark and B. Grapengieser in 1960[2], in most alkali halides, saccharides and some organic compounds. A more detailed study of this effect was made by Ahnström in 1962[3], who proposed a mechanism for the light emission when irradiated NaCl is dissolved in water. On dissolution, the excited chloride ion $(Cl^-)^*$ is formed and light emission takes place either in the solvent or at the solid-solvent interface. A more detailed study of lyoluminescence has recently been carried out by N. A. Atari[4] in NaCl, KCl and KBr halides irradiated by γ radiation from a ^{60}Co source to a dose of 6 kGy. He finds that the presence of both F and V_2 centres (see Chapter 2) is essential for exhibiting lyoluminescence phenomena. Neither centre is capable of producing lyoluminescence on its own. Lyoluminescence is mainly due to the liberation of trapped electrons (from F centres) by the hydration process to form hydrated electrons (e^-_{aq}) followed by recombination with V_2 centres at the water-solid interface. The basic process can be expressed as

$$F \text{ centre} \xrightarrow{\text{hydration}} e^-_{aq}$$

$$e^-_{aq} + V_2 \text{ centre} \xrightarrow{\text{recombination}} h\nu$$

(Hydrated electron: After an ionisation process following irradiation of water, the freed electron is thermalised and captured in a molecular cage of a few water molecules. This configuration is called a hydrated electron.)

1.6.   Current trends in thermoluminescence

Because of the ability to glow upon heating possessed by a large variety of substances, minerals, chemical compounds in crystalline and amorphous forms and biological materials, extensive research and development work concerning the theory and applications of TL has been carried out. Ultraviolet and nuclear radiation incident on a TL sample produces mobile electrons and holes, which are caught in their respective trap states within the band gap of the sample. The population of these occupied traps is proportional to the incident radiation dose[5]. As TL intensity is also proportional to the occupied trap density, the sample can act as a TL dosemeter; therefore a large amount of work in TL relates to radiation dosimetry[6,7]. Several of the TL materials, e.g. $CaSO_4$:Dy, Mg_2SiO_4:Tb, CaF_2:Mn, $Li_2B_4O_7$:Mn and LiF:Mg,Ti, have been found especially valuable in this work. Because of the high susceptibility of certain TL materials to both low and highly penetrating radiations of various types (γ, X, e π, n, and heavy ions) these materials are being employed in space flight experiments for cosmic ray dose measurement in the upper atmosphere[8]. An interesting example is the measurement of cosmic ray exposure of passengers in high altitude commercial aircraft[9,10].

Another large and productive area of activity is the field of archaeological and geological dating, in which the TL technique is used for a reasonably accurate determination of the age of samples[11-13]. Here, the (valid) assumption is that the high temperatures of kiln firing at the time of formation of an archaeological material (pottery, ceramic, artefacts etc.) would have cleared all the existing TL signal, thereby setting the TL clock to zero and all the present day observed TL signal has built up during the life of the sample. In the case of geological materials (sedimentary rocks and ocean sediments) exposure to sunlight, mechanical breaking of grains, surface temperatures during the weathering of rocks, all provide for the zeroing mechanism (Chapter 8). The TL dating technique has also been applied to fossil bones[14,15] and prehistoric human teeth. Such samples, however, present many complexities and these studies have not progressed much.

While the basic features of TL instrumentation remain standard efforts towards automation and miniaturisation with a view to developing reliable and portable systems (for space applications) have been made[16a].

Better reliability and precision are being achieved by incorporating microcomputer controlled operations in the TL equipment[16b,c].

The use of TL techniques for the evaluation of environmental radiation and continuous monitoring of cosmic rays is growing. TL dosemeters are being used with a good degree of reliability for making estimates of the exposure of a population from the walls and floors of dwellings[17].

Currently, interest is developing[18a,b] in the TL dosimetry of UV radiation, particularly in the wavelength range 200-320 nm, because of its biological significance. This radiation can cause erythema (skin reddening), skin burns, and inflammation of the cornea, and has a potential industrial application in the sterilisation of enclosures, containers, and surgical instruments. Efforts to develop efficient and sensitive TL phosphors for UV dosimetry are being made; Al_2O_3:Si,Ti phosphor, among others, has been found to be efficient for this purpose.

TL glow curves of many materials are unique and characteristic. This feature has allowed the use of TL in the study of the residual compressive strength of concrete in fire damaged buildings[19], with a view to establish criteria for the acceptance or rejection of the salvaged building.

Further exploitation of this feature of TL glow curves in forensic applications has also been suggested[20], where a 'find' (e.g. automobile headlamp glass recovered from a hit-and-run accident site) can be compared and identified with suspect materials. Because of the high demands for reliability in forensic work and the legal implications of the use of such results, the TL characterisation technique, in its current state of the art, will require elimination of chance conclusions.

Some work, mainly experimental, is being done to study the correlation of TL with other accompanying thermally stimulated processes, e.g. conductivity (TSC), exoelectron emission (TSEE), and ionic thermo-current* (ITC) or depolarisation current in

---

*Bucci *et al*[21] first studied the phenomenon of thermally stimulated depolarisation (TSD) currents in insulators containing dipolor ionic defects. They called the phenomenon 'ionic thermo-current' (ITC). This term has been widely used for TSD currents due to ionic motion. The current relative to temperature due to dipole relaxation can be described by an equation similar to that of a first order TL process.

samples. For example, the occurence of TL and TSEE peaks at the same sample temperature[22] can testify that the origin of TL glow peaks is due to the involvement of electron, and not hole, traps. Similarly, TL and ESR can both be applied in parallel to a sample to draw some conclusions about the nature of defects produced during irradiation of the sample[23]. ESR data offer quite definite information on the state of charge and aggregation of impurity, particularly the paramagnetic impurity. Takeuchi *et al*[24], for instance, have shown that in single crystals of NaCl:Cu, the prominent TL peak at 67°C in the X irradiated sample at room temperature, arises from the Cu^+ ion. ESR studies of this sample shows that this glow peak is caused by the capture of thermally released electrons (from the F centres) by luminescence centres.

Efforts are currently being made to find still more suitable materials whose TL study can increase understanding of the different features of these various solids, and to see what specific applications in dosimetry and luminescence can be developed. BeO has been found to possess many desirable features[25,26], particularly the strong TL yield and simple glow curves. Its fair sensitivity to γ rays and negligible response to thermal neutrons allow it to be used as a dosemeter in the vicinity of a nuclear reactor facility. The low atomic number ($Z_{eff} \approx 7$) compares favourably with that of water and soft tissue ($Z_{eff} \approx 7.4$), which makes BeO a useful TL dosemeter in medical applications.

TL methodology is constantly being improved. As an example, the LiF:Mg,Ti phosphor, which is an important TL dosemeter material, is required, like many others, to be annealed to reset the dosemeter for the next use. The programme was to anneal it at 400°C for one hour to clear thermally any remaining TL signal, followed by a 24 hour anneal at 80°C to erase the shallow traps that contribute to error and instability of performance due to fading. Regulla[27] achieved the same results by using a quicker annealing procedure – 5 min at 400°C followed by 15 min at 100°C. This modification in the annealing procedure can obviously shed light on the nature of traps and the mechanism of their thermal ionisation because the different annealing procedures adopted also influence the shape of the glow curves. Recently Lau *et al*[28] have reported that the irradiation of LiF:Mg,Ti chips by a thermal neutron fluence of about 10^{16} cm^{-2} in a reactor core causes an increase in their TL output by a factor of 8000. The increase is nearly proportional to the thermal neutron fluence and the TL output from the irradiated phosphors is stable. It is postulated that irradiation increases the number of luminescent centres in the phosphors. (See also Lakshmanan *et al*[29] who have contradicted the conclusion of Lau *et al*, stating that the large enhancement is due to the TL signal from thermal neutron induced ^{3}H beta particles in LiF.)

TL study has been further exploited to investigate other physical processes such as those involving diffusion activation energy and anion Frenkel defect formation energy. In CaF_2:Pr phosphor, the diffusion controls the high temperature glow peaks[30] so that the resulting TL data can give estimates of the energies mentioned.

References

1. Harvey, E. N. *A History of Luminescence – from the Early Times until 1900.* American Philosopical Society, 1045, Fifth Street, Philadelphia, PA 19106, USA (1957).
2. Grapengiesser, B. Nature **188**, 395 (1960).
3. Ahnström, G. Acta Chem. Scand. **16**, 2113 (1962).
4. Atari, N. A. J. Lumin. **21**, 305, 387 (1980).
5. Mahesh, K. *Thermoluminescence in Radiation Protection and Control.* (review) Joint Symp. on Radiation Safety and Dosimetry, 6-8 June, 1984, University of Rome, Italy. Nucl. Sci. J. (Taiwan) **22**, 211 (1985).
6. Mckinlay, A. F. *Thermoluminescence Dosimetry.* (Bristol: Adam Hilgar) (1981).
7. Horowitz, Y. S. (ed) *Thermoluminescence and Thermoluminescent Dosimetry.* Vols I-III (Florida: CRC Press) (1984).
8. Akatov, Yu. A. *et al*, Adv. Space. Res. **1**, 67 (1981).
9. Kramer, R., Regulla, D. F. and Drexler, G. IN Proc. Fifth Int. Conf. on Luminescence Dosimetry, Sao Paulo, Brazil. Ed. A. Scharmann, p. 298 (1977).
10. Feher, I. *et al*, Adv. Space Res. **1**, 61 (1981).
11. McDougall, D. J. (ed) *Thermoluminescence of Geological Materials,* (London: Academic Press) (1968).
12. Aitken, M. J. *Physics and Archaeology* (Oxford: Clarendon Press) (1974).
13. Mejdahl, V., Bowman, S. G. E., Wintle, A. G. and Aitken, M. J. IN Third Proc. Specialist Seminar on TL and ESR Dating, 26-31 July 1982, Helsingor, Denmark. PACT, Vol. **9** (Parts I and II) (1983).
14. Jasinska, M. and Niewiadomski, T. Nature **227**, 1159 (1970).
15. Christodulides, C. and Fremlin, J. H. Nature **232**, 257 (1971).
16. a. Vij, D. R. and Mahesh, K. IN *Techniques of Radiation Dosimetry,* eds. K. Mahesh and D. R. Vij (Chichester: John Wiley) (1985).
16. b. Davis, J. E. and Ramsay, P. Radiat. Prot. Dosim. **4**(3-4), 177 (1983).
16. c Takamiya, H. and Nishimura, S. Nucl. Tracks Radiat. Meas. **11**, 251 (1986).
17. Niewiadomski, T., Jasinska, M., Koperski, J., Rybe, E. and Schwabeuthan, J. *A Comparative Study of Different Methods used to Assess Population Exposure to Terrestrial Radiation.* Report No. 1117/D (Institute of Nuclear Physics, Krakow, Poland) (1980).

18. a. Lakshmanan, A. R. Bull. Radiat. Prot. **2**, 67 (1979).
18. b. Sweet, M. A. S. and Rennie, J. Phys. Status Solidi a **100**, K67 (1987).
19. Placido, F. Mag. Concr. Res. **32**, 112 (1980). Also and FIRE, Feb. 1981, p. 464.
20. Ingham, J. D. and Lawson, D. D. J. Forensic Sci. **18**, 217 (1973).
21. Bucci, C., Fieschi, R. and Guidi, G. Phys. Rev. **148**, 816 (1966).
22. Tomita, A., Hirai, N. and Tsutsumi, K. Proc. 5th Int. Symp. on Exoelectron Emission and Dosimetry, Zvikov, Sept. 1976, p. 157.
23. Lopez, F. J., Jaque, F., Forte, A. J. and Agullo-Lopez, F. J. Phys. Chem. Solids **38**, 1101 (1977).
24. Takeuchi, N., Adachi, M. and Inabe, K. J. Lumin., **18/19**, 897 (1979).
25. Scarpa, G. Phys. Med. Biol. **15**, 667 (1970).
26. Scarpa, G., Benincasa, G. and Ceravolo, L. IN Dosimetry in Agriculture, Industry, Biology and Medicine. STI/PUB/311 (Vienna: IAEA) p. 207 (1973).
27. Regulla, D. F. IN Instrumentation in Radiotherapy. Report GSF-S 471 (Gesellschaft für Strahlen-und Umweltforschung mbH, Munchen, West Germany) (1978).
28. Lau, H. M., Gui, A. A. and Harasawa, S. Nucl. Instrum. Methods Phys. Res. **B17**, 170 (1986).
29. Lakshmanan, A. R., Chandra, B. and Bhatt, R. C. Nucl. Instrum. Methods Phys. Res. **B26**, 557 (1987).
30. Mukherjee, M. L. and Sinha, R. K. J. Lumin. **21**, 435 (1980).

Bibliography on historical development of luminescence.

General

Garlick, G. F. J. IN Handbuch der Physik **26**, 1-128 (1958).

Henderson, S. T. and Schulman, J. H. *et al*, Br. J. Appl. Phys. Suppl. 4, 1 (1955).

Hercules, D. M. (ed) *Fluorescence and phosphorescence Analysis* (New York: Interscience) (1966).

Kallmann, H. P. and Marmor Spruch, G. (eds) *Luminescence of organic and inorganic materials* (New York: John Wiley) (1962).

Kroger, F. A. *Some Aspects of the Luminescence of Solids.* (New York: Elsevier) (1948).

Leverenz, H. W. IN Encyclopaedia Britannica, Vol. 14.

Leverenz, H. W. *An Introduction to Luminescence of Solids.* (New York: John Wiley) (1950).

Phillips, Tech. Rev. **31** (10) (1970).

Pringsheim, P. *Fluorescence and Phosphorescence* (New York: Interscience) (1949).

The Story of Fluorescence, Raytech Equipment Company, USA (1965). (Available from Carolina Biological Supply Company, 2700 York Road, Burlington, NC 27215, USA)

Bioluminescence

Baruah, G. D. and Dewri, R. P. IN Proc. Symp. Luminescence and Allied Phenomena, 14-16 December 1978, Indian Institute of Technology, Kharagpur, India.
Baruah, G. D. Ind. J. Pure Appl. Phys. 24, 47 (1986).

DeLuca, M. A. (ed.) *Bioluminescence and Chemiluminescence.* IN Methods in Enzymology, Vol. 57 (New York: Academic Press) (1978).

Harvey, E. N. Am. Sci. **45**, 372 (1957).

Harvey, E. N. *Bioluminescence* (New York: Academic Press) (1952).

Hastings, J. W. *Bioluminescence.* IN Ann. Rev. Biochem. **37**, 597 (1968).

Kricka, L. J. (ed.). J. Biolumin. Chemilumin. (starting June 1986, Wiley, New York).

Proceedings of the 4th International Symposium on Bioluminescence and Chemiluminescence, 8-10 September, 1986, University of Freiburg, FRG (New York: Wiley).

Ruby, E. G. and Nealson, K. H. Science **196**, 432 (1977).

Stone, H. J. *Bioluminescence.* Carolina Tips **49**, 13 (1986). (Carolina Biological Supply Co., Burlington, NC 27215, USA).

Van Dyke, K. (Ed.) Bioluminescence and Chemiluminescence: Instruments and Applications, Vols 1 and 2 (Florida, USA: CRC Press) (1985).

Cathodoluminescence

Bröcker, W. and Pfefferkorn, G. E. Bibliography on Cathodoluminescence. Scanning Electron Microscopy, Vol. 1 (SEM Inc., AMF O'Hare, Il 60666, USA) (1978).

Garlick, G. F. J. *Cathodoluminescence.* IN Advances in Eelectronics, ed. L. Marton, Vol. 2 (New York: Academic Press) (1950).

Leverenz, H. W. R.C.A. Rev. **5**, 131 (1940).

Leverenz, H. W. R.C.A. Rev. **7**, 199 (1946).

Phillips Research Reports **7**, 401-432 (1952).

Electroluminescence

Diemer, G. Phillips Research Reports **11**, 353 (1956).

Destriau, G. Phil. Mag. **38** 700, 774, 880 (1947).

Destriau, G. and Ivey, H. F. *Electroluminescence and Related Topics.* Proc. I.R.E. **43**, 1911 (1955).

Henisch, H. K. *Electroluminescence* (Oxford: Pergamon) (1962).

Ivey, H. F. *Electroluminescence and Field Effects in Phosphors.* Trans. Electrochem. Soc. **104** 740 (1957).

Ivey, H. F. *Electroluminescence and related effects* (New York: Academic Press) (1963).

Pankove, J. I. (Ed.) *Electroluminescence* (New York: Springer Verlag) (1977).

Proc. IEEE **61** (1973) (also December 1955).

Shrader, R. E. and Kaisel, S. F. J. Opt Soc. Am. **44** 135 (1954).

CHAPTER 2

General Features of Luminescence

K. Mahesh

2.1. Introduction

Many solids luminesce when heated to incandescence; this, however, constitutes a different phenomenon and is not included in the category of either luminescence or thermoluminescence. Luminescence in solids can be caused chiefly by ultra-violet radiation, energetic electrons, nucleons and pions, γ or X radiation, and also by an electric field. Some chemical and biological processes can also give rise to the glow in solids. The emission depends greatly on the characteristic features of the impurity atom which acts as an activator.

The ability of some materials to emit light in the visible or near visible range when the exciting radiation ceases is the phenomenon of luminescence. The duration for which this afterglow persists can be very short (10^{-8} to 10^{-7} s) while in many cases there may be an additional component of afterglow persisting for much longer. This component, termed phosphorescence, usually arises from the role attributed to the metastable states of the activator in the host material.

The use of zinc sulphide phosphor for counting alpha particles in Rutherford's famous alpha ray scattering experiment, and for detecting nuclear particles, especially γ rays, by using a single crystal of NaI (activated with thallium) in scintillation counter experiments, are well known examples of the application of the luminescence process. Fluorescent mercury vapour tubes (for household lighting) utilise electrical energy to produce a mercury discharge which gives out UV radiation; this strikes the coating of calcium halo-phosphate (containing suitable activator impurities) on the outer glass tube and is converted to visible light. The cathode ray tube screens used in a television or an oscilloscope have a coating of sulphide and silicate phosphors which convert the kinetic energy of electrons into light.

For a considerable time, luminescence remained an empirical technique, and quantitative development was possible only after the advent of electrodynamics and quantum mechanics. These theories provided the appropriate basis for an understanding of luminescence phenomena, leading to considerable progress in technique and methodology.

The following table (Table 2.1) gives the various luminescence processes and the corresponding exciting modes.

2.2. Defects and colour centres

As defects and colour centres are central to any luminescence process, it is pertinent to introduce these concepts here, in an elementary way[1]. Among the simple defects that are found in a crystal structure and which are easily amenable to mathematical formulation are the point defects. These are illustrated in Fig 2.1. The introduction of defects and impurities creates states within the energy gap of the host material and these states can be of electron and hole type. As these new states can absorb light from some part of the visible spectrum, the crystals are coloured and,

Table 2.1. Luminescence processes and the exciting modes.

Process	Exciting mode
1. Photoluminescence	Light (ultraviolet)
2. Cathodoluminescence	Electrons
3. Radioluminescence	X rays, γ rays; α and β particles, protons, neutrons, pions, fission fragments, etc.
4. Electroluminescence	Application of an electric field.
5. Triboluminescence	Mechanical/frictional forces (e. g. grinding sugar causes luminescence).
6. Chemiluminescence	Chemical reaction; e.g. oxidation of phosphorous in a humid atmosphere causes light emission.
7. Bioluminescence	Biochemical action; e.g. in glow-worms and fireflies, the light is produced by oxidation of a biochemical substance, luciferrin.
8. Thermoluminescence	Heating, though it is not the exciting mode, rather a stimulating cause. To this extent the term is rather a misnomer. This is a phenomenon of light emission due to heating of a solid following its excitation by radiation. More appropriately, the term should be called thermally stimulated luminescence (TSL) which is now in use in some current literature.

therefore, these states are also termed colour centres. It is also common to refer to the defect states as carrier traps.

Configurationally, the states lying close to the bottom of the conduction band (CB) possess a greater probability of trapping electrons, while those lying near the top of the valence band (VB) will trap holes with greater ease.

A colour centre designated by F is shown in Fig 2.2. The constitution of some colour centres is shown in Fig 2.3. F centres correspond to electrons trapped at negative ion vacancies. M and R centres are aggregates of two or three F centres. Hole centres are denoted by V. A V_K centre is a hole trapped by two adjacent negative ions; the resulting V_K centre is therefore a molecular ion of the type X_2^-.

The thermal detrapping of electrons from the occupied traps is possible if we supply thermal energy, which we call the activiation energy E, to the occupied trap. The probability per unit time for the electron to escape from the trap to the bottom of CB is governed by the Boltzmann factor $\exp(-E/k_BT)$. One way to create and populate traps is by irradiation of the sample with ionising

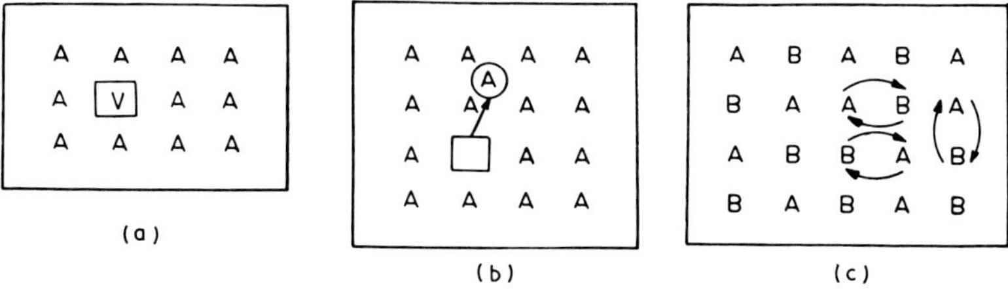

Figure 2.1. Point defects. (a) Vacancy (Schottky defect). (b) Interstitial (Frenkel defect). (c) Disorder.

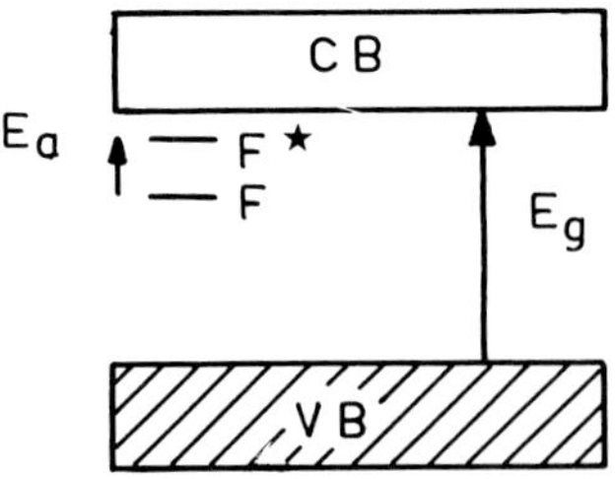

Figure 2.2. Ground and excited states of the colour centre F in an ionic crystal. E_a is the optical absorption and emission energy which gives colour to the crystal.

radiation. As the lifetimes of the occupied traps may be long enough (months and years), and if a crystal initially had no occupied traps, then over the period of irradiation the traps will be populated proportionally. This period is its 'memory' (hence applications in dosimetry and dating). By heating, a thermal activation energy is supplied to the filled traps which may then be emptied. This process can result in the emission of photons (thermoluminescence) or electrons (thermally stimulated exoelectron emission; TSEE).

2.3.   Glossary of impurity centres

In an ionic solid, a negative ion vacancy, represented as $\square$ is called an α centre. If an alpha centre traps an electron, then it is an F centre. Some of these centres are illustrated in Figure 2.3.

In addition, we have the so-called N centre which is a cluster of four F centres. Similarly, M′ and R′ centres denote M and R centres with an extra electron in each. An F′ centre, likewise, is an F centre with one extra electron. A V_t centre describes a hole trapped at a complex of positive and one negative vacancies. A hole trapped at a positive vacancy is a V_F centre. A V_3 centre is a negative ion vacancy with two trapped holes. In the Nink and Kos model[2,3] of LiF:Mg,Ti

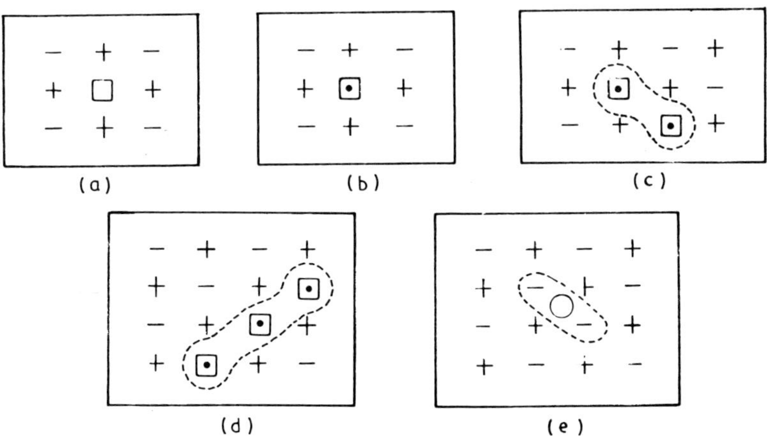

Figure 2.3. Schematic representation of colour centres in an ionic solid. (a) Negative ion vacancy (α-centre) denoted by $\square$. (b) Trapped electron at negative ion vacancy (F centre), denoted by $\boxdot$. (c) Aggregate of two neighbour F centres (M centre). (d) Aggregate of three neighbour F centres (R or H centre). (e) A hole trapped between two adjacent negative ion sites (V_k centre).

phosphor, some other centres have been defined. For instance, a magnesium ion plus an alpha centre (Mg^{2+}:α) is called a Z_0 centre. (Mg^{2+}:F′) is called a Z_1 centre, which is evidently an electrically neutral centre. A Z_3 centre is Mg^{2+}:F.

2.4.   Radiation emission and absorption

In the classical sense, a moving or an oscillating charge constitutes a system that is a source of radiation. An electric dipole can generate radiation having a definite intensity, polarisation and angular distribution.

If an element of charge δe is confined in an element of volume $\delta\upsilon$ ($\delta e = \rho\delta\upsilon$; where ρ is the charge density), then classically the retarded potentials ϕ and A of the radiation field can be expressed by (see Fig 2.4 and Appendix III).

$$\phi = (1/4\pi\varepsilon_0) \int \frac{1}{R} \rho\,(t')\,d\upsilon \qquad \text{(scalar potential)}$$

and

$$A = \frac{\mu_0}{4\pi} \int \frac{1}{R} j\,(t')\,d\upsilon \qquad \text{(vector potential)}$$

$$\left. \right\} (2.1)$$

where $t' = t-(R/c)$, c being the velocity of light and j the current density.

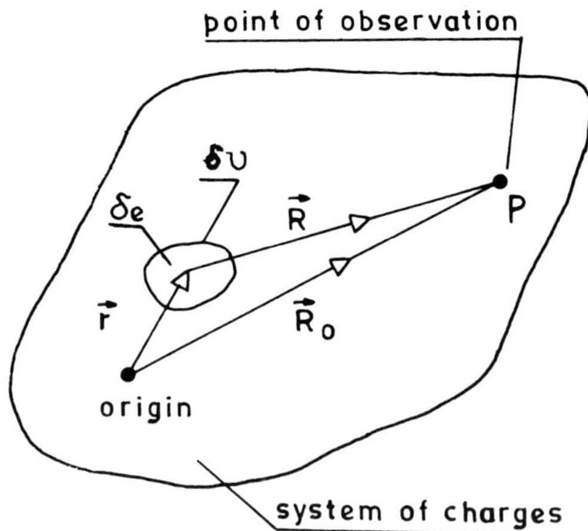

Figure 2.4. Retarded potentials at point P of the system of charges; $R = R_0 - r$, where R_0 is position vector of the observation point and r is position vector of the elementary charge (δe).

It is evident that the radiation field at the point P at time t is determined by the state of the system of charges at an earlier time t', since the radiation, moving with speed c, takes time R/c to reach P. The electric and magnetic field vectors of the radiation (electromagnetic wave) are then given by

$$E = -\frac{\partial A}{\partial t'} - \text{grad } \phi$$

(2.2)

$$H = \text{curl } A$$

2.5.   Oscillating dipole

If one* or both the charges of a dipole execute simple harmonic oscillations along its axis (see Figure 2.5), which may be along the x-axis, then the relative displacement x may be expressed as

$$x = x_0 \cos \omega t \qquad (2.3)$$

where x_0 is the equilibrium separation between the two charges and ω is the angular frequency of the harmonic oscillator. We may then represent a source of radiation by a set of harmonic oscillators in which electric dipoles execute simple harmonic motion. Under these conditions the potential energy of the dipole is

$$V(x) = \tfrac{1}{2}\,k\,x^2 = \tfrac{1}{2}\,m\,\omega_0^2\,x^2 \qquad (2.4)$$

where $k\ (= m\,\omega_0^2)$ is the force constant (force per unit displacement

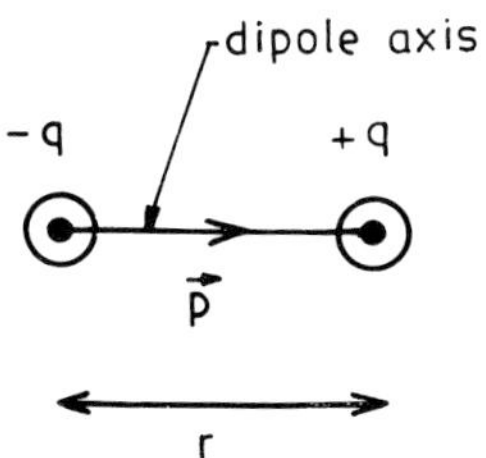

Figure 2.5. The electric dipole. p (electric dipole moment) $= q\,r$.

---

* As in the hydrogen atom the nucleus is stationary and the electron is orbiting around (we can call the system the electric rotator) making a uniform rotational motion. This motion is equivalent to two oscillatory linear motions along mutually perpendicular directions.

from the equilibrium position). m is the mass of the oscillator and ω_o ($= 2\pi \nu_o$) is the natural frequency of the oscillator.

The force F acting on the dipole is

$$F = - dV/dx = - kx$$

or
$$m(d^2x/dt^2) = - kx$$

or
$$(d^2x/dt^2) + \omega_0^2 x = 0 \tag{2.5}$$

The solution of the above equation is
$$x = x_0 \cos(\omega_0 t + \theta) \tag{2.6}$$

where θ is the initial phase of the oscillator.

The oscillations considered here are undamped. In reality this is not so, for even if other damping factors (such as collision with other oscillators and the decelerating effect of the surrounding medium) are ignored, the radiation emitted from the oscillating dipole reacts with the charges of the dipole itself and damps its oscillations (the damping being termed radiative friction or resistance). When damping is also considered, the amplitude x at time t is expressed by including a damping term in Equation 2.5, i.e.

$$(d^2x/dt^2) + 2\gamma (dx/dt) + \omega_0^2 x = 0$$

and its solution is

$$x = x_0 (e^{-\gamma t}) \cos(\omega_1 t + \theta) \tag{2.7}$$

where $\omega_1^2 = \omega_0^2 - \gamma^2$ and $\gamma = \gamma_r + \gamma_0$ is the damping constant in which γ_r refers to the radiative and γ_0 to every other kind of damping constant. In the absence of a medium, γ_0 will be zero. As compared to $t = 0$, the amplitude of oscillations will nearly vanish when $t > 1/\gamma$.

2.6. Spectral line shape

The radiation emitted from a freely oscillating dipole is monoenergetic (within the limits of natural line broadening). The electric field intensity associated with this monoenergetic radiation is given by

$$E(t) = E_0 \exp(i\, \omega_0 t)$$

If the amplitude of emitted radiation is damped

$$E(t) = E_0 e^{-\gamma t} \exp(i\, \omega_1 t)$$

The monoenergetic component $E(\omega)$ of the spectral distribution of the line of finite width, using a Fourier integral, is

$$E(\omega) = (1/2\pi) \int_0^\infty E(t) \exp(-i\,\omega\,t)\,dt$$

$$= (E_0/2\pi) \int_0^\infty \exp(-\gamma t) \exp[it(\omega_1-\omega)]dt$$

$$= (E_0/2\pi) \int_0^\infty \exp\{-t[\gamma+i(\omega-\omega_1)]\}dt$$

$$= (E_0/2\pi) / [i(\omega - \omega_1) + \gamma] \tag{2.8}$$

The intensity $I(\omega)$ of the component ω of the spectral line, which has a distribution of its intensity with ω, is proportional to the square of the electric field intensity $E(\omega)$;

$$K\,|E(\omega)|^2 = (K/4\pi^2)\,\{|E_0|/[\gamma+i(\omega-\omega_1)]\}\,\{|E_0|/[\gamma-i(\omega-\omega_1)]\}$$

$$I(\omega) = (K/4\pi^2)\,|E_0|^2/[\gamma^2 + (\omega - \omega_1)^2]$$

The total intensity I_0 of the spectral line profile (area under the curve) is given by

$$I_0 = \int_0^\infty I(\omega)\,d\omega = (K/4\pi^2)\,|E_o|^2 \int_0^\infty d\omega/[\gamma^2 + (\omega - \omega_1)^2]$$

$$= K\,|E_0|^2/(4\pi\gamma) \text{ (for } (\omega_1/\gamma > 1))$$

This gives $K = 4\pi\gamma I_0 / |E_0|^2$

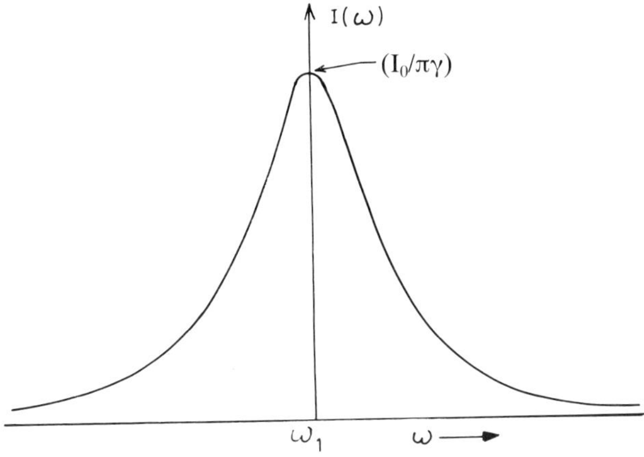

Figure 2.6. Spectral line profile of radiation emitted from a damped oscillator.

Hence $I(\omega) = (\gamma/\pi) \{I_0 / [(\omega - \omega_1)^2 + \gamma^2]\}$

The maximum in $I(\omega)$ will result when $(\omega - \omega_1)^2$ is a minimum, i.e. when it is zero. Since $\omega_1 = (\omega_0^2 - \gamma^2)^{1/2}$, and

$$\omega - \omega_1 = 0$$

we have $\omega = (\omega_0^2 - \gamma^2)^{1/2} = \omega_0[1-(\gamma^2/\omega_0^2)]^{1/2}$

since usually $(\gamma / \omega_0) < 1$, we can use this to obtain

$$\omega \approx \omega_0[1 - (\gamma^2/2\,\omega_0^2)] \tag{2.9}$$

For the maximum value of $I(\omega)$, we have

$$I(\omega)_{max} = I_0 /\pi\gamma$$

This is shown in Fig 2.6. In passing, we may note that a free oscillator will, strictly, give out monoenergetic radiation of frequency ω_0. The intensity $I(\omega)$ is zero for frequencies $\omega \neq \omega_0$.

2.7. Doppler broadening

This is a temperature dependent process in which the increase in temperature of the free emitting atoms in a gas broadens the spectral profile. The effect arises from the presence of relative motion between the spectrograph (at rest) and the radiation-emitting atom (making a random motion) giving rise to the Doppler effect. The frequency of the emitted radiation is therefore correspondingly modified. Considering the simple case of free atomic oscillators in which the emitted frequency of zero relative motion ($\upsilon = 0$) is monoenergetic, ω_0, the instrument still records a spectrum of frequencies. We have, from the Doppler effect,

$$\omega(\upsilon) = \omega_0 (1 \pm \upsilon/c) \tag{2.10}$$

The positive sign applies when the oscillator is moving towards the spectrograph while the minus sign indicates its motion away.

Using Maxwell's velocity distribution law for molecules in a gas at temperature T, the number (dn) lying between the velocity interval υ and $\upsilon + d\upsilon$ is given by

$$dn = n(\upsilon)\, d\upsilon = n\left(\frac{\mu}{2\,\pi k_B T}\right)^{1/2} \exp(- \mu\upsilon^2/2k_B T)d\upsilon \tag{2.11}$$

where υ is the velocity of the molecule along the axis of the spectrograph and it is assumed that each one of the molecules vibrates with the same frequency and the same amplitude. μ is the mass of the molecule and n is the total number of molecules. From Equation 2.10 we introduce the modulation of the frequency ω_0 of the emitted light as a result of the molecular velocity υ

$$\upsilon = (c/\omega_0)(\omega - \omega_0)$$
$$d\upsilon = (c/\omega_0)d\omega$$

Hence

$$(dn/n) = f(\omega)\,d\omega$$
$$= (\mu/2\pi k_B T)^{1/2}(c/\omega_0)\exp\left[-\mu c^2(\omega - \omega_0)^2/2\omega_0^2 k_B T\right]d\omega$$

where $f(\omega)d\omega$ represents the fraction of radiation in the frequency interval ω and $\omega + d\omega$ received by the spectrograph. Also, since $I(\omega)d\omega = Af(\omega)d\omega$

$$I(\omega)/I(\omega_0) = \exp\left[-\mu c^2(\omega - \omega_0)^2/2\omega_0^2 k_B T\right]$$

$$= \exp\left[-(\omega - \omega_0)^2/\delta^2\right]$$

Thus $I(\omega) = I(\omega_0)\exp\left[-(\omega - \omega_0)^2/\delta^2\right]$ \hfill (2.12)

where $\delta = (\omega_0/c)(2k_B T/\mu)^{1/2}$ is defined by Equation 2.13, below.

The frequency $\omega_{1/2}$ at which the intensity $I(\omega) = \frac{1}{2}I(\omega_0)$ is given by

$$\ln 2 = (\omega_{1/2} - \omega_0)^2/\delta^2$$

$$2|(\omega_{1/2} - \omega_0)| = \triangle\omega_D = 2\,\delta\,(\ln 2)^{1/2} \hfill (2.13)$$

This is shown in Figure 2.7. The width $\triangle\omega_D$ (Doppler broadening) as

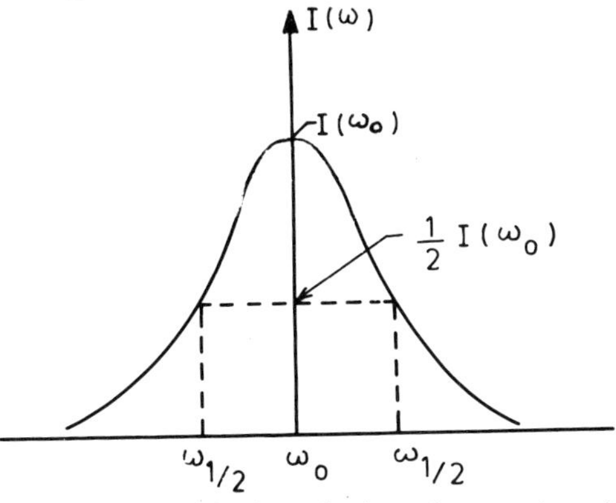

Figure 2.7. Doppler broadening of spectral profile.

seen from Equation 2.13 increases with temperature. The concept of Doppler broadening outlined here for gaseous molecules does not apply to the atoms and molecules in solids because the equilibrium position of atoms or molecules in a solid remains unaltered relative to the position of the spectrometer.

2.8.   Configuration coordinate model

The luminescent or emission centre interacts with its neighbours in the lattice. This interaction results in the emission of luminescence peaked at a lower frequency compared with that of the incident radiation which excited the centre (this is Stoke's law). Physically it means that when the centre is excited its electron moves to outer orbits of larger size which can overlap with the electron orbits of the neighbouring atoms. In this process a fraction of energy is transferred from the excited centre. When the centre returns to the ground state, the emitted energy is now less than that received. This concept is mathematically treated in terms of the coordinates of configurations (ions) around the centre. This model was first considered by von Hippel in 1936 and was applied to the description of luminescence by Seitz in 1939[4,5]. It affords a convenient way of describing the processes of optical absorption and luminescence emission. The two curves in Fig 2.8 show the energy of interaction of the centre as a function of movement of ions relative to their equilibrium position as if the host atom had occupied the centre-site.

The minimum, point A, in the potential energy curve represents the ground state equilibrium of the centre. After absorption of incident radiation, the centre goes over to the excited state at point B and comes to point C through emission of phonons, from whence it makes a transition to point D and again, through phonon emission, returns to the ground state equilibrium position A. It is obvious from Fig 2.8, that

$$AB(= \hbar\omega_a) > CD(= \hbar\omega_e)$$

where ω_a and ω_e refer to the absorbed and emitted photon frequencies. If $U_0 = \hbar\omega_0$, and U_0 is the energy by which the ground state and excited state of the centre are separated (see Fig 2.9), then:

For absorption

$$E_a = U_0 + U_{vib}$$

$$= U_0 + n_e \hbar \, \Omega_e$$

For emission

$$E_e = U_0 - n_a \hbar \, \Omega_a$$

If the potential energy curves in the ground state and the excited state are reasonably parabolic (true for small amplitudes), then for phonon frequencies we may take $\Omega_e \approx \Omega_a = \Omega$ (say), and

$$E_a - E_e = \hbar\Omega \, (n_a + n_e) \approx 2n\hbar\Omega = S \text{ (Stokes shift)}$$

(2.14)

In the above we have used the quantum mechanical expression $(U_{vib} = n\hbar\Omega)$ for the vibrational energy. If we treat this problem

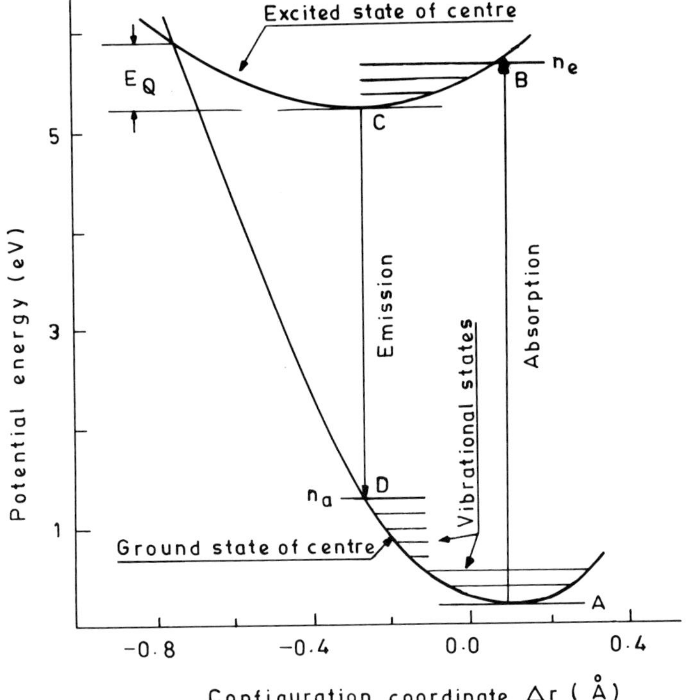

Figure 2.8. Configuration coordinate diagram; vertical arrows show electronic transition; $\triangle r = r - r_0$; see section 2.11.

classically, then we may write, for the displacement of the centre from its equilibrium position,

$$U_g = 0 + \tfrac{1}{2} k_g x^2$$

and

$$U_e = U_0 + \tfrac{1}{2} k_e(x - x_0)^2$$

$$\left. \right\} \qquad (2.15)$$

2.9. Emission and absorption bands

If m and n denote vibrational levels in the ground and excited states of the emission centre, then the emission energy on account of transition from the nth vibrational level to the mth level in the ground state is given by

$$\hbar\omega_{nm} = E_n^{\,e} - E_m^{\,g} \qquad (2.16)$$

Corresponding to the nth state of vibration, the emission centre can be in any state of displacement along the whole width of the nth level (its amplitude) (shown in Figure 2.9) at the time of transition

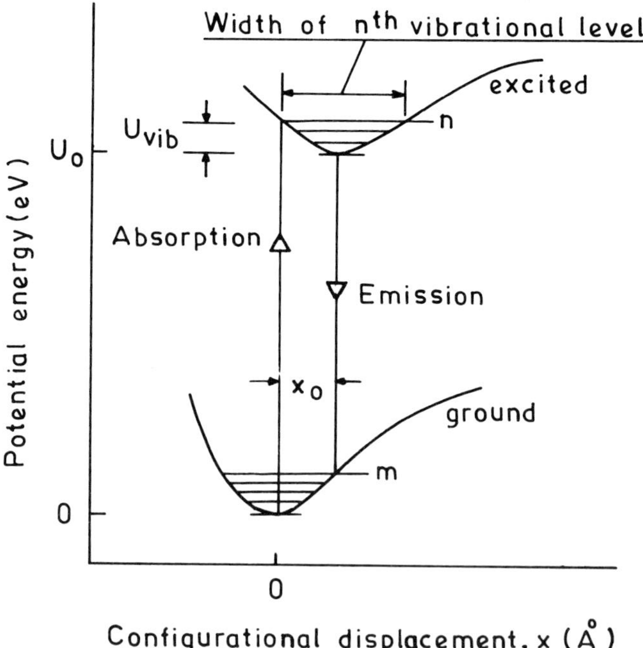

Figure 2.9. Ground and excited states of the emission centre.

to the ground state. Since each transition from different points on the *n*th level will give rise to a different energy of the emission line, an emission band is obtained having maxima corresponding to the maximum probability of the state of displacement of centre in the *n*th vibrational level. In the same manner one obtains the absorption band.

It is found that the emission and absorption bands (distribution of intensity of the line as a function of frequency) are reasonably Gaussian in shape (Fig 2.10) which is expressed as

$$I(\hbar\omega) = (1/2\pi)^{1/2} \exp\left[-(\hbar\omega - \hbar\omega_0)^2/2\,\sigma^2\right] \tag{2.17}$$

where ω_0 is the peak frequency and σ is the standard deviation

$$\sigma = \left[\int (\hbar\omega - \hbar\omega_0)^2\, I(\hbar\omega)\, d(\hbar\omega)\right]^{1/2}$$

and

$$\hbar\omega_0 = <\hbar\omega> = \int I(\hbar\omega)\hbar\omega\, d(\hbar\omega)/\int I(\hbar\omega)\, d(\hbar\omega)$$

where

$$\int I(\hbar\omega)\, d(\hbar\omega) = \text{area under the curve}$$

which is normalised to unity, i.e. it is equal to 1.

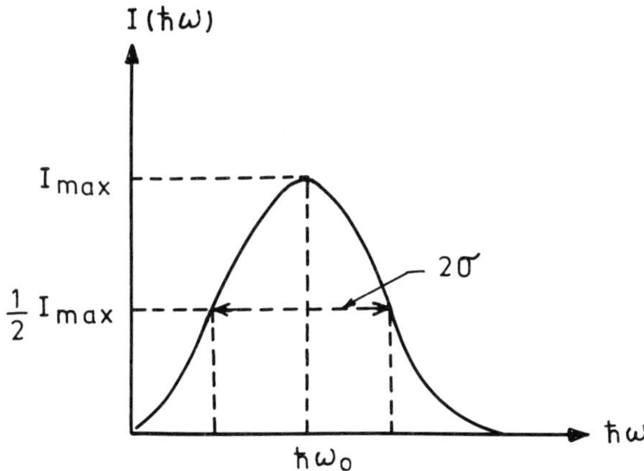

Figure 2.10. Emission and absorption band.

2.10. Temperature dependence of bandwidth

The dependence of bandwidth on temperature arises from the variation of the amplitude of vibration with temperature.

The amplitude increases with increase in temperature and the activator ion will have a correspondingly larger span of transition to excited or from excited to the ground state.

Consider the case of an excited activator ion at 0 K. The zero point vibration amplitude $A_e(0)$ of the excited ion may be expressed as

$$\tfrac{1}{2}\, k_e\, A_e^2(0) = \tfrac{1}{2}\, \hbar\Omega_e \qquad (2.18a)$$

while

$$\tfrac{1}{2}\, k_e A_e^{\,2}(n) = (n + \tfrac{1}{2})\hbar\Omega_e \qquad (2.18b)$$

where $\hbar\Omega_e$ is the energy separation of the consecutive vibrational levels of the ion in the excited state. $A_e(n)$ is the amplitude at an elevated temperature where n represents the nth vibrational state of the ion at this temperature. In determining the temperature variation of bandwidth, $A_e(n)$ is replaced by its average value $<A_e(n)>$ for all possible values of n. Since A_e is both positive and negative with respect to the equilibrium position, the average is taken to be the root mean square value.

$$<A_e\,(n)> = [\sum_n P_n\, A_e^{\,2}(n)]^{1/2} = \sum_n A_e(n)\,(P_n)^{1/2}$$

P_n is the probability that the emission centre will be found in the nth vibrational state at temperature T, i.e.

$$P_n = \exp[-\,n\,\hbar\,\Omega_e/k_B T]/\sum_{n=0}^{\infty} \exp\,(-n\hbar\Omega_e/k_B T)$$

Using Equations 2.18a, b, $\langle A_e(n)\rangle = A_e(0)\sum_n [P_n(2n + 1)]^{1/2}$

where $A_e(0)$, from Equation 2.18b, is equal to $(\hbar\Omega_e/k_e)^{1/2}$

$$A_e(n) = A_e(0)\Big[\sum_{n=0}^{\infty} (2n+1)\,\exp(-n\hbar\Omega_e/k_B T)/\sum_{n=0}^{\infty} \exp(-n\hbar\Omega_e/k_B T)\Big]^{1/2}$$

$$= A_e(0)(S_n)^{1/2}$$

where, by putting $(\hbar\Omega_e/k_B T) = x$, we have

$$S_n = \sum_{n=0}^{\infty} (2n+1)\exp(-nx)/\sum_{n=0}^{\infty} \exp(-nx)$$

or

$$S_n = (1+3e^{-x}+5e^{-2x}+7e^{-3x}+\dots\dots)/(1+e^{-x}+e^{-2x}+\dots\dots) = (N/D)\ \text{(say)}$$

Thus

$$N = 1+3e^{-x}+5e^{-2x}+7e^{-3x}+\ldots$$

or

$$Ne^{-x} = e^{-x}+3e^{-2x}+5e^{-3x}+7e^{-4x}+\ldots.$$

$$N(1-e^{-x}) = 1 + 2(e^{-x}+e^{-2x}+e^{-3x}+\ldots.) = 1+[2e^{-x}/(1-e^{-x})]$$

$$N = [1/(1-e^{-x})] + [2e^{-x}/(1-e^{-x})^2]$$

and $D = 1 +e^{-x}+e^{-2x}+\ldots. = 1/(1-e^{-x})$

Thus

$$(N/D) = S_n = 1+[2e^{-x}/(1-e^{-x})] = (1+e^{-x})/(1-e^{-x}) = \coth(x/2)$$

So $$\langle A_e(n)\rangle = A_e(0) \left[\coth\left(\frac{\hbar\Omega_e}{2k_BT}\right) \right]^{1/2}$$

Since A_e has already been averaged over n values, the only variable is temperature. Thus

$$\langle A_e(T)\rangle = A_e(0) \left[\coth\left(\frac{\hbar\Omega_e}{2k_BT}\right) \right]^{1/2} \tag{2.19}$$

Through Equation 2.19, the emission (as well as absorption) bandwidth depends upon temperature. $[\mathrm{Coth}(\hbar\Omega_e/2k_BT)]^{1/2}$ dependence of $A_e(T)$ has, for instance, been demonstrated in CaF_2:Mn by Abbundi *et al*[6].

2.11. **Example of KCl(Tl) phosphor**

KCl with Tl activator is a popular TL material which has been extensively studied, both experimentally and theoretically (see Figure 2.11). The interaction potential $U_{e,g}$ for the Tl^+ ion may be expressed as

$$U_{e,g}(\triangle r) = \alpha_{e,g}(r - r_0)^2 \tag{2.20}$$

where the subscripts e,g, refer to the excited and ground states of the ion or, more appropriately, of the $TlCl_6$ group and

r_0 = equilibrium position of nearest Cl⁻ ions in a perfect KC1 lattice.

r = displacement of the Cl⁻ ions when K⁺ ion is replaced by Tl⁺ ion.

α = force constant

Further details regarding emission and absorption spectra, excited and ground states of the luminescent centre in the KCl(Tl) phosphor are discussed in Chapter 3.

2.12. Luminescence efficiency; non-radiative transition

An important feature of luminescence is its efficiency η. It is the ratio of the total energy emitted $(h\nu)$ in the form of light to the energy absorbed $(h\nu_o)$ by the specimen in the process of excitation.

$$\eta = (h\nu/h\nu_o) = (\lambda_o/\lambda)$$

Thus, η is proportional to the exciting wavelength λ_o. This behaviour is illustrated in Figure 2.12(a). At long enough wavelengths, at which luminescence cannot be excited, η drops to

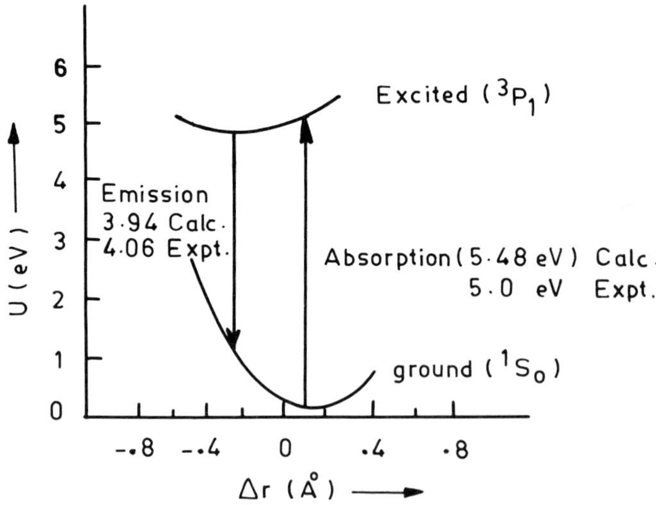

Figure 2.11. Configurational coordinate diagram for KCl(Tl).

zero. Figure 2.12(b) shows the experimentally observed efficiency for ZnS.

Referring to Figure 2.8, if the point B is high enough to be in line with Q, then the emission centre may have a definite probability of reaching the point A at the ground state curve without any significant emission of radiation, but, mainly through lattice vibrational transition, accompanied only by the phonon emission. The same situation can obtain if the electron trapped at a depth E is thermally activated to reach the point Q (energy of activation required is $E + E_Q$). Such events constitute the non-radiative transition. If q is the probability per second for the non-radiative transitions, it is expressed as

$$q = \nu \exp \left[-(E + E_Q)/k_B T\right] \qquad (2.21)$$

where $\nu\ (= 10^{12} - 10^{13}\ \mathrm{s}^{-1})$ is the escape probability or the frequency factor (on account of its dimensionality) for the process. If $p\ [= s\ \exp\ (-\ E/k_B T)]$ is the probability per second for the radiative transition (i.e. the transition results in the emission of detectable

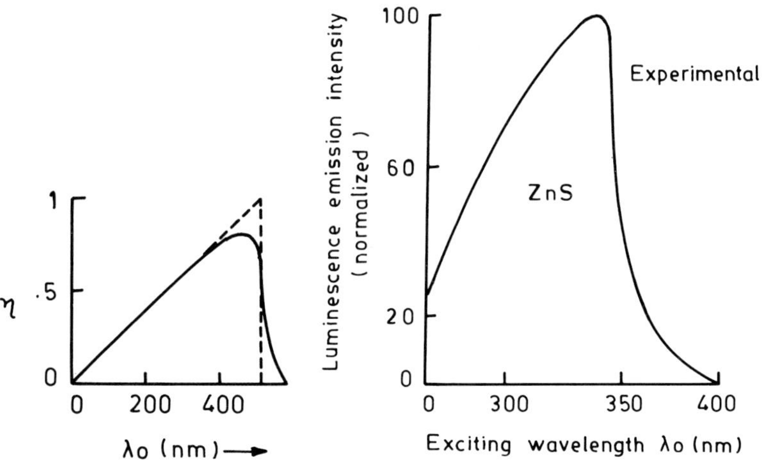

Figure 2.12.(a) Schematic variation of the luminous efficiency (η) as a function of exciting wavelength λ_o. (b) Luminescence intensity in ZnS as a function of exciting wavelength [7].

radiation), then the luminescence efficiency η is expressed by

$$\eta = \frac{\text{radiative events}}{\text{total events}}$$

$$= p/(p+q) = 1/[1 + (v/s) \exp(-E_Q/k_B T)] \qquad (2.22)$$

Equation 2.22, by parameterising E_Q for a given value of (v/s), is fitted with the experiment for the Mg_2TiO_4:Mn phosphor in Figure 2.13. Thus E_Q can be determined.

2.13.   **Duration of luminescence**

The time between the instant at which the exciting radiation is switched off and the instant at which complete disappearance of afterglow takes place is called duration of luminescence τ, and

$$\tau = \frac{\int_0^\infty t\, W_1(t)\, dt}{\int_0^\infty W_1(t)\, dt} = \frac{t_1 W_1 + t_2 W_2 + t_3 W_3 + \ldots}{W_1 + W_2 + W_3 + \ldots = W} \qquad (2.23)$$

where $W_1(t)$ = luminescent yield per unit time at time t. The time is counted after the cessation of excitation.

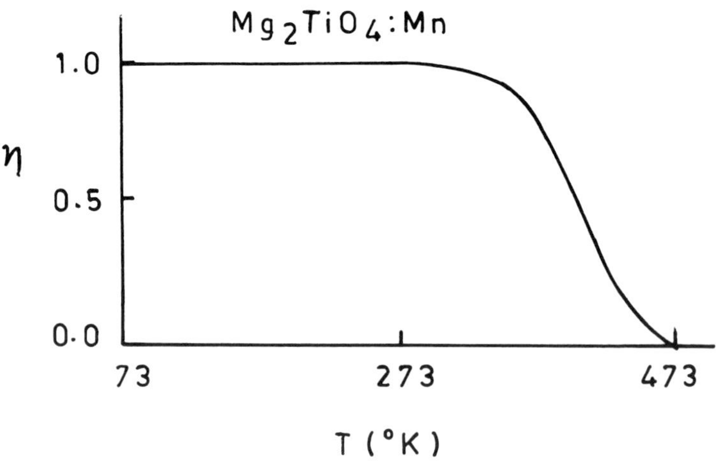

Figure 2.13. Luminescent efficiency η as a function of temperature. F. A. Kröger *et al*, Physica, **14**, 81 (1948).

2.14.   Lifetime of the excited state

Let $n_i(0)$ be the population in the ith excited level at an instant when the exciting radiation is switched off (taken as the zero time). Levels higher than ith are supposed to be empty (see Figure 2.14). The number $n_i(0)$ begins to decrease after the excitation is switched off, and gives rise to luminescence. Let $n_i(t)$ be the level population at time t. The decrease $(-dn_i)$ in the population level in the next time interval dt is given as

$$(-dn_i) \propto n_i(t)\, dt$$

or

$$dn_i = -\left(\Sigma_j A_{ij} + \Sigma_j B_{ij}\right) n_i(t)\, dt$$

where $\Sigma_j A_{ij}$ and $\Sigma_j B_{ij}$ are radiative and non-radiative transition probabilities from the ith to the set of j levels. Integrating, we get

$$\ln n_i(t) = -\left(\Sigma_j A_{ij} + \Sigma_j B_{ij}\right) t + \text{constant}$$

At $t = 0$, $n_i(t) = n_i(0)$. Hence

$$\ln\left[n_i(t)/n_i(0)\right] = -\left(\Sigma_j A_{ij} + \Sigma_j B_{ij}\right) t$$

or

$$n_i(t) = n_i(0) \exp(-t/\tau) \tag{2.24}$$

where

$$\tau = \left(\Sigma_j A_{ij} + \Sigma_j B_{ij}\right)^{-1}$$

is called the lifetime of the excited state. It is defined as the time in

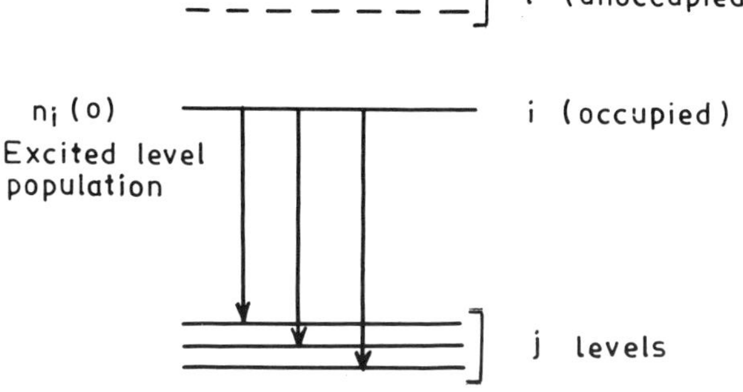

Figure 2.14. Lifetime of the excited state.

which the population of the excited state becomes (l/e) of the initial value. As we see below, the lifetime of the excited state is the mean lifetime of the particle in that state. Differentiating Equation (2.24) we get

$$\mathrm{d}n_i\,(t) = (1/\tau)\,n_i\,(0)\,\exp(-t/\tau)(-\mathrm{d}t) \qquad (2.25)$$

The mean lifetime $\bar{t}$ per particle is

$$\bar{t} = \left(\int_0^\infty t\,\mathrm{d}n_i\right)/\left(\int_0^\infty \mathrm{d}n_i\right) = \left(\int_0^\infty t\,\mathrm{d}n_i\right)/\,n_i(0)$$

where $\int_0^\infty \mathrm{d}n_i = n_i(0)$ is the normalisation condition. Using Equation (2.25)

$$\bar{t} = (1/\tau)\int_0^\infty t\exp(-t/\tau)\,\mathrm{d}t = \tau \qquad (2.26)$$

Typically the quantity A_{ij} is of the order of 10^8 s^{-1} and therefore the lifetime τ of the excited state in the absence of non-radiative transition (termed quenching) is of the order of 10^{-8} s. The levels with this order of magnitude of τ are known as labile states in contrast to metastable states for which the lifetime is greater by several orders of magnitude.

The rate of luminescence decay (energy output) as a function of time t with $\hbar\omega_{ij}$ as the energy of the emitted photon is given by

$$W_l^{ij}\,(t) = A_{ij}n(t)\,\hbar\,\omega_{ij}$$
$$= A_{ij}\,n_i(0)\,\exp(-t/\tau)\,\hbar\,\omega_{ij}$$

or

$$W_l^{ij}\,(t) = W_l^{ij}\,(0)\,\exp(-t/\tau) \qquad (2.27)$$

In the luminescence decay process we can also include the effect of internal conversion, i.e. a part of the emitted light may be absorbed by the particles in j levels which will then be pushed up to the i_{th} level. The rate of change of n_i will then be given by

$$\mathrm{d}n_i/\mathrm{d}t = -\,n_i\sum_j A_{ij} + \sum_j n_j A_{ji}$$

where $n_j = n_o\,\exp(-j\hbar\omega/k_B T)$

and $j\hbar\omega$ is the position of the jth level as compared with the ground state and n_0 is the total number of particles in the system.

2.15. Temperature dependence of lifetime of an electron trap

If an electron trap is occupied and is at an energy depth E

below the bottom of the conduction band then the probability per second of release of an electron from the trap to the conduction band (inverse of this probability is the lifetime of the trap) is given by the Boltzmann relation. If p and τ denote the probability and lifetime respectively, we have

$$p = (1/\tau) = s \exp(-E/k_B T)$$

where s is the frequency factor; thus

$$\ln \tau = (E/k_B T) - \ln s \qquad (2.28)$$

Equation (2.28) gives the temperature variation of the trap lifetime. This is shown in Figure 2.15 for the CaS:Bi phosphor.

2.16.   Luminescence decay kinetics
(i)   *Fluorescence decay*

In the fluorescence process the energy of excitation is absorbed directly by the luminescent centre* and re-emitted as light within about 10^{-8} s. Thus, the electron in the centre, in this process,

---

* A quantum state within the band gap will act as a centre of recombination of carriers if it captures a carrier and holds it until such time as another carrier of opposite sign is captured. The two excited carriers combining at the centre cause the release of excess energy as photons/phonons. The centre at which the recombination results in the emission of light is called a luminescent centre.

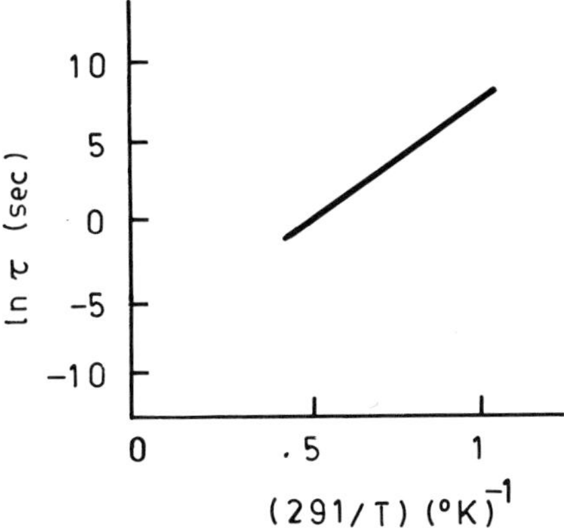

Figure 2.15. Temperature variation of trap lifetime in CaS:Bi phosphor. (See D. Curie in Bibliography)

is excited to a higher quantum state from which it eventually returns to the ground state accompanied by the emission of light. The fluorescent materials are characterised by the presence in them of the large number of luminescent centres as compared with that of the traps. In other words, the role of traps in the fluorescence mechanism is considered to be negligible. The decay of the excited luminescent centres, after the excitation has ceased, is essentially similar to that of radioactive nuclei. If $n(t)$ is the number of excited centres (i.e. electrons) at any instant of time t, the rate of decay $(-dn/dt)$ at time t is proportional to the number $n(t)$.

$$(-dn/dt) = \alpha n \tag{2.29}$$

where α is the probability of decay (decay constant of the process). Also $\alpha = 1/\tau$ where τ is the lifetime of the excited state of the centre. The emission intensity I is given by

$$I = -c(dn/dt) = c\alpha n \tag{2.30}$$

where c is a constant of proportionality. Since each decay may not necessarily lead to the emission of a photon, $c<1$. Sometimes, however, in the literature c is taken as unity as no generality is lost in doing so. Solution of Equation (2.29) gives

$$n = n_{o}e^{-\alpha t} \tag{2.31}$$

which using Equation (2.30) gives

$$I = I_{o}e^{-\alpha t} \tag{2.32}$$

where I_{o} is the initial luminescence intensity just when the excitation ceased ($t = 0$). Equation (2.30) is characteristic of monomolecular or first order decay kinetics. Here α has been taken to be temperature independent. One can verify this assumption experimentally by plotting ln I against t at different temperatures and determining α from the slopes of the straight lines.

If the luminescent centres, following the absorption of energy, are ionised so that the electrons cross over to the conduction band leaving behind, on an average, an equal number of empty centres, then the problem essentially becomes one of combining n electrons with the equal number of holes (empty centres). The rate of loss of conduction electrons at time t is then given by

$$(-dn/dt) = rn^2 \tag{2.33}$$

where r here means the probability of recombination ($=$ inverse of the 'recombination lifetime' of conduction band electrons). Equation (2.33) upon integration gives

$$(-1/n) = (rt) + (1/n_o)$$

where n_o is the initial electron concentration in the conduction band at $t=0$.

$$n = n_o (1 + n_o rt)^{-1}$$

and since $I = crn^2$, we get

$$I = I_o/ (1 + n_o rt)^2 = I_o / (1 + at)^2 \tag{2.34}$$

where $a = (I_o r/c)^{1/2}$. Equation (2.34) illustrates the hyperbolic decay of fluorescence emission and is characteristic of bimolecular or second order decay kinetics. In general the recombination lifetimes ($\sim 10^{-5}$s) can be larger than the excited state lifetimes ($\sim 10^{-8}$s).

(ii)   ***Phosphorescence decay***
Phosphorescence emission involves the role of traps along with that of the activator levels (luminescent centres). The trap levels can belong to either the host atoms themselves under the perturbing influence of neighbouring activator atoms or to the lattice defects whose presence is also linked to the incorporation of activator atoms. If the trapping levels are within close proximity of the excited states of the centres, the electrons released from the traps can 'get over' to the excited states of the centres. Since the excited state lifetimes are shorter than the trap lifetimes, the luminescence process will be governed by the lifetime of the traps. If $n(t)$ is the concentration of occupied traps at time t and τ is the mean trap lifetime, then the rate of release of electrons from the traps ($=$ rate of decrease of the occupied trap concentration) is given by

$$(-dn/dt) = n(t)/\tau$$

This gives

$$n(t) = n_o \exp(-t/\tau) \tag{2.35}$$

where n_o is the initially occupied trap concentration (at $t = 0$).

Thus

$$I = -c(\mathrm{d}n/\mathrm{d}t) = (cn_\mathrm{o}/\tau)\,\exp(-t/\tau) \qquad (2.36)$$

Equation (2.36) represents the first order kinetics of the phosphorescence decay mechanism.

In many phosphors, such as the thallium activated alkali halides, the trap lifetime is found to be considerably larger and also temperature dependent. These factors require the consideration of traps lying lower (on the energy scale) than the excited states of the centres. Let us, for simplicity, consider these traps to be all at a single level. The probability per unit time of the thermal excitation of electrons from these traps lying at a position lower by an amount of energy ε as compared with the excited state of a centre is, at temperature T, given by

$$p = (1/\tau) = s\,\exp(-\varepsilon/k_\mathrm{B}T)$$

Using this value of τ in Equation (2.36) we get

$$I(t,T) = (csn_\mathrm{o}\,[\exp(-\,\varepsilon/k_\mathrm{B}T)]\exp[-st\,\exp(-\,\varepsilon/k_\mathrm{B}T)] \quad (2.37)$$

We find from Equation (2.37) that the fluorescence decay is rapid at lower temperatures. This is as expected because the thermal detrapping of electrons at low temperatures is small. Analysis of luminescence decay at different temperatures can, therefore, be used to find the value of ϵ [8].

In the second order decay process of phosphorescence [9]*, the escaping electron from the trap is assumed to have equal probability of either being retrapped or of recombining with the empty centre. If N is the concentration of traps (all of a single kind) of which, at any instant of time t, n are occupied and m is the concentration of empty centres (charge neutrality requires that $m = n$), then the probability that an electron escaping from the trap will recombine (and not be retrapped) at the centre is given by

$$\frac{m}{(N-n)+m} = \frac{n}{(N-n)+n} = \frac{n}{N}$$

The phosphorescence intensity I is given by the rate of decrease of the occupied trap density resulting in the recombination of released electrons with the empty centres, i.e.

---

* Equation (5) of Garlick and Gibson, i.e. $I = n_\mathrm{o}^2\,\tau/N[\,1 + (n_\mathrm{o}t/N\tau)]^2$ (they assumed $c = 1$) has an algebric error. Equation (2.39) in the present text is correct.

$$I = -c(\mathrm{d}n/\mathrm{d}t) = c(n/N)(n/\tau)$$

or

$$\mathrm{d}n/n^2 = -(1/N\tau)\mathrm{d}t \qquad (2.38\text{a})$$

Integrating Equation (2.38a) and incorporating $n = n_\mathrm{o}$ at $t = 0$, we get

$$n = n_\mathrm{o}N\tau/(n_\mathrm{o}t + N\tau) = n_\mathrm{o}/[1 + (n_\mathrm{o}t/N\tau)] \qquad (2.38\text{b})$$

and

$$I = cn^2/N\tau = (cn_\mathrm{o}^2/N\tau)\{1/[1 + (n_\mathrm{o}t/N\tau)]^2\} \qquad (2.39)$$

Equation (2.39) describes the hyperbolic form of the phosphorescence decay and arises from the inclusion of the role played by the retrapping mechanism. Figure 2.16 shows the observed phosphorescence decay in ZnS:Cu and Zn$_2$SiO$_4$:Mn

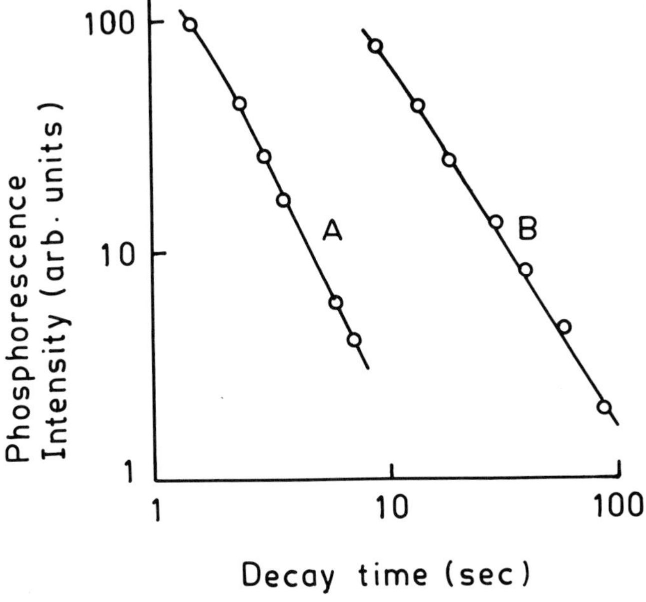

Figure 2.16. Phosphorescence decay of UV excited ZnS:Cu (A) and Zn$_2$SiO$_4$:Mn (B) phosphors at room temperature (291 K). The curves show hyperbolic decay forms at longer decay times[9].

phosphors at room temperature (291 K). The curves show hyperbolic decay (second order decay kinetics) forms at longer decay times.

(iii)   ***Thermoluminescence decay***
    The luminescence intensity of a pre-irradiated phosphor is recorded as a function of the rising temperature. The resulting curve between the luminescence intensity and temperature, known as the glow curve, first increases and then, after reaching a maximum, decreases.

    If n represents the occupied trap density at time t, then

$$(-\mathrm{d}n/\mathrm{d}t) = (n)(1/\tau) = ns\,\exp(-E/k_\mathrm{B}T)$$

gives the rate of thermal release of electrons at temperature T. If the phosphor is heated at a constant rate of increase of temperature, i.e. $\mathrm{d}T/\mathrm{d}t$ = constant ($=q$), the above equation can be written as

$$(\mathrm{d}n/n) = -(s/q)\,\exp(-E/k_\mathrm{B}T)\,\mathrm{d}T \qquad (2.40)$$

Assuming that initially at the start of heating, i.e. when $T = T_i$, $n = n_\mathrm{o}$, the integration of Equation (2.40) yields

$$n = n_\mathrm{o}\,\exp\left[-(s/q)\int_{T_i}^{T}\exp(-E/k_\mathrm{B}T')\,\mathrm{d}T'\right] \qquad (2.41)$$

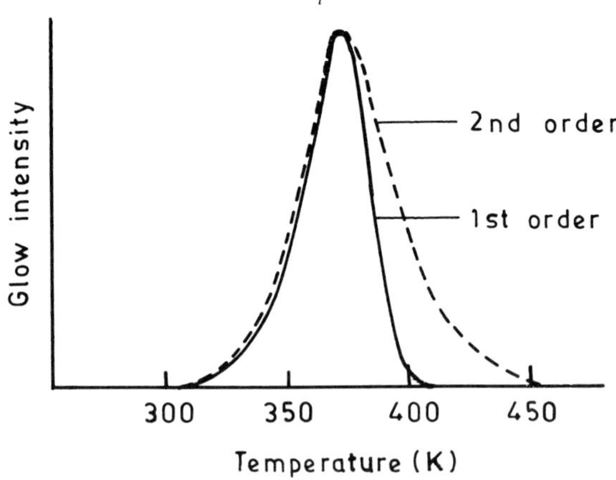

Figure 2.17. Computed glow curves corresponding to first and second order kinetics. The parameters chosen are: $q = 5$ K.s^{-1}, electron trap depth $E = 0.8$ eV and $s = 5\times 10^{11}$s^{-1}. The initial occupied trap densities for the first and second order kinetics are taken as 10^{10} and 10^{9} cm^{-3} respectively. The two curves are normalised to the same peak-height[10].

The thermoluminescence intensity I as a function of temperature is as follows:

$$I = -c(dn/dt) = cn_o s \left[\exp(-E/k_B T)\right]\exp\left[-(s/q) \int_{T_i}^{T} \exp(-E/k_B T')dT'\right]$$

$$(2.42)$$

Equation (2.42) describes the first order kinetics of the glow curve.
If, however, one uses Equation (2.38a) for $(-dn/dt)$, i.e.

$$-(dn/dt) = (n^2/N)(1/\tau) = (n^2 s/N)\exp(-E/k_B T)$$

the second order kinetics equation is obtained, which is

$$I = (csn_o^2/N)\left[\exp(-E/k_B T)\right]\left[1 + (sn_o/qN)\int_{T_i}^{T} \exp(-E/k_B T')dT'\right]^{-2}$$

$$(2.43)$$

Second order kinetics will be exhibited by those phosphors in which the concentrations of traps (say 10^{16}) and recombination centres almost match. For a larger concentration of the latter (say by a factor of about 100), the glow curve shape is rather asymmetric (falls off rapidly beyond the maxima) conforming to the first order kinetics. Figure 2.17 depicts these two cases of the order of kinetics.

References

1. Townsend, P. D. and Kelly, J. C. *Colour Centres and Imperfections in Insulators and Semiconductors* (London: Chatto and Windus) (1973).
2. Nink, R. and Kos, H. J. Phys. Status Solidi a **35**, 121 (1976).
3. Nink, R. and Kos, H. J. Nucl. Instrum. Methods **175**, 15, 24 (1980).
4. von Hippel, A. Z. Phys. **101**, 680 (1936).
5. Seitz, F. Trans. Faraday Soc. **35**, 79 (1939).
6. Abbundi, R. J., Cooke, D. W., Mathur, V. K., Royce, G. A. and Brown, M. D. Radiat. Prot. Dosim. **6**, 329 (1984).
7. Bube, R. H. *Photoconductivity of Solids* (Chichester: Wiley) p. 164 (1960).
8. Teh, C. K., Tin, C. C. and Weichman, F. L. J. Lumin. **35m** 17 (1986).
9. Garlick, G. F. J. and Gibson, A. F. Proc. Phys. Soc. **60**, 574 (1948).
10. Bull, R. K. Nucl. Tracks Radiat. Meas. **11**, 105 (1986).

Bibliography

Curie, D. *Luminescence in Crystal* (London: Methuen) (1963).
Garlick, G. F. J. *Luminescent Materials* (Oxford: Clarendon Press) (1949).
Garlick, G. F. J. *Cathodoluminescence*. IN Advances in Electronics, Vol. II, ed. L.

Marton (New York: Academic Press) (1950).

Garlick, G. F. J. *Luminescence.* IN Handbuch der Physik, XXVI, 1-128 (1958).

Kallmann, H. P. and Spruch, G. M. (eds) *Luminecence of Organic and Inorganic Materials* (Conf. Proc.) (Chichester: John Wiley) (1962).

Klick, C. C. and Schulman, J. H. *Luminescence in Solids.* IN Solid State Phys. **5**, 97-172 (1957).

Kroger, F. A. *Some Aspects of Luminescence of Solids* (Amsterdam: Elsevier) (1948).

Leverenz, H. W. *An Introduction to Luminescence of Solids.* (New York: John Wiley) (1950).

Lim, E. C. (ed.) *Molecular Luminescence* Proc. Int. Conf. (New York: Benjamin) (1969).

Marfunin, A. S. *Spectroscopy, Luminescence and Radiation Centres in Minerals* (New York: Springer Verlag) (1979).

Pringsheim, P. *Fluorescence and Phosphorescence* (New York: Interscience) (1949).

Przibram, K. *Irradiation Colours and Luminescence* (New York: Pergamon Press) (1956).

Stepanov, B. I. and Gribkovskii, V. P. *Theory of Luminescence* (London: Iliffe Books) (1968).

Williams, E. W. and Hall, R. *Luminescence and the Light Emitting Diode* (Oxford: Pergamon Press) (1978).

Williams, F. *Theoretical Basis for Solid State Luminescence.* IN *Luminescence of Inorganic Solids,* ed. P. Goldberg (New York: Academic Press) (1966).

Williams, F. E. *Solid State Luminescence.* IN Advances in Electronics. Vol. V, ed. L. Marton (New York: Academic Press) (1953).

CHAPTER 3

Principles and Methods of Thermoluminescence

K. Mahesh and C. Furetta

3.1.   Introduction

Following excitation by radiation of insulators containing electron as well as hole trapping levels, carriers of both types become mobile in their respective bands, which are then eventually trapped into these levels. These trapped carriers can be released if thermal energy is supplied; this stimulates the carriers to cross the potential barrier of these traps and, in the process, allows them to move to the suitable recombination centres that contain a hole or an electron (as the case may be), resulting in the emission of light. This process of light emission by thermal stimulation is called 'thermally stimulated luminescence' (TSL), although the older but less justified name 'thermoluminescence' (TL) still persists and is more often used. If the probability of transition from a trap to the band is high even at room temperature, the well known phosphorescence process of crystals is observed.

The study of these phenomena is based on models that describe the number of charge carriers present in the traps and centres relative to time and temperature. In order to know the charge carrier population, Boltzmann's statistics are usually applied and probabilities of all possible transitions are deduced. Various models will be discussed in Chapter 7. A simple approximate model is that of Randall and Wilkins[1a] which assumes that all traps are located at the same level.

At temperature T, the de-trapping of the carriers depends upon the occupied trap density n and the potential barrier (known as the thermal activation energy) E of these traps. The rate of de-trapping, $-dn/dt$, can be expressed as

$$-dn/dt \propto n \exp(-E/k_B T)$$

or

$$-dn/dt = s\, n \exp(-E/k_B T) \tag{3.1}$$

where s is a constant known as the frequency factor and k_B is the

Boltzmann constant ($=0.862 \times 10^{-4}$ eV.K^{-1}). Note the dimensionality of s which is time^{-1}; hence the name frequency factor.

The plot of the luminescence intensity as a function of the rising temperature of the TL phosphor being heated (at constant rate of increase of temperature, $dT/dt = q =$ constant) exhibits one or more peaks, known as glow peaks. The analysis of these glow curves provides information on:
 (i) trap depths (i.e. thermal activation energy corresponding to each glow peak);
 (ii) electron/hole trap densities;
(iii) frequency factor, s;
(iv) carrier mobility, and
 (v) capture cross section of traps.

In addition to the glow curves, if an emission spectrum (TL intensity versus wavelength) is also recorded, then the nature and characteristics of luminescent centres can also be studied. Simultaneous measurement of thermally stimulated conductivity (TSC) can give information on the type of carriers (i.e. electrons or holes) responsible for the luminescence emission.

In this chapter our aim is to describe the fundamentals of the TL process.

3.2.   The Randall-Wilkins band model

This model[1a], briefly outlined in ch.2(2.16(iii)), assumes the presence of a single defect (trap) level within the forbidden gap between the valence and conduction bands of the phosphor. When it is assumed that the thermally stimulated electron from the trap has a negligible chance of re-trapping and can go straight to the luminescent centre, it constitutes the basis of the first order kinetics. Figure 3.1 shows the effect following the excitation and subsequent thermal stimulation of a TL phosphor. As the temperature increases at a constant rate, q, the traps lying close to the main trap at depth E below CB become vacated. The peak in the glow curve thus relates to the maximum in the occupied trap density distribution around the main trap of activation energy E. The peak temperature, through its relation to E, is characteristic of the trapping states.

3.3.   Determination of trap depths

Methods for determining trap depths and other TL

parameters have been reviewed recently by Azorin[1b].

From Equation 3.1

$$dn/n = -s \exp(-E/k_B T)\, dt$$
$$= (-s/q) \exp(-E/k_B T)\, dT$$

where $q = dT/dt$. Integration gives

$$\ln(n) = \int -(s/q) \exp(-E/k_B T')\, dT' + \text{constant}$$

When $t = 0$, let $T = T_i$ (some initial temperature) and $n = n_o$ then the

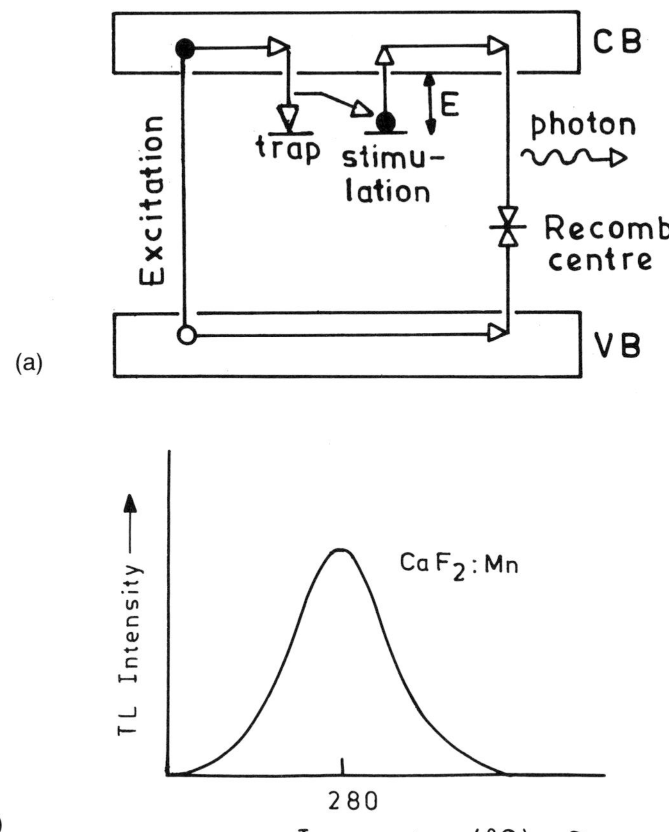

Figure 3.1. (a) Schematic diagram of excitation and thermal stimulation in a TL phosphor. (b) Typical TL glow curve of CaF$_2$:Mn phosphor.

constant is equal to $\ln n_o$

$$\ln(n/n_o) = \int_{T_i}^{T} -(s/q) \exp(-E/k_B T') \, dT'$$

or

$$n = n_o \exp[-\int_{T_i}^{T} (s/q) \exp(-E/k_B T') \, dT'] \tag{3.2}$$

The prime in T' is used merely to distinguish between the variable and the upper limit of the integral. The thermoluminescence intensity I that results when electrons, after having been released from traps by heating, recombine with holes in the recombination centres, is given by the following relation in which c is a constant:

$$I = -c \, (dn/dt)$$

$$= c \, n \, s \, \exp(-E/k_B T)$$

or

$$I = c \, n_o \, s \, \exp(-E/k_B T) \exp[-\int_{T_i}^{T} (s/q) \exp(-E/k_B T') \, dT'] \tag{3.3}$$

For low temperatures $(T \geq T_i)$ the second exponential term in Equation 3.3 approaches unity. Hence, the initial rise of the thermoluminescence intensity may be expressed by

$$I \approx c \, n_o \, s \, \exp(-E/k_B T) \tag{3.4}$$

Equation 3.4 is the basis of the initial rise method for determining E.

At high temperatures, the second exponential term becomes more important and the thermoluminescence decreases. Thus a maximum is observed at a certain temperature T_m. For given values of s and q, $E \propto T_m$. Thus for a maximum in the I, T curve, $dI/dT = 0$ when $T = T_m$. Noting that $(d \ln I/dT) = (1/I)(dI/dT) = 0$ (at $T = T_m$), and setting the logarithmic derivative of I equal to zero at $T = T_m$, Equation 3.3 yields

$$(-s/q) \exp(-E/k_B T_m) + (E/k_B T_m^2) = 0$$

or

$$\frac{qE}{k_B T_m^2} = s \exp(-E/k_B T_m) \tag{3.5}$$

Equation 3.5 gives the value of E from the measured values of T_m and q, provided s is known.

3.4. Second order kinetics

Garlick and Gibson[2] introduced the possibility of second order glow peaks arising from the equal probability of the thermally stimulated electron being either re-trapped or recombined at the luminescent centre (more precisely, the model supposes re-trapping to be predominant compared with recombination; however, it is customary to take re-trapping and recombination as equal). The rate equation, following Equation 2.43, is

$$-dn/dt = s' \, n^2 \exp(-E/k_B T) \tag{3.6}$$

where n is the concentration of trapped electrons per unit volume at time t and s' $(= s/N)$, known as the pre-exponential factor, is a constant with the dimensions of $cm^3.s^{-1}$. If q is the constant heating rate

$$-dn/n^2 = (s'/q) \exp(-E/k_B T) \, dT$$

$$\int_{n_o}^{n} (-dn'/n'^2) = (s'/q) \int_{T_i}^{T} \exp(-E/k_B T') \, dT'$$

$$(1/n) - (1/n_o) = (s'/q) \int_{T_i}^{T} \exp(-E/k_B T')dT'$$

$$n = n_o \left[1 + (s'n_o/q) \int_{T_i}^{T} \exp(-E/k_B T')dT'\right]^{-1} \tag{3.7}$$

Using Equation 3.6,

$$I = - c \, dn/dt = [c \, s' \exp(-E/k_B T)]n_o^2$$
$$\times \left[1 + (s'n_o/q) \int_{T_i}^{T} \exp(-E/k_B T')dT'\right]^{-2} \tag{3.8}$$

3.5. Method of two heating rates

By employing two different warming rates (q_1 and q_2), the need to know the value of s for determining E can be avoided. The maximum for the given glow peak is found to yield two different T_m values, T_{m1} and T_{m2}, corresponding to heating rates q_1 and q_2. It is also seen (Figure 3.2) that if $q_1 > q_2$ then $T_{m1} > T_{m2}$. From Equation 3.5

$$\left.\begin{aligned}
E \exp(E/k_B T_{m1}) &= k_B T_{m1}^2 \, s/q_1 \\
E \exp(E/k_B T_{m2}) &= k_B T_{m2}^2 \, s/q_2
\end{aligned}\right\} \tag{3.9}$$

Equations 3.9 yield

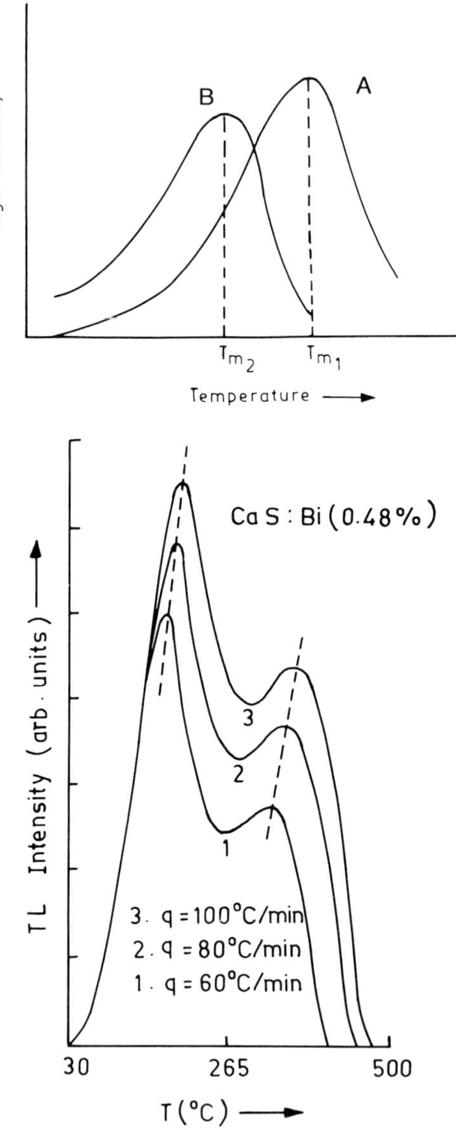

Figure 3.2. (a) Glow curves at two heating rates: A, q_1=2.5°C.s^{-1}; B, q_2=0.5°C.s^{-1}. E=0.67 eV and s=2.9×10^9 s^{-1}. (Reference 1, page 365). (b) Glow-peak shift due to change in heating rate (broken lines) (K. Mahesh and R. Pal, 1985, unpublished).

$$\ln(q_1 T^2_{m2}/q_2 T^2_{m1}) = (E/k_B)\,[(1/T_{m2}) - (1/T_{m1})] \tag{3.10}$$

Mutual subtraction of Equations 3.9 can also give the value of s

$$s = (T^2_{m1} - T^2_{m2})^{-1}\,(E/k_B)[q_1 \exp(-E/k_B T_{m1}) - q_2 \exp(-E/k_B T_{m2})] \tag{3.11}$$

It may be emphasised that since the value of T_m is not very sensitive to changes in q, this method is only approximate, giving a possible error of 20-30%.

3.6. Initial rise method

When sample temperatures are much below the T_m value ($T_i \lesssim T \ll T_m$), Equation 3.4 applies to the variation of intensity in this range of temperature. This procedure for determining the trap depths is known as the initial rise method, proposed by Garlick and Gibson[2]. This method can be applied when the sample has a single glow curve or, if more than one, non-overlapping glow curves. If s is assumed to be independent of temperature, we may write Equation 3.4 as

$$\ln(I) = \ln\,(csn_o) - (E/k_B T) \tag{3.12}$$

A graph of $\ln(I)$ against $(1/T)$ will give a straight line whose slope is equal to E/k_B from which the value of E can be found.

The initial rise method is applicable to first, second, and also the general order cases. This method has also been successfully used to determine activation energy in some other phenomena; for example, the thermally stimulated conductivity (TSC) and thermally stimulated depolarisation current (TSDC).

The initial rise method is based on the fact that the density of unoccupied recombination centres, f, and the density of trapped electrons, h, remain approximately constant as the glow peak begins to rise. Braunlich[3] has shown that the values of E obtained will be smaller than the true ones unless the condition $R[h\,(T_i)/f(T_i)] \ll 1$ is satisfied, where R is the re-trapping factor,

$$R = \frac{\text{probability of re-trapping}}{\text{probability of recombination}}$$

and $h(T_i), f(T_i)$ are h and f values evaluated at the initial temperature T_i, the temperature at which the glow curve begins to rise.

Christodoulides[4] has analysed the initial rise method and has

49

estimated that if the signal levels used are higher than a few per cent of the peak height, the correct value of the activation energy, E_c is given by

$$E_c = (1 + 0.74a_1 + 0.082a_2)E_i - (2a_1 + 0.22a_2)(T_m/11605)$$

where E_i is the activation energy measured by the initial rise method, T_m is the peak-temperature (K) and a_1, a_2 are the fractions of the peak value of the TL intensity observed at temperatures T_1, T_2 (both $<$ T_m).

3.7. Half-width method

The method suggested by Halperin and Braner[5] uses the shape of the glow curves* for determining E. Chen[6,7] has shown empirically from analysis of the experimental data that in the case of first order kinetics (referring to Figure 3.3)

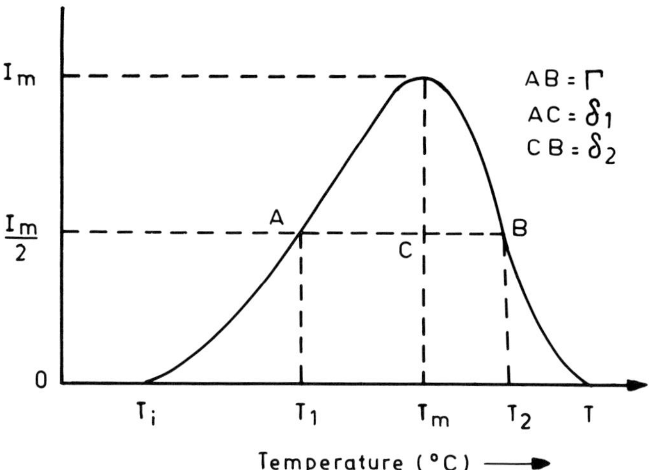

Figure 3.3 A typical glow curve shape. $\Gamma = (T_2 - T_1)$, $\delta_1 = (T_m - T_1)$, $\delta_2 = (T_2 - T_m)$. The ordinate shows TL intensity.

---

*First order and second order expressions for E_1 and E_2 respectively, are $E_1 =$ (1.71) $(k_B T_m^2/\delta_1)$ $(1-1.58\,\Delta_1)$ and $E_2 = 2k_B T_m^2 (1-\Delta_2)(1-2\,\Delta_2)$ $(1/\delta_1)$ where $\Delta_{1,2} = 2k_B T_m/E_{1,2}$. This is an iterative method in which one has to estimate E_1 assuming $\Delta=0$, then calculate Δ and then re-evaluate E_1. Equations 3.13 and 3.14 are convenient and more accurate.

$$E_1 = \frac{(1.52)\, k_B T_m^2}{\delta_1} - (1.58)\,(2k_B T_m) \tag{3.13}$$

and for the second order kinetics

$$E_2 = \frac{(1.81)\, k_B T_m^2}{\delta_1} - 2k_B T_m \tag{3.14}$$

Equations 3.13 and 3.14 are related to the case of temperature independent frequency factor. If s depends upon the temperature as $T^{\,b}$, then Chen has shown[6] that the correct result will be obtained by subtracting $b k_B T_m$ from the values of E obtained from Equations 3.13 and 3.14. The form factor μ_g introduced by Halperin and Braner[5] and defined by $\mu_g = (\delta_2/\Gamma)$ has also been shown by Chen[6] to be equal to 0.42 for first order peaks and 0.52 for the second order case. A slightly different version of Equation 3.13 has been given by Chen[8]

$$E_1 = \frac{(2.52)\, k_B T_m^2}{\delta_1} - 2k_B T_m \tag{3.15}$$

(and a general description of different formulae for E and the frequency factor has been given by Braunlich[9]). We may also use a more recent approach given by Balarin[10] which allows the determination of E with only a negligible error of about 0.5%. The analytical relationship obtained by Balarin assumes a constant rate of heating. The significant feature of these deductions is that they are valid for TL, TSC, TSEE, TSD and other thermal processes for which a plot of the intensity of the effect as a function of temperature is a curve with a maximum. Balarin's equations for E in eV are:

$$E = (T_m^2/4998)\ (1/\Gamma),\ \text{1st order kinetics}$$
$$= (T_m^2/3542)\ (1/\Gamma),\ \text{2nd order kinetics} \tag{3.16}$$
$$= (T_m^2/2872)\ (1/\Gamma),\ \text{3rd order kinetics}$$

where both Γ and T_m are in Kelvin. Half-width parameters can be used differently to assign the order of kinetics. By determining the form factor μ_g, the corresponding order of kinetics can be predicted from the calibration curve given by Chen[11] and shown in Figure 3.4. μ_g depends almost wholly on the order of kinetics, with little dependence on E and s.

It is pertinent to mention here the observation of Stoebe and Watanabe[12] that the half-width method, though easier, is less accurate in determining E and s. The initial rise method is considered to be a most satisfactory experimental method as the initial portion of the glow curve is presumably unaffected by the individual details of the luminescent process. Isothermal decay results (see section 3.18), however, seem to be generally consistent.

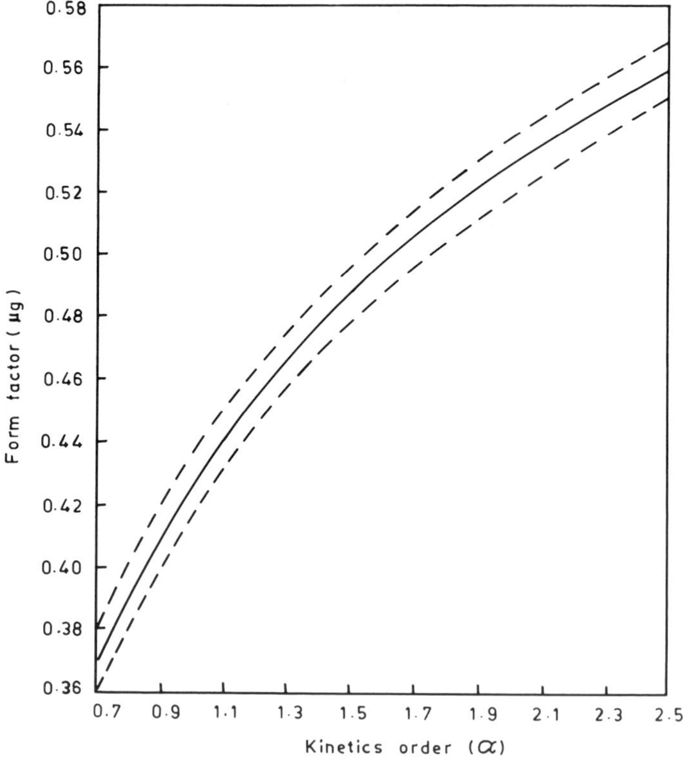

Figure 3.4. Variation of μ_g with order of kinetics α. Solid line gives the average values and dashed lines give the largest variations in α for the given μ_g value for different E and s values[11].

3.8.   Capture cross section of a recombination centre

If u is the thermal velocity of a carrier, ρ the number of recombination centres per cm^3 and σ the capture cross section, then the free carrier lifetime may be expressed as

$$\tau = (\rho\,\sigma\,u)^{-1}$$

where

$$u \simeq (2\,k_\mathrm{B}T/m)^{1/2}$$

where m is the carrier mass.

The charged centres interact with carriers and will attract or repel them depending upon their respective signs. The carrier will be captured by (or will be bound to) the oppositely charged centre within the approach distance r for which the interaction binding energy is more than the thermal energy $k_\mathrm{B}T$ at temperature T of the sample, i.e.

$$e^2/\varepsilon r \gtrsim k_\mathrm{B}T$$

where ε is the dielectric constant of the material. Now

$$\sigma = \pi r^2 \approx \frac{\pi e^4}{(\varepsilon k_\mathrm{B}T)^2} \tag{3.17}$$

and for $\varepsilon = 10$ and $T = 300$ K, $\sigma \approx 10^{-12}$ cm^{2}[13]. If carrier and centre both have the same charge then Coulomb repulsion takes place and this opposes the capture. In this situation the capture cross section is reduced to about 10^{-22} cm^2. An uncharged centre has a cross section of about 10^{-17} cm^2, which may be imagined as the geometric mean of the two values, i.e. 10^{-12} cm^2 and 10^{-22} cm^2.

A recombination centre captures a carrier (hole or electron) that stays with the centre until it captures another carrier of opposite sign whereupon the two recombine at the site of the recombination centre releasing the excitation energy as photons (luminescence), phonons (non-radiative recombination) or through the three-body event in which a third electron or hole takes up the energy. Among the centres distributed within the energy band gap, the ones near the top of the VB will most likely capture holes; those near the bottom of the CB will capture electrons. For centres lying within the central region of the gap the electron and hole capture probabilities are comparable and these are the centres which act as recombination centres. The value of ρ varies from 10^{12} cm^{-3} in the purest and most perfect crystals to 10^{19} cm^{-3} for crystals with a high level of imperfections. Inserting these values of ρ and σ, the value of τ varies from 10^{-14} to 10^3 s, at room temperature.

3.9.   KCl(Tl) phosphor

3.9.1.   *Introduction*
KCl(Tl) is a material widely studied for its absorption and emission characteristics, TL output[14] and, to some extent, its dosimetric response. Its value as a TL material for practical applications is small but the various simple theoretical models fit this phosphor well and thus provide a valuable insight into the phenomena of luminescence. KCl, being an alkali halide, is also better understood because of its cubic crystal structure.

The phosphor is prepared by adding a dilute quantity (e.g. 100 ppm) of Tl in the KCl lattice. Tl sits in the KCl system at some of the vacant K sites, and gives one of its outer electrons to the $3p^5$ orbital of Cl atom.

3.9.2.   *Conduction and valence bands in KCl*
The outer electronic configuration of Cl$^-$ ions is $3p^6$, thus it constitutes the filled band which is the valence band. The K$^+$ ions with $3p^6\,4s^0$ configuration are deficient in electrons and can, therefore, accept them, and these ions form the empty band which

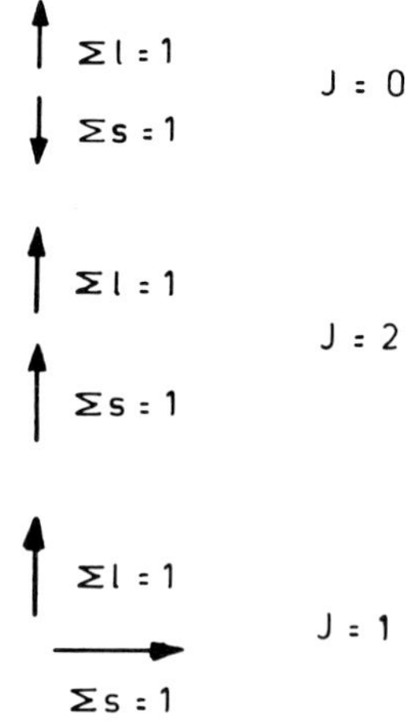

Figure 3.5. Spin orbit coupling.

is the conduction band.

3.9.3. *Level structure of the Tl$^+$ ion in the KCl lattice*

The outer configuration of the Tl atom is $6s^2 6p^1$. The configuration of the Tl$^+$ ion can, therefore, be either $6s^1 6p^1$ or $6s^2$. In the former configuration there is an angular momentum contribution to the energy, due to the presence of $6p^1$ orbital, while this contribution is absent in the latter configuration ($6s^2$). Hence, obviously, $6s^2$ forms the ground state of the Tl$^+$ ion while $6s^1 6p^1$ is the excited state. Since the resultant electronic spin on account of pairing in the $6s^2$ orbital is zero, it is a singlet with designation ^1S_o ($\Sigma l=0$, $\Sigma s=0$; hence $\Sigma j=0$). For the excited state configuration $6s^1 6p^1$, the spins may arrange in a parallel or antiparallel manner due to spin-exchange interaction. The parallel arrangement of spins gives a triplet $3P$ ($\Sigma l=1$, $\Sigma s=1$; for J, see Figure 3.5) whereas the antiparallel spin arrangement gives a singlet ^1P-state ($\Sigma s=0$, $J=\Sigma l=1$). The level scheme of the Tl$^+$ ion in the KCl lattice is shown in Figure 3.6.

3.9.4. *Absorption and emission wavelengths in KCl (Tl)*

In the absorption process the electric dipole transitions $^1S_o \rightarrow {}^3P_o$ ($\triangle J=0$) and $^1S_o \rightarrow {}^3P_2$ ($\triangle j=2$) are forbidden. This means that ^3P_o and $^3P_2$ are electron-trapping states. The other allowed transitions are

$$^1S_o \rightarrow {}^1P_1 \ (1960 \ \text{Å})$$

$$^1S_o \rightarrow {}^3P_2 \ (2100 \ \text{Å})$$

$$^1S_o \rightarrow {}^3P_1 \ (2500 \ \text{Å})$$

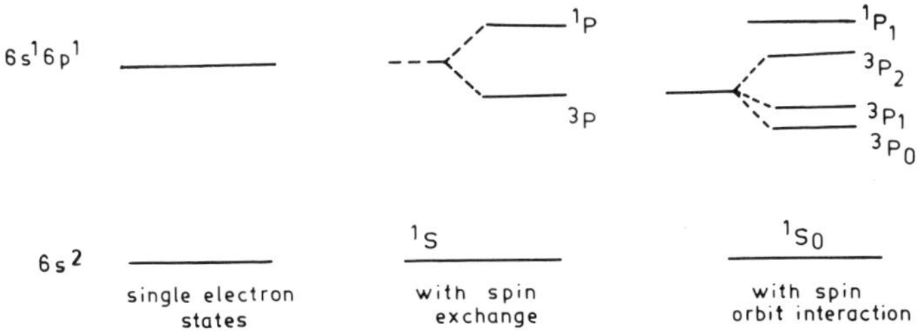

Figure 3.6. Level structure scheme of Tl$^+$ ion.
D. Curie, *Luminescence in crystals*, (Methuen, London) (1963).

(The second of these transitions is a weak electric quadrupole transition which the selection rule $\triangle J = \pm\, 2$ permits.)

The absorption spectrum is depicted in Figure 3.7. The emission spectrum of the phosphor corresponds to the transitions

$$^1P_1 \rightarrow {}^1S_o \; (2470 \text{ Å})$$

$$^3P_1 \rightarrow {}^1S_o \; (3050 \text{ Å})$$

The intensity of the 2470 Å line is very weak in the emission spectrum as this wavelength, being close to 2500 Å, is strongly re-absorbed by the $^1S_o \rightarrow {}^3P_1$ transition, which then appears as a strong emission line.

3.9.5.   TL of KCl(Tl) phosphor

The states $^3P_1$ and $^1P_1$ serve as the absorption and emission bands, as shown above, while the states ^3P_o and $^3P_2$ are metastable and act as trapping levels. For low Tl^+ ion concentrations (0.05%) we observe two peaks at 205K and 300K respectively as shown in Figure 3.8. The activation energy E for the second peak ($T_m = 300$K) is found to be 0.67 eV ($s = 2.9 \times 10^9 \, \text{s}^{-1}$) while for the first peak ($T_m = 205$K), E is 0.4 eV. It is also observed that when the Tl^+ ion concentration

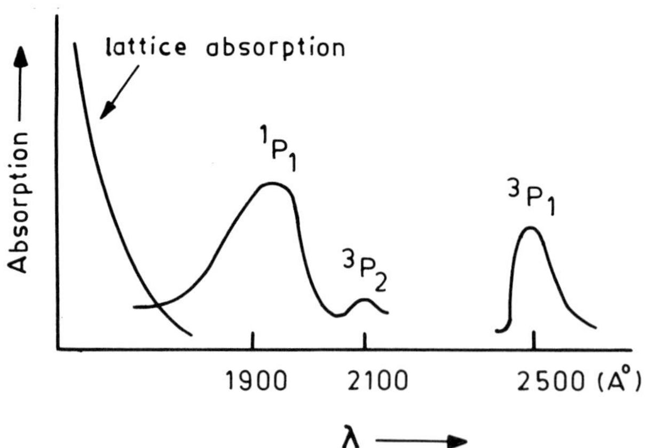

Figure 3.7. Absorption spectrum of KCl(Tl) phosphor.
F. Seitz, Trans. Faraday Soc. **35**, 79 (1939).

increases, additional peaks also appear. For the KCl(Tl) system the analytical formula for E (in eV) due to Urbach[14] is given as:

$$E = T_m/500 \tag{3.18}$$

where T_m is in K.

3.10. General order kinetics

3.10.1. *Constant s'*

The general order kinetics equation is expressed as

$$-dn/dt = p' \, n^\alpha \exp(-E/k_B T) \tag{3.19}$$

where α is the order of kinetics and p' is a pre-exponential factor, except for $\alpha = 1$ in which case $p' \equiv s$, is the frequency factor. (Note that earlier we had used s' for what we are using here as p'. In what follows, s' will be used differently – paucity of symbols!). The value of α is essentially a measure of the affinity of the released carrier to its trap. $\alpha = 1$, which defines the first order kinetics, gives, for example, the fact that the released carrier does not return to the trap. $\alpha = 2$ (second order) gives an equal probability for the released carrier (electron) to get trapped again or to recombine with a hole in the recombination centre during its transition from the CB. The physical and chemical characteristics of TL phosphors actually determine α.

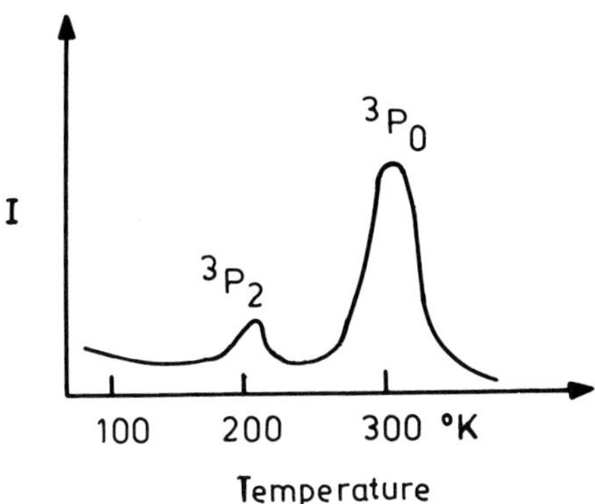

Figure 3.8. TL curve for KCl(Tl) (Tl=0.05% and q=0.83 K.s^{-1}). P. D. Johnson and F. E. Williams, J. Chem. Phys. **21**, 125 (1953).

From Equation 3.19

$$\int_{n_o}^{n} (n')^{-\alpha}\, dn' = - \int p'\, \exp(-E/k_B T)\, dt$$

Using $q = dT/dt$ and integrating the left hand side,

$$n^{(1-\alpha)} = n_o^{(1-\alpha)} \left[1 + \frac{p'(\alpha-1)(n_o)^{\alpha-1}}{q} \int_{T_i}^{T} \exp(-E/k_B T')\,dT'\right]$$

or

$$n = n_o \left[1 + \frac{s'\,(\alpha-1)}{q} \int_{T_i}^{T} \exp(-E/k_B T')\, dT'\right]^{1/(1-\alpha)}$$

where $s' = p'\, n_o^{(\alpha-1)}$.

The intensity $I(T)$ of TL emission is given by Equation 3.19 as

$$I(T) = -c\,(dn/dt) = cp'\, n^{\alpha} \exp(-E/k_B T)$$

$$= c\,p'\, \exp(-E/k_B T)\, n_o^{\alpha} \left[1 + \frac{s'\,(\alpha-1)}{q} \int_{T_i}^{T} \exp(-E/k_B T')\,dT'\right]^{\alpha/(1-\alpha)}$$

and since $p'\, n_o^{\alpha} = s'n_o$

$$I(T) = c\, s'\, n_o \exp(-E/k_B T) \left[1 + \frac{s'\,(\alpha-1)}{q} \right.$$
$$\left. \times \int_{T_i}^{T} \exp(-E/k_B T')\,dT'\right]^{\alpha/(1-\alpha)} \qquad (3.20)$$

It is seen from Equation 3.20 that $I(T)$ contains two factors. The first, the exponential factor, increases with temperature and the second, inside the brackets, decreases as T increases. The interplay of these two factors gives rise to the bell shape of the glow curve $I(T)$ that is observed experimentally.

3.10.2.   *Variable s′*

In certain cases the factor s' is found to be temperature dependent[5,6,15]. Let us assume that s' depends upon T according to the following relation

$$s' = s'_o T^b \qquad (3.21)$$

where b has a value between -2 and $+2$[16] (these values result from different dependencies of recombination cross section on temperature). Considering, for simplicity, the case of first order kinetics, $\alpha = 1$, we may write for the intensity $I(T)$, using Equation 3.3 in which $s' = p' \equiv s$ (the frequency factor), and putting $s = s_0 T^b$

$$I(T) = c\, n_\mathrm{o}\, s_\mathrm{o}\, T^b \exp(-E/k_\mathrm{B}T)\, \exp\!\left[-\int_{T_i}^{T} (s_\mathrm{o} T'^{\,b}/q)\exp(-E/k_\mathrm{B}T')\,\mathrm{d}T'\right]$$

or $\qquad\qquad\qquad\qquad\qquad\qquad\qquad\qquad\qquad\qquad\qquad$ (3.22)

$$\ln I(T) = \ln(c n_\mathrm{o} s_\mathrm{o}) + b\ln T - (E/k_\mathrm{B}T) - \int_{T_i}^{T} (s_\mathrm{o} T'^{\,b}/q)\exp(-E/k_\mathrm{B}T')\,\mathrm{d}T'$$

The condition of maximum TL emission is given by

$$[\mathrm{d}(\ln I(T))/\mathrm{d}T]_{T=T_\mathrm{m}} = 0$$

The resulting equation that can be solved either numerically or graphically is

$$\exp(-E/k_\mathrm{B}T_\mathrm{m}) = (qb/s_\mathrm{o})T_\mathrm{m}^{-(b+1)} + (qE/s_\mathrm{o}k_\mathrm{B})\,T_\mathrm{m}^{-(b+2)}$$

or

$$qT_\mathrm{m}^{-(b+2)} = s_\mathrm{o}\left[(E/k_\mathrm{B}) + bT_\mathrm{m}\right]^{-1}\exp(-E/k_\mathrm{B}T_\mathrm{m})$$

or, putting $\triangle_\mathrm{m} = 2k_\mathrm{B}T_\mathrm{m}/E$ $\qquad\qquad\qquad\qquad\qquad\qquad$ (3.23)

$$q/T_\mathrm{m}^{b+2} = \left[k_\mathrm{B}s_\mathrm{o}/E\left(1 + \frac{b\triangle_\mathrm{m}}{2}\right)\right]\exp(-E/k_\mathrm{B}T_\mathrm{m}) \qquad\qquad (3.24)$$

As will be explained later, $\triangle_\mathrm{m}$ is very important for obtaining approximate solutions of $\int\exp(-E/k_\mathrm{B}T)\mathrm{d}T$. Since $(b/2)$ is not greater than 1, the quantity $(b\triangle_\mathrm{m}/2)$ is very small and can be regarded as a constant without introducing any appreciable error. A procedure for an experimental use of Equation 3.24 for determining E has been proposed by Schön[17]. When two heating rates q_1 and q_2 are employed and a special case of $b = 1.5$, as considered by Schön, his equation for E is obtained from Equation 3.24:

$$E = \left[k_\mathrm{B}\, T_\mathrm{m1}\, T_\mathrm{m2}/(T_\mathrm{m1}-T_\mathrm{m2})\right]\ln\left(q_1 T_\mathrm{m2}^{7/2}/q_2 T_\mathrm{m1}^{7/2}\right) \qquad (3.25)$$

The derivation of Equation 3.25 will be taken up later in this chapter.

3.11.   Condition of maximum emission for general order kinetics

3.11.1.   *Constant s′*

For the maximum in the $I(T)$ curve in the case of general order kinetics ($\alpha \neq 1$), we take Equation 3.20 and equate its logarithmic derivative to zero:

$$\left[\frac{d\,(ln\,I(T)}{dT}\right]_{T=T_m} = 0$$

$$= (E/k_B T_m^2) - \left(\frac{\alpha}{\alpha-1}\right)\left[1 + \frac{s'\,(\alpha-1)}{q}\int_{T_i}^{T}\exp(-E/k_B T')dT'\right]_{T=T_m}^{-1}$$

$$\times\left(\frac{s'\,(\alpha-1)}{q}\exp(-E/k_B T_m)\right)$$

This gives

$$\left(\frac{s'\,\alpha k_B T_m^2}{qE}\right)\exp(-E/k_B T_m) = 1 + \frac{(\alpha-1)\,s'}{q}\int_{T_i}^{T_m}\exp(-E/k_B T)\,dT$$

$$(3.26)$$

Equation 3.26 has to be solved experimentally by the 'two heating rates' method.

3.11.2. *Variable s'*

Let $s' = s'_o T^b$ so that Equation 3.20 will be modified to the following form

$$I(T)=c\,s'_o\,n_o T^b \exp(-E/k_B T)\left[1+\frac{s'_o\,(\alpha-1)}{q}\int_{T_i}^{T} T'^b \exp(-E/k_B T')\,dT'\right]$$

$$(3.27)$$

Maximising $I(T)$ as in section 3.11.1 above,

$$(s'_o\alpha T_m^b/q)\left[\exp\left(\frac{-E}{k_B T_m}\right)\right]\left(\frac{b}{T_m} + \frac{E}{k_B T_m^2}\right)^{-1}$$

$$= \left(1 + \frac{s'_o\,(\alpha-1)}{q}\int_{T_i}^{T_m} T^b \exp(-E/k_B T)dT\right)$$

$$(3.28)$$

As expected, when $b = 0$, Equation 3.28 reverts to Equation 3.26.

3.12. **Physical interpretation of TL kinetics**

3.12.1. *Kinetics of the first order*

A satisfactory physical intrepetation of the first order kinetics is possible if we assume that the system of TL centres in the forbidden gap is completely known, in terms of energy levels, population size, and transition probabilities (see Figure 3.9). For this purpose, the following parameters are introduced

α = probability for thermal excitation of trapped electrons into CB

β = capture coefficient for conduction electrons into traps

β^* = capture coefficient for holes into activator levels

γ = capture coefficient for conduction electrons into activator levels

γ^* = capture coefficient for holes from the valence band into traps

δ = probability for thermal excitation of holes into VB

n = concentration of conduction electrons in CB

p = concentration of holes in VB

h = concentration of trapped electrons

f = concentration of trapped holes

H = concentration of traps

A = concentration of recombination centres (activators)

q = dT/dt (rate of heating)

The population of free and trapped electrons, and free and trapped holes, can be described by four differential equations, assuming that there exists an absolute symmetry between the action of activators and VB on holes, on the one hand, and the action of traps and CB on electrons on the other. The first equation is

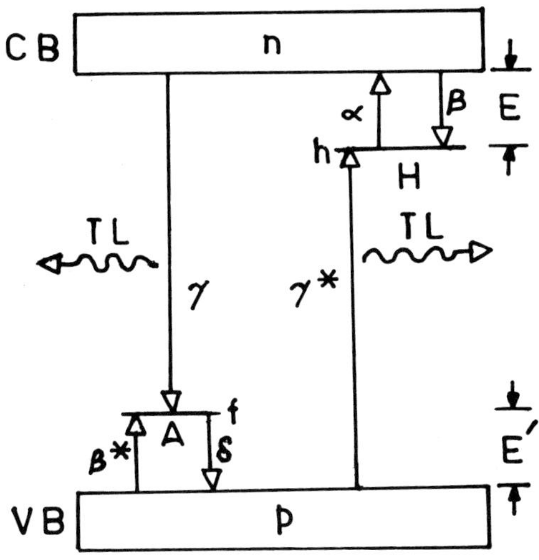

Figure 3.9. Band model of TL and TSC processes.

$$\mathrm{d}n/\mathrm{d}t = \alpha h - \beta n \, (H{-}h) - \gamma \, nf \tag{3.29a}$$

which results from the following balance: (i) the input rate of conduction electrons in CB is proportional to the concentration of trapped electrons and probability of thermal excitation, (ii) the re-trapping rate is proportional to the number of free traps and the capture coefficient β, and (iii) the radiative recombination is proportional to the concentration of conduction electrons and empty activators. From similar balances we get the remaining three equations:

$$\mathrm{d}h/\mathrm{d}t = -\alpha h + \beta n \, (H{-}h) - \gamma^* \, p \, h \tag{3.29b}$$

$$\mathrm{d}p/\mathrm{d}t = \delta f - \beta^* \, p \, (A{-}f) - \gamma^* \, p \, h \tag{3.29c}$$

$$\mathrm{d}f/\mathrm{d}t = -\delta f + \beta^* \, p \, (A{-}f) - \gamma \, n \, f \tag{3.29d}$$

As the crystal is electrically neutral, the concentration of free and trapped holes must be equal to the concentration of electrons in CB and in traps; i.e.

$$p + f = n + h \tag{3.30}$$

In the case of first order kinetics, the re-trapping rate for both electrons and holes is so small that in the former equations, the terms with β or β^* can be neglected. In addition, it can be realistically assumed that $\mathrm{d}n/\mathrm{d}t = 0$, i.e. there is no net increase in the CB electron concentration as those excited into the conduction band recombine almost instantaneously with holes in activation centres. Hence the set of Equations 3.29 becomes

$$\mathrm{d}n/\mathrm{d}t = \alpha h - \gamma nf \tag{3.31a}$$

$$\mathrm{d}h/\mathrm{d}t = -\alpha h - \gamma^* \, ph \tag{3.31b}$$

$$\mathrm{d}p/\mathrm{d}t = \delta f - \gamma^* \, ph \tag{3.31c}$$

$$\mathrm{d}f/\mathrm{d}t = -\delta f - \gamma \, nf \tag{3.31d}$$

Also, as most of the electrons and holes are trapped, i.e.

$$n \ll h \text{ and } p \ll f$$

we find that Equation 3.30 gives $f \approx h$. Thus, from Equations 3.31a and c

$$n \approx \alpha/\gamma$$
$$p \approx \delta/\gamma^* \tag{3.32}$$

Equation 3.31b, using Equations 3.32, reduces to

$$\mathrm{d}h/\mathrm{d}t = -\alpha h - \gamma^* ph \approx -(\alpha + \delta) h \qquad (3.33)$$

showing that $\mathrm{d}h/\mathrm{d}t$ depends on h according to the first order process. The solution of Equation 3.33 leads to a generalised form of the Randall-Wilkins relation, i.e. taking into account a constant heating rate q $(= \mathrm{d}T/\mathrm{d}t)$,

$$\mathrm{d}h/h = -\frac{(\alpha + \delta)}{q} \, \mathrm{d}T \qquad (3.34)$$

Applying Boltzmann's statistics to α and δ, the final result is obtained

$$h(T) = h(T_i) \exp\left\{-\frac{1}{q} \int_{T_i}^{T} [\alpha_o \exp(-E/k_{\mathrm{B}}T') + \delta_o \exp(-E'/k_{\mathrm{B}}T')]\mathrm{d}T'\right\} \qquad (3.35)$$

where E is the electron trap activation energy (Figure 3.9) and E' is the hole trap activation energy (distance between activator levels and the top of VB).

Similar considerations (negligible retrapping) lead to the relation describing the intensity of TL emission which, in its general form, is given by

$$I = (-\mathrm{d}n/\mathrm{d}t) + (-\mathrm{d}h/\mathrm{d}t) = \gamma nf + \gamma^* h p \qquad (3.36)$$

or, using the assumption defined in Equation 3.32

$$I \approx \alpha f + \delta h$$

and using $f \approx h$ from Equation 3.30 with the assumption of $n \ll h$ and $p \ll f$, we get

$$I \approx (\alpha + \delta) h$$

or, using Equation 3.35,

$$I \approx h(T_i) [\alpha_o \exp(-E/k_{\mathrm{B}}T) + \delta_o \exp(-E'/k_{\mathrm{B}}T)]$$

$$\times \exp\left[-\frac{1}{q} \int_{T_i}^{T} (\alpha_o \exp(-E/k_{\mathrm{B}}T') + \delta_o \exp(-E'/k_{\mathrm{B}}T'))\mathrm{d}T'\right] \qquad (3.37)$$

By neglecting transitions from the forbidden band to VB, i.e.

assuming $\delta = 0$, the Randall-Wilkins relation is obtained

$$I(T) \approx h(T_i) \, \alpha_o \, \exp(-E/k_B T) \, \exp\left(\frac{1}{q} \int_{T_i}^{T} \alpha_o \exp(-E/k_B T') \mathrm{d}T'\right)$$

(3.38)

3.12.2. *Kinetics of the second order*

For physical interpretation of second order kinetics, we must start again from the general case and assume that re-trapping of charge carriers prevails over the recombination transitions, i.e. β^*, $\beta \gg \gamma^*$, γ. Equations 3.29 will then become

$$\mathrm{d}n/\mathrm{d}t = \alpha h - \beta \, n \, (H-h) \tag{3.39a}$$

$$\mathrm{d}h/\mathrm{d}t = -\alpha h + \beta n \, (H-h) - \gamma^* \, ph \tag{3.39b}$$

$$\mathrm{d}p/\mathrm{d}t = \delta f - \beta^* \, p \, (A-f) \tag{3.39c}$$

$$\mathrm{d}f/\mathrm{d}t = -\delta f + \beta^* \, p \, (A-f) - \gamma nf \tag{3.39d}$$

Further, as we had assumed that concentrations n and p correspond to the quasi-equilibrium condition, i.e. $\mathrm{d}n/\mathrm{d}t = \mathrm{d}p/\mathrm{d}t \approx 0$, we have, from Equation 3.30, $\mathrm{d}f/\mathrm{d}t \approx \mathrm{d}h/\mathrm{d}t$, and from Equations 3.29 we get

$$\mathrm{d}n/\mathrm{d}t + \mathrm{d}h/\mathrm{d}t \approx \mathrm{d}h/\mathrm{d}t = -\gamma \, nf - \gamma^* \, ph \tag{3.40}$$

or, using $f \approx h$,

$$\mathrm{d}h/\mathrm{d}t \approx -(n\gamma + \gamma^* p) \, h \tag{3.41}$$

also, Equation 3.39a gives

$$\mathrm{d}n/\mathrm{d}t \approx 0 = \alpha h - \beta n \, (H-h) \tag{3.42}$$

In experimental situations, the actual initial conditions lead to a further realistic approximation, namely $h(t=0) \ll H$, so that relation 3.42 becomes

$$n \approx \alpha h/\beta H \tag{3.43}$$

Similar considerations apply to p, with $f \ll A$, we have

$$p \approx \delta f/\beta^* A \tag{3.44}$$

Substituting relations 3.43 and 3.44 into Equation 3.41, we obtain the final relation

$$\mathrm{d}h/\mathrm{d}t \approx -[(\gamma\alpha/\beta H) + (\gamma^* \delta/\beta^* A)] \, h^2 \tag{3.45}$$

from which, applying Boltzmann's statistics to α and δ, we get

$$(dh/h^2) = -(1/q)\,[(\gamma/\beta H)\,\alpha_o\,\exp(-E/k_B T)$$
$$+ (\gamma^*\delta_o/\beta^* A)\,\exp(-E'/k_B T)]\,dT \qquad (3.46)$$

By integration we get

$$h = 1\Big/\Big[\frac{1}{h_o} + \frac{1}{q}\int_{T_i}^{T}\Big(\frac{\gamma\alpha_o}{\beta H}\exp(-E/k_B T') + \frac{\gamma^*\delta_o}{\beta^* A}\exp(-E'/k_B T')\Big)dT'\Big]$$
$$(3.47)$$

where h_o is the value of h at time $t = 0$ (i.e. when $T = T_i$). In conclusion, using Equations 3.47 and 3.45, the TL intensity is obtained as (taking $c = 1$)

$$I(T) = -\frac{dh}{dt} = \Big(\frac{\gamma}{\beta H}\alpha_o\exp(-E/k_B T) + \frac{\gamma^*\delta_o}{\beta^* A}\exp(-E'/k_B T)\Big)$$

$$\times \Big[\frac{1}{h_o} + \frac{1}{q}\int_{T_i}^{T}\Big(\frac{\gamma\alpha_o}{\beta H}\exp(-E/k_B T') + \frac{\gamma^*\delta_o}{\beta^* A}\exp(-E'/k_B T')\Big)dT'\Big]^{-2}$$
$$(3.48)$$

It must be noted that, by neglecting the probability for transitions from activators into VB, i.e. by supposing $\delta_o = 0$, relation 3.48 becomes Garlick and Gibson's equation[2] for second order kinetics.

3.13. Experimental situations

In the previous sections we have seen that first order kinetics correspond to situations in which the probability for electron-hole recombinations is very large and, consequently, the probability for re-trapping is negligible. In the case of second order kinetics there is, on the contrary, a very large re-trapping factor. Of course, these are limiting situations obtained from radical assumptions. It must be taken into account that most often TL experimental results indicate intermediate kinetics processes of order $1 < r < 2$, or even $0 < r < 1$ (in these sections, since α denotes probability of thermal excitation, we use r to denote order of kinetics.) Nevertheless, by adding to the limit situations a further factor independent of h, i.e. corresponding to kinetics of order zero, intermediate situations can be directly interpreted.

Let us consider the case $1 < r < 2$; it can be treated as one resulting from superimposition of first and second order kinetics, and explained by assuming that capture coefficients for traps are prevailing over the capture coefficients for activators for both CB electrons and VB holes ($\beta \gg \gamma$, $\beta^* \ll \gamma^*$), or vice versa. Therefore, Equation 3.29 reduces to

$$\mathrm{d}n/\mathrm{d}t = \alpha h - n\beta\,(H{-}h) \qquad (3.49\mathrm{a})$$

$$\mathrm{d}h/\mathrm{d}t = -\alpha h + n\beta\,(H{-}h) - \gamma^* ph \qquad (3.49\mathrm{b})$$

$$\mathrm{d}p/\mathrm{d}t = \delta f - \gamma^* ph \qquad (3.49\mathrm{c})$$

$$\mathrm{d}f/\mathrm{d}t = -\delta f + \beta^*\, p(A{-}f) - \gamma\, nf \qquad (3.49\mathrm{d})$$

Assuming that: (i) $n(t)$ is in quasi-equilibrium conditions; (ii) $H \gg h$; and (iii) n and p are very small, i.e. $f \approx h$; the following relations are valid:

$$\alpha h - n\beta\,(H{-}h) \approx 0 \qquad (3.50)$$

or

$$n \approx \alpha h/\beta H \qquad (3.51)$$

$$\delta f - \gamma^*\, ph \approx 0 \qquad (3.52)$$

or

$$p \approx \delta/\gamma^* \qquad (3.53)$$

Equations 3.49 then yield

$$\mathrm{d}h/\mathrm{d}t = -\alpha h + n\beta\,(H{-}h) - \gamma^*\, ph = -(\mathrm{d}n/\mathrm{d}t) - \gamma^*\,(\delta/\gamma^*)\, h = -\delta h \qquad (3.54)$$

and a relation analogous to that of first order kinetics is obtained:

$$h = h_{\mathrm{o}} \exp\!\left[-(1/q) \int_{T_{\mathrm{i}}}^{T} \delta_{\mathrm{o}} \exp(-E'/k_{\mathrm{B}} T')\, \mathrm{d}T'\right] \qquad (3.55)$$

The TL emission intensity is given by

$$I = \gamma\, nf + \gamma^*\, ph \qquad (3.56)$$

whence, from relations 3.51 and 3.52, along with the assumption (iii) above

$$I \approx \gamma h\,(\alpha h/\beta H) + \delta h \approx (\alpha\gamma h^2/\beta H) + \delta h \qquad (3.57)$$

using relation 3.55

$$I \approx [h_o^2(\gamma\alpha_o/\beta H)\,\exp(-E/k_B T]$$

$$\times \exp[-\frac{2}{q}\int_{T_i}^{T}\delta_o\,\exp(-E'/k_B T')\,dT'] + h\delta_o\,\exp(-E'/k_B T) \qquad (3.58)$$

When describing an intermediate situation of the type $1 < r < 2$ by means of relation 3.57, we see that it is formally possible, with proper re-definitions of coefficients, to find an equivalent single-term expression proportional to h^r for a given T:

$$(\alpha(T)\,\gamma h^2/\beta H) + \delta\,(T)\,h = A(T)\,h^r$$

If all coefficients are known and r is experimentally assessed, $A(T)$ can be obtained and a qualitative and quantitative description of the process completed.

3.14. General features of the two heating rates method

3.14.1. *First order kinetics*

With the assumption of first order kinetics and a constant frequency factor, the activation energy E can be determined experimentally by using Equation 3.10. However, if the frequency factor s depends upon the temperature according to Equation 3.21

$$s = s_o\,T^b$$

then E can be evaluated from Equation 3.24 by means of Schön's method[17] in which the corresponding peak shift of the glow curve is observed by employing two heating rates q_1 and q_2. Schön took $b = 1.5$. Equation 3.24 then yields

$$(q_1/T_{m1}^{7/2}) = \left[k_B s_o/E\left(1 + \frac{3\triangle_{m1}}{4}\right)\right]\exp(-E/k_B T_{m1})$$

and

$$(q_2/T_{m2}^{7/2}) = \left[k_B s_o/E\left(1 + \frac{3\triangle_{m2}}{4}\right)\right]\exp(-E/k_B T_{m2})$$

By dividing these two equations we get,

$$(q_1/q_2)\,(T_{m2}/T_{m1})^{7/2} = \left(1 + \frac{3\triangle_{m2}}{4}\right)\left(1 + \frac{3\triangle_{m1}}{4}\right)^{-1}$$

$$\times \left[\exp(-E/k_B)\left(\frac{1}{T_{m1}} - \frac{1}{T_{m2}}\right)\right]$$

As mentioned earlier, $\triangle_m$ is a small quantity which can be disregarded in comparison with unity. We then find

$$E = [k_B\,T_{m1}\,T_{m2}/(T_{m1}-T_{m2})]\,\ln\,[(q_1/q_2)\,(T_{m2}/T_{m1})^{7/2}] \qquad (3.60)$$

which is Schön's equation for determining E.

In the general case, for any value of b,

$$E = [k_B\,T_{m1}\,T_{m2}/(T_{m1}-T_{m2})]\,\ln\,[(q_1/q_2)\,(T_{m2}/T_{m1})^{b+2}] \qquad (3.61)$$

Note that if E is already known by other methods, Equation 3.61 can be used to evaluate b and hence s_o from Equation 3.24. Thus a complete knowledge of the trap level under investigation is possible.

3.14.2.   *Kinetics of any order, constant s'*

In the general case of r order kinetics, Equation 3.26 can be used for the two heating rates method. Thus

$$s'k_B r\,T_{m1}^2/q_1 E)\,\exp(-E/k_B T_{m1})$$
$$= 1 + [s'(r-1)/q_1]\int_{T_i}^{T_{m1}} \exp(-E/k_B T)\mathrm{d}T$$

and

$$(s'k_B r\,T_{m2}^2/q_2 E)\,\exp(-E/k_B T_{m2})$$
$$= 1 + [s'(r-1)/q_2]\int_{T_i}^{T_{m2}} \exp(-E/k_B T)\mathrm{d}T \qquad (3.62)$$

By subtraction we get

$$[rk_B/E(r-1)]\,[(T_{m1}^2/q_1)\,\exp(-E/k_B T_{m1}) - (T_{m2}^2/q_2)\,\exp(-E/k_B T_{m2})]$$

$$=\left\{\left[(1/q_1)\int_{T_i}^{T_{m1}} \exp(-E/k_B T)\,\mathrm{d}T\right]\right.$$
$$\left. - \left[(1/q_2)\int_{T_i}^{T_{m2}} \exp(E/k_B T)\,\mathrm{d}T\right]\right\} \qquad (3.63)$$

The solution of Equation 3.63 is not easy. It can, however, be greatly simplified if we know the approximate analytical form of the integral. This is given as (Ref. 24, pp. 12 and 322)

$$\int_{T_i}^{T} \exp(-E/k_B T')\,\mathrm{d}T' \approx (k_B T^2/E)\,(1-\triangle)\,\exp(-E/k_B T) \qquad (3.64)$$

where $\triangle = 2k_B T/E <1$ (as $E \approx 1$ eV and $k_B T \lesssim 0.05$ eV) so that $\triangle^2$ and $\triangle^2/2$ can be neglected in comparison with 1. Substituting

Equation 3.64 in Equation 3.63 we get

$$\frac{T_{m1}^2}{q_1}\left(\frac{1}{r-1} + \triangle_{m1}\right) \exp(-E/k_B T_{m1})$$

$$= \frac{T_{m2}^2}{q_2}\left(\frac{1}{r-1} + \triangle_{m2}\right) \exp(-E/k_B T_{m2}) \tag{3.65}$$

From Equation 3.65 E can be found.

As a crude approximation, in those samples in which glow peak temperatures are not very high, $\triangle_{m1}$ and $\triangle_{m2}$ may also be disregarded in comparison with $1/(r-1)$ whose value is typically $\gtrsim 1$, and Equation 3.65 transforms to a simpler form that can be used to find E independently of the order of kinetics, i.e.

$$E \approx \frac{k_B T_{m1} T_{m2}}{(T_{m2} - T_{m1})} \ln (q_2 T_{m1}^2/q_1 T_{m2}^2) \tag{3.66}$$

3.14.3. *Kinetics of any order, variable s′*

We can determine E in this case by using Equation 3.28 and specialising it for two heating rates, i.e.

$$(s_o' r k_B T_{m1}^{b+2}/q_1)(k_B b T_{m1} + E) \exp(-E/k_B T_{m1})$$
$$= 1 + [s_o'(r-1)/q_1] \int_{T_i}^{T_{m1}} T^b \exp(-E/k_B T) \, dT$$

and $\hspace{9cm}$ (3.67)

$$(s_o' r k_B T_{m2}^{b+2}/q_2)(k_B b T_{m2} + E) \exp(-E/k_B T_{m2})$$
$$= 1 + [s_o'(r-1)/q_2] \int_{T_i}^{T_{m2}} T^b \exp(-E/k_B T) \, dT$$

where $s' = s_o' T^b$ has been used. Subtraction of one from the other in Equations 3.67 eliminates s_o' and we get

$$rk_B\{[T_{m1}^{b+2}/q_1 (k_B T_{m1} b + E)] \exp(-E/k_B T_{m1})$$
$$- [T_{m2}^{b+2}/q_2(k_B T_{m2}b+E] \exp(-E/k_B T_{m2})\}$$
$$= (r-1) \{[(1/q_1) \int_{T_i}^{T_{m1}} T^b \exp(-E/k_B T) \, dT]$$
$$- (1/q_2) \int_{T_i}^{T_{m2}} T^b \exp(-E/k_B T) \, dT]\} \tag{3.68}$$

For this we use

$$\frac{\partial}{\partial T}\,[(k_B T^n/E)\,(1-n\,k_B T/E)\,\exp(-E/k_B T)] = T^b\,\exp(-E/k_B T)$$
(3.69)

where $n = b + 2$. (The validity of Equation 3.69 can be checked by differentation, in which $\triangle^2$ and $\triangle^2/2$ are ignored in comparison with unity.) Thus

$$\int T^b\,\exp(-E/k_B T)\,dT = (k_B/E)\,T^{b+2}\,[1-(b+2)\,k_B T/E]\,\exp(-E/k_B T)$$
(3.70)

Note that for $b = 0$, Equation 3.70 transforms to Equation 3.64. Substitution of Equation 3.70 in Equation 3.68 gives an expression from which E can be found.

In all the cases that we have considered so far, the method of evaluation of the activation energy consists of the following steps:

(i) Assumption of a suitable equation for E, with r and b as known parameters. This assumption is rather critical since the equations used for evaluation of E depend upon r and b.

(ii) Determination of the peak current I_m and the corresponding peak temperature T_m.

(iii) Observation of the shift in T_m and I_m for a known change in q.

(iv) Evaluation of E.

(v) Evaluation of s by substitution of E in the equation for $I_m(E)$.

3.15. Variable heating rate method

3.15.1. *Constant s'*

A different experimental method for the evaluation of E, and possibly r, is also used[18]. This is based on observation of I_m and T_m by employing a changing but linear heating rate. Let us assume $b = 0$, giving the condition of constant s'. If we maximise Equation 3.20 by letting $[d(\ln I)/dT]_{T=T_m} = 0$, we get (replacing α in Equation 3.26 by r)

$$(s'\,k_B r\,T_m^2/qE)\,\exp(-E/k_B T_m) = 1 + [s'\,(r-1)/q]\int_{T_i}^{T_m}\exp(-E/k_B T)\,dT$$
(3.71)

Substituting in Equation 3.20

$$I_m = c\,s'n_o\,\exp(-E/k_B T_m)\,[(s'\,k_B\,r\,T_m^2/qE)\,\exp(-E/k_B T_m)]^{r/(1-r)}$$
(3.72)

Equation 3.72 can be written alternatively as

$$\ln\left[I_m^{r-1}\,(T_m^2/q)^r\right] = r\ln E + \ln\left[(cs'n_o)^{r-1}/(s'k_B r)^r\right] + (E/k_B T_m) \tag{3.73}$$

A plot of the term on the left hand side of Equation 3.73 against $(1/T_m)$ will be a straight line. The slope of the line will give E/k_B. The value of r has to be parameterised for a best fit with the experimental data, (I_m, T_m), obtained at various q values.

3.15.2. *Variable s'*

Using Equations 3.27 and 3.28 which utilise the dependence of s' on $T(s' = s'_o T^b)$ and proceeding in the same manner as outlined in section 3.15.1 we get

$$I_m = c\,s'_o n_o\,T^b\left[\exp(-E/k_B T_m)\right]$$

$$\{s'_o k_B\,r\,T_m^{b+2} \times \left[\exp(-E/k_B T_m)\right]/q(b\,k_B T_m + E)\}^{r/(1-r)}$$

Equation 3.74 can alternatively be expressed as

$$\ln\left[I_m^{r-1}\,T_m^{(b+2)r}/q^r\right] = (E/k_B T_m) + \text{other terms} \tag{3.75}$$

from which E can be determined in the usual manner. Sufficient care should be exercised in choosing the gap between different q values. In any case it should be well above the instrumental error in the value of q as specified by the manufacturer of the temperature programmer.

3.16. **Analysis of the initial rise method**

Evaluation of E in this method is based on the use of the low temperature portion of the glow curve and it is assumed that the concentration of trapped electrons is constant in this region. Furthermore, the low temperature portion is taken to be short enough to regard the lower and upper limits of the temperature, T_i and T in the integral to be nearly the same in Equation 3.3. The second factor in Equation 3.3, therefore, becomes unity, and the TL intensity is adequately described by Equation 3.4. We may rewrite Equation 3.3 as

$$I(T) = c\,s\,n_o\,Z\,\exp(-E/k_B T) \tag{3.76}$$

where

$$Z = \exp\left[-(s/q)\int_{T_i}^{T}\exp(-E/k_B T')\,dT'\right] \tag{3.77}$$

is close to unity in the approximation outlined above, i.e. $T_i \lesssim T$.

If we define the upper limit of T by T_c, at which no appreciable error is introduced by taking $Z \approx 1$, we may write

$$Z(T_c) = 1 - \varepsilon \tag{3.78}$$

where ε is the maximum permissible error. Thus when $T < T_c$

$$I(T) \propto \exp(-E/k_B T)$$

and when $T > T_c$, Z cannot be assumed to be constant as it starts decreasing. In order to find T_c by taking into account the maximum permissible error level, we first evaluate E by means of a linear plot (sometimes known as the Arrhenius plot) between $\ln I$ and $(1/T)$, Equation 3.4, and s by means of Equation 3.5. If the resolution of the TL peak is not good, i.e. if there is a second minor peak superimposed, the method of initial rise is not applicable. In order to find T_c, we make use of the following approximation:

$$\exp(-\varepsilon) \approx 1 - \varepsilon$$

since $\varepsilon < 1$

$$Z(T_c) = \exp[-(s/q) \int_{T_i}^{T_c} \exp(-E/k_B T)\, dT] = 1 - \varepsilon = \exp(-\varepsilon)$$

or

$$(s/q) \int_{T_i}^{T_c} \exp(-E/k_B T)\, dT = \varepsilon \tag{3.79}$$

Using Equation 3.64 for the left hand side of Equation 3.79

$$(k_B T_c^2/E)\,[\exp(-E/k_B T_c)]\,(1 - \triangle) = q\varepsilon/s \tag{3.80}$$

where $\triangle = (2k_B T_c/E)$.

From Equation 3.80, T_c can be easily calculated.

3.17   Thermal quenching of luminescence

Equation 2.22 (Chapter 2) describes luminescence efficiency η by accounting for the thermal quenching (expressed by the probability factor, q) of luminescent centres. At low temperatures the thermal quenching is negligible ($q \approx 0$) and, therefore $\eta \approx 1$; but above a certain temperature T^*, say, the thermal quenching of luminescence assumes significance, thus making η smaller than unity and also decreasing as the temperature increases. At these temperatures, therefore, the thermoluminescence intensity $I(T)$ in the initial rise method will be adequately described by including the efficiency factor η (T) in its expression

$$I(T) = csn_o \, \eta \, (T) \exp(-E/k_B T)$$

where

$$\eta \, (T) = p/(p+q) = 1/[1 + (v/s) \exp(-E_Q/k_B T)]$$

At temperatures $T \leq T^*$, as the thermal quenching is negligible, we have $\eta \, (T \leq T^*) = 1$. Generally, however, $(v/s) \exp(-E_Q/k_B T) \gg 1$, so that we can write

$$\eta \, (T=T^*) = 1 = 1/ \, (v/s) \exp(-E_Q/k_B T^*)$$

This gives

$$T^* = E_Q/k_B \ln \, (v/s)$$

If the temperature dependence of both v and s is taken to be identical so that the ratio (v/s) is independent of temperature then T^* can be regarded as a constant for a given phosphor.

From this it can be concluded that for $T \leq T^*$, the slope of the Arrhenius plot will read (E/k_B), while for $T > T^*$, the temperature dependence of η is also to be considered. Using the inequality mentioned above

$$\eta \, (T) \approx 1/(v/s) \exp(-E_Q/k_B T)$$

hence

$$I(T) = csn_o \, (s/v) \exp[-(E-E_Q)/k_B T]$$

The slope of the Arrhenius plot will now be equal to $(E-E_Q)/k_B$. Delunas *et al*[19] have stated that the onset of thermal quenching of luminescence at T^* will show by a sharp decrease of the slope of the Arrhenius plot. This is expected because at $T=T^*$, the slope E/k_B will turn to $(E-E_Q)/k_B$ as the temperature of the phosphor increases.

3.18   Error in E and s

It may be emphasised here that any error in the evaluation of E introduces a much higher error in the determination of s. For this, we refer to Equation 3.5 from which

$$\ln(qE/k_B T_m^2) = \ln(s) - (E/k_B T_m) \tag{3.81}$$

If we assume that measurements of q and T_m involve no error, we can differentiate Equation 3.81 to give

$$(\triangle s/s) = (\triangle E/E) + (\triangle E/k_{\mathrm{B}}T_{\mathrm{m}}) = (\triangle E/E)\,(1+E/k_{\mathrm{B}}T_{\mathrm{m}})$$

Thus

$$\triangle s/s >> \triangle E/E \qquad\qquad (3.82)$$

and we find that any error in the evaluation of E introduces a much larger error in the determination of s. A method to determine s, independently of E, has been suggested by Moharil[20] and will be described in section 3.20.

3.19. **Analysis of the total glow peak**

Assuming first order kinetics and constant s, the Randall-Wilkins relation is

$$I = -c\,(\mathrm{d}n/\mathrm{d}t) = csn\,\exp(-E/k_B T)$$

$$\int I\,\mathrm{d}t = \int c\,\mathrm{d}n = cn = I/[s\,\exp(-E/k_{\mathrm{B}}T)]$$

the negative sign is dropped as only the magnitudes are relevant here)

$$I/\int I\mathrm{d}t = s\,\exp(-E/k_{\mathrm{B}}T)$$

or

$$I/\int_{T_{\mathrm{i}}}^{T_{\mathrm{f}}} I\,\mathrm{d}T = (s/q)\,\exp(-E/k_{\mathrm{B}}T)$$

where T_{f} refers to the final temperature at which the glow curve terminates

$$\ln[I(T)/\int_{T_{\mathrm{i}}}^{T_{\mathrm{f}}} I\mathrm{d}T] = \ln\,(s/q) - (E/k_{\mathrm{B}}T) \qquad (3.83)$$

A plot of the function on the left hand side of Equation 3.83 against $1/T$ is a straight line whose slope gives E/k_{B} and whose intercept gives $\ln\,(s/q)$.

3.20. **Isothermal decay method**
3.20.1. *Determination of E and r*

We will now discuss a method for determining E which does not employ any particular heating cycle. The TL material is kept at a constant temperature and the light emission, which decays as a function of time, is monitored. This is known as the isothermal decay method.

Assuming first order kinetics (for simplicity, in the first instance) and using the Randall-Wilkins relation

$$dn/dt = - n\, s\, \exp(-E/k_{\rm B}T)$$

$$\int_{n_{\rm o}}^{n} \frac{dn}{n} = - s\, \exp(-E/k_{\rm B}T) \int_{\rm o}^{t} dt$$

$$n = n_{\rm o}\, \exp[-st\, \exp(-E/k_{\rm B}T)] \qquad (3.84)$$

$$I = - c(dn/dt)$$

$$= cs\, n_{\rm o}\, [\exp(-E/k_{\rm B}T)]\, [\exp\{- st\, \exp(-E/k_{\rm B}T)\}] \qquad (3.85)$$

$$\ln I = \ln(csn_{\rm o}) - [st\, \exp(-E/k_{\rm B}T)] - E/k_{\rm B}T$$

A plot of $\ln I$ against time t is linear and the slope m is given by

$$m = s\, \exp(-E/k_{\rm B}T) \qquad (3.86)$$

If the experiment is carried out at two different constant temperatures T_1 and T_2; two different slopes, m_1 and m_2, are obtained and are given by

$$\ln(m_1/m_2) = (-E/k_{\rm B})\, [(1/T_1) - (1/T_2)] \qquad (3.87)$$

Once E is known from Equation 3.87, s can be determined from Equation 3.86.

In the more general case of kinetics of any order r, we can write (using p' here for the pre-exponential factor)

$$I = c\, p'\, n^{r}\, \exp(-E/k_{\rm B}T) = - c(dn/dt)$$

$$\int_{n_{\rm o}}^{n} dn/n^{r} = \int -p'\, \exp(-E/k_{\rm B}T)\, dt$$

$$(r-1)^{-1}[n_{\rm o}^{(1-r)} - n^{(1-r)}] = -p'\, \exp(-E/k_{\rm B}T)\, t \qquad (3.88)$$

Putting $A = n_{\rm o}^{(1-r)}$
$$B = -(1-r)\, p'\, \exp(-E/k_{\rm B}T)$$

we get

$$n = (A + Bt)^{1/(1-r)}$$

$$I = - c\frac{dn}{dt} = - c\,\frac{B}{(1-r)}\,(A + Bt)^{r/(1-r)} \qquad (3.89)$$

or

$$I^{(1-r)/r} = [c\, p'\, \exp(-E/k_{\rm B}T)]^{(1-r)/r}\, (A + Bt) \qquad (3.90)$$

or

$$I = (a + bt)^{r/(1-r)} \qquad (3.91)$$

where

$$a = A \, [c \, p' \, \exp(-E/k_\mathrm{B}T)]^{(1-r)/r}$$
$$b = B \, [c \, p' \, \exp(-E/k_\mathrm{B}T)]^{(1-r)/r}$$

Equation 3.91 can be used to determine the order of kinetics parameter[21]. We differentiate this equation to get

$$dI/dt = [rb/(1-r)] \, (a+bt)^{(2r-1)/(1-r)} = [rb/(1-r)] \, I^{(2r-1)/r}$$

$$\ln(dI/dt) = \ln[rb/(1-r)] + [(2r-1)/r] \ln I$$

Thus a plot of $\ln(dI/dt)$ against $\ln I$ is a straight line whose slope m is

$$m = 2 - (1/r)$$

or

$$r = 1/(2-m) \tag{3.92}$$

Equation 3.92 can predict fractional order kinetics quite reliably. This matter has been discussed recently for TLD-100 (LiF:Mg,Ti) by Moharil[22].

Experimentally I is seen to decay as a function of time at a sample temperature held fixed at glow peak temperature ($T \lesssim T_\mathrm{m}$). The experimental (I,t) curve is differentiated to read dI/dt at different values of t (t is the mid-point of equal intervals dt on the time axis taken as abcissa). From these data one may finally plot $\ln(dI/dt)$ against $\ln I(t)$. Mahesh *et al*[21] have recently used this procedure to determine r in CaS:Bi phosphors. It is found that the glow curve of this phosphor has two peaks. The earlier one corresponds to the first order and remains unaffected by changing the Bi concentration, while the other, higher temperature peak, changes its order from first to second as Bi concentration increases from low to higher values.

3.20.2. *Determination of s*

As explained in section 3.16.1 any error involved in the estimation of E causes a much larger ambiguity in the subsequent determination of s. Isothermal decay procedures, as proposed by Moharil[20], allow s estimation without the need to know E. A considerable spread is found in the value of E using the same experimental data when different methods of calculation are followed. As a consequence, the reported values of s, e.g. for peak 5 of TLD-100 (LiF:Mg,Ti) phosphor, vary from 10^{12} to 10^{42}.

Isothermal decay curves are obtained at two temperatures T_1 (just below T_m) and $T_2(>T_m)$ for a given glow curve of the TL material. The area under the isothermal decay curve is proportional to the initial number of filled traps n_o. The intensity at any time t will be proportional to the number of filled traps ionised during this time interval (filled traps that have been emptied). Thus the total area minus the area between $t = 0$ and $t = t$, will be proportional to the number of populated traps n remaining at time t. We denote these two areas by A_o and δA_t respectively, as shown in Figure 3.10.

For general order kinetics and using Equation 3.88

$$(n_o/n)^{r-1} = 1 + [(r-1)\, p't\, n_o^{(r-1)} \exp(-E/k_B T)]$$

Putting $s' = p'\, n_o^{(r-1)}$

$$(n_o/n)^{r-1} = (A_o/\delta A_t)^{r-1} = 1 + [(r-1)\, s't \exp(-E/k_B T)] \qquad (3.93)$$

The plot between $(A_o/\delta A_t)^{r-1}$ (where r is pre-determined) and t is a straight line with slope m given by

$$m = (r-1)\, s'\, \exp(-E/k_B T)$$

Thus if m_1 and m_2 are the slopes at temperatures T_1 and T_2, then

$$m_1 = (r-1)\, s'\, \exp(-E/k_B T_1)$$

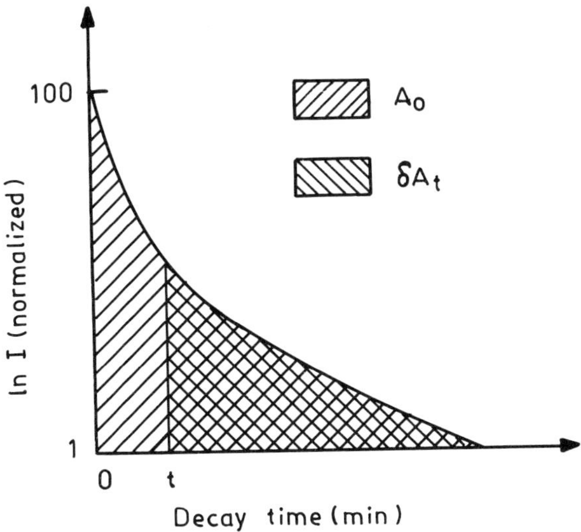

Figure 3.10. Isothermal decay characteristics of a phosphor.

and

$$m_2 = (r-1)\, s'\, \exp(-E/k_B T_2)$$

Eliminating E and solving for s' we find

$$s' = \frac{1}{(r-1)}\, \frac{(m_2)^{T_2/(T_2-T_1)}}{(m_1)^{T_1/(T_2-T_1)}} \tag{3.94}$$

Thus it is possible to find s' without using E. For first order kinetics, we use Equation 3.86

$$m = s\, \exp(-E/k_B T)$$

and hence

$$s = \frac{(m_2)^{T_2/(T_2-T_1)}}{(m_1)^{T_1/(T_2-T_1)}} \tag{3.95}$$

Peak 12 of TLD-100 was used by Moharil for this purpose and the isothermal decay data for this peak at two temperatures (370°C and 380°C) obtained by Kathuria and Sunta[23] was employed to determine slopes m_1 and m_2. The plots are shown in Figure 3.11. $r = 1.5$, as determined for peak 12 by Kathuria and Sunta was used. In

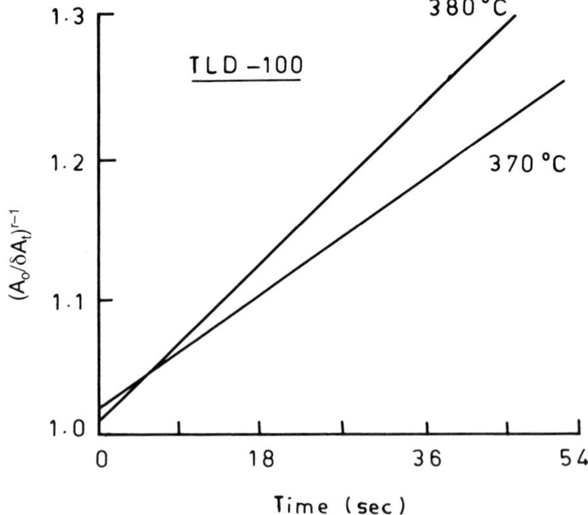

Figure 3.11. Isothermal decay processing for determining s'[20].

this manner, Moharil found $s' = 3.2 \times 10^7$ which is in sharp contrast to 5.2×10^{10} determined by Kathuria and Sunta[23] using the initial rise method. A 1% error in the ratio m_1/m_2, however, affects the s' value by a factor of 2[20].

3.21.   Analysis of glow curve shape

Another experimental procedure for the evaluation of E and s (or s') is by using the features of the glow curve shape. Various parameters in this regard have been defined in section 3.7.

Chen[11,24] has discussed these procedures, which are based on the experimental determination of Γ, δ_1, δ_2 and T_m. As the glow peaks are not sharp enough to allow unambiguous determination of T_m as compared with T_1 and T_2, the procedures requiring a knowledge of Γ $(= T_2-T_1)$ yield E and s values which are less ambiguous in comparison with those requiring the use of either or both δ_1 and δ_2. It is possible to express a relationship for E in a general form[11,24] as

$$E_\alpha = [C_\alpha (k_B T_m^2)/\alpha]-2k_B T_m\, a_\alpha \tag{3.96}$$

where α stands for δ_1, δ_2 or Γ. The values of C_α and a_α for the three methods in the first and second order kinetics are given in Table 3.1.

In the table b is the power of the temperature dependence of the pre-exponential factor, $s' =s'_o T^{\,b}$. The value of b usually lies in the range -2 to $+2$. Chen[6] has also suggested a method for finding C_α and a_α for the case of general order kinetics, i.e. when $r \neq 1,2$. Equation 3.96 holds here also, but the coefficients are now as follows

$$\left. \begin{aligned}
C_{\delta 1} &= 1.51 + 3\,(\mu_g - 0.42) \\
C_{\delta 2} &= 0.976 + 7.3\,(\mu_g - 0.42) \\
C_\Gamma &= 2.52 + 10.2\,(\mu_g - 0.42)
\end{aligned} \right\} \tag{3.97}$$

and

$$\left. \begin{aligned}
a_{\delta 1} &= 1.58 + 4.2\,(\mu_g - 0.42) + b/2 \\
a_{\delta 2} &= b/2 \\
a_\Gamma &= 1 + (b/2)
\end{aligned} \right\} \tag{3.98}$$

3.22.   Analysis of complex thermoluminescence
3.22.1.   *Sweet-Urquhart method*

The method of Haering and Adams[25] for TSC

measurements has been generalised by Sweet and Urquhart[26] with a view to apply it to find E in the case of the glow peak arising from the 'thermally connected' (having close energy difference) trapping states. This method allows us to determine separately the two activation energies characterising these two thermally connected traps that generate the overlapping glow peaks. The procedure is based upon measurements for a set of glow curves generated with different linear heating rates. The significance of this approach is that the knowledge of s values for these two traps is not required.

For different heating rates q_i, the glow intensity $I(T_{xi})$ at the temperature T_{xi} is given by

$$I(T_{xi})/\exp(-E/k_B T_{xi}) = \text{constant, for all } i$$

where T_{xi} is the temperature of the glow curve obtained at the heating rate q_i such that the area of the portion of the curve on the lower side of this temperature, i.e. A_{xi}, divided by the total area A_{ti} is equal to the fixed fraction x chosen arbitrarily, i.e.

Table 3.1. Coefficients of Equation 3.96.

	First order			Second order		
	δ_1	δ_2	Γ	δ_1	δ_2	Γ
C_α	1.51	0.976	2.52	1.81	1.71	3.54
a_α	$1.58+(b/2)$	$b/2$	$1+(b/2)$	$2+(b/2)$	$b/2$	$1+(b/2)$

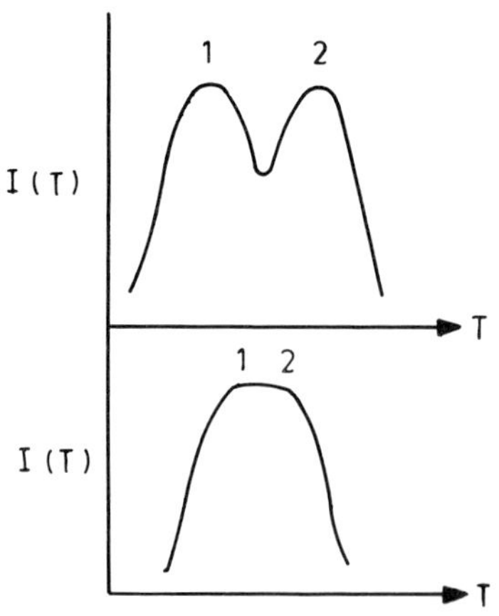

Figure 3.12. Schematic view of separated and overlapping glow peaks[27].

$$x = A_{xi}/A_{ti}$$

Thus, for different glow curves i, the temperature T_{xi} is so chosen that the value x is the same. A set of T_{xi} values and corresponding $I(T_{xi})$ values is then found. A plot between ln $I(T_{xi})$ and $1/T_{xi}$ is a straight line whose slope is E/k_B from which E can be found. When x is taken to be small (up to about 20%), the plot gives the E_1 value accurately and when x is large ($\approx 80\%$), the slope then gives E_2. The E_2 value, so determined, is not very accurate and can be regarded only as an estimate. An example of overlapping and separated glow peaks (by Sweet-Urquhart method) is illustrated in Figure 3.12[27].

3.22.2.   *McKeever's method*

This method, suggested by McKeever[28] enables us to separate peaks that overlap (but not too closely) when there are more than two and determine the T_m value for each of them. The method, like that of Sweet and Urquhart, is applicable to both TL and TSC glow peaks. McKeever's method consists in heating at a linear rate a pre-irradiated specimen, to some temperature T_s lying on the low temperature tail (see Figure 3.13(i)) of the first glow peak. The sample is then cooled rapidly to room temperature and then re-heated at the same rate in order to record all of the remaining glow curve. The position of the first maximum only in the glow curve is noted. The whole process is repeated on a freshly irradiated sample using a different value of T_s (in small increments of between 2 to 5K) and the corresponding value of T_m noted every time. A plot of T_m values and T_s values is made (shown in Figure 3.13 (ii)). The plot is a straight line for a single peak glow curve and exhibits first order kinetics. For a glow curve containing several well separated first order peaks, the T_m-T_s curve shows a staircase structure, each flat portion in the curve indicating the presence of an individual peak. When peaks are too close, corresponding to quasi-distribution of trapping centres, the T_m-T_s plot becomes a continuous line, of slope ≈ 1.0. This sets a limit to the adopted technique as peaks can no longer be resolved. This resolution is typically of the order of 5K. It should be noted, however, that this technique only highlights the most prominent peaks in the glow curve and the smaller peaks may get suppressed by their larger neighbours so that they do not show as a flat region in the T_m-T_s plot.

In the case of second order TL peaks, the T_m-T_s curves are characteristically different in shape compared with first order peaks. This is depicted in Figure 3.13 (iii). In curve *a*, T_m increases at higher values of T_s ($T_s > T_m$). However, the *E* value can be represented by a constant value of T_m (i.e. for $T_s \ll T_m$). In curve *b*, the staircase structure is rather smoother, while in *c*, the straight line behaviour is evident.

An example of TL glow peaks for LiF(TLD-100) and the corresponding T_m–T_s plot is shown in Figure 3.14. Here the TLD-100 sample was annealed in air at 400°C for 1 h and then quickly cooled to room temperature. The specimen was then irradiated with 255 cGy of ^{90}Sr/^{90}Y β particles of energies 0.54 MeV (from ^{90}Sr) and 2.27 MeV (from ^{90}Y decay).

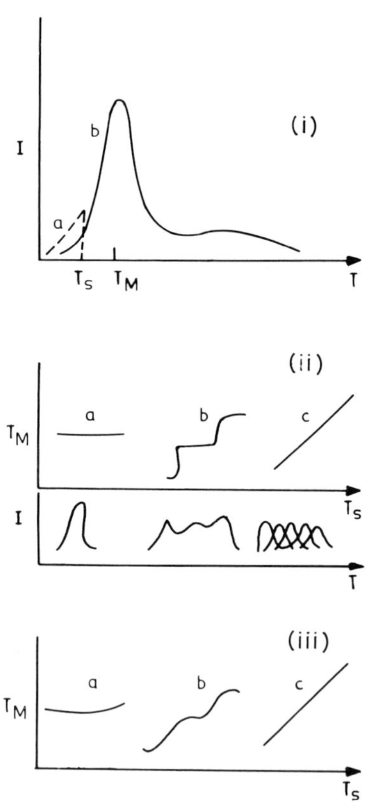

Figure 3.13. Separation of overlapping glow peaks, McKeever's method[28].

The heat treatment given to the sample eliminates one peak at about 80°C and the TL process in LiF follows first order kinetics. This technique of partial heating is of special significance in the analysis of glow in meteorite samples. In meteorites the TL peaks overlap closely and individual peaks are not easily detectable. However, only an estimate of peak positions can be made as the T_m values are usually higher than the actual peak positions due to overlapping.

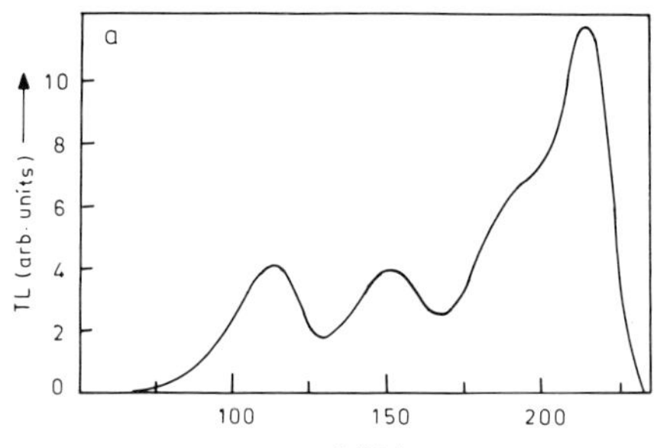
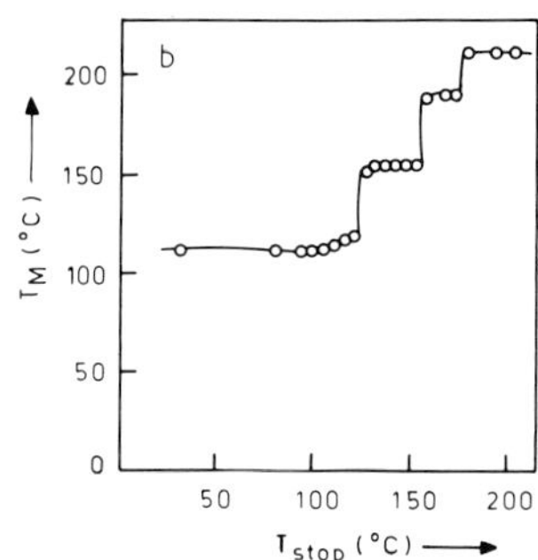

Figure 3.14. (a) Glow curve for LiF(TLD-100) and (b) its T_m–T_{stop} curve. LiF is a material showing a series of well separated, first order glow peaks[28].

References

1. (a)Randall, J. T. and Wilkins, M. H. F. Proc. R. Soc. A **184**, 347, 366, 390 (1945).
1. (b)Azorin, J. Nucl. Tracks Radiat. Meas. **11**, 159 (1986).
2. Garlick, G. F. J. and Gibson, A. F. Proc. R. Soc. A **60**, 574 (1948).
3. Braunlich, P. J. Appl. Phys. **38**, 2516 (1967).
4. Christodoulides, C. J. Phys. D: Appl. Phys. **18**, 1665 (1985).
5. Halperin, A. and Braner, A. A. Phys. Rev. **117**, 408 (1960).
6. Chen, R. J. Appl. Phys. **40**, 570 (1969).
7. Chen, R. J. Mater. Sci. **9**, 345 (1974).
8. Chen, R. J. Electrostatics **3**, 15 (1977).
9. Braunlich, P. IN *Thermoluminescence of Geological Materials*, ed. D. J. McDougall (New York: Academic Press) p. 61 (1968).
10. Balarin, M. J. Therm. Anal. **17**, 319 (1979).
11. Chen, R. J. Electrochem. Soc. **116**, 1254 (1969).
12. Stoebe, T. G. and Watanabe, S. Phys. Status Solidi a **29**, 1 (1975).
13. Mattern, P. L., Lengweiler, K., Levy, P. W. and Esser, P. D. Phys. Rev. Lett. **24**, 1287 (1970).

14. Urbach, F. *Cornell Symposium* (New York: Wiley) p. 115 (1948).
15. Lax, M. Phys. Rev. **119**, 1502 (1960).
16. Keating, P. N. Proc. Phys. Soc. **78**, 1408 (1961).
17. Schön, M. Tech. Wiss. Abh. Osram Studienges. **7**, 175 (1958).
18. Hoogenstraaten, W. Philips Res. Rep. **13**, 515 (1958).
19. Delunas, A., Maxia, V., Ortu, A. and Spano, G. J. Lumin. **36**, 373 (1987).
20. Moharil, S. V. J. Phys. D: Appl. Phys. **14**, 1677 (1981).
21. Mahesh, K., Vij, D. R., Singh, N., Lal, N. and Nagpaul, K. K. Nucl. Tracks Radiat. Meas. **9**, 139 (1984).
22. Moharil, S. V. J. Phys. D: Appl. Phys. **17**, L107 (1984).
23. Kathuria, S. P. and Sunta, C. M. J. Phys. D: Appl. Phys. **12**, 1573 (1979).
24. Chen, R. and Kirsh, Y. *Analysis of Thermally Stimulated Processes* (Oxford: Pergamon) p. 159 (1981).
25. Haering, R. R. and Adams, E. N. Phys. Rev. **117**, 451 (1960).
26. Sweet, M. A. S. and Urquhart, D. J. Phys. C: Solid State Phys. **14**, 773 (1981).
27. Sweet, M. A. S. and Urquhart, D. Phys. Status Solidi a **59**, 223 (1980).
28. McKeever, S. W. S. Phys. Status Solidi a **62**, 331 (1980).

CHAPTER 4

Thermoluminescent Phosphors

P. S. Weng

4.1.   Introduction

There are many materials which exhibit TL: the problem is, therefore, not to find such a material but rather to select the most appropriate TL phosphor for a purpose from the many available. In this chapter the important characteristics and features of TL phosphors are examined, reviewed and compared. The materials considered here have been chosen on the basis of popularity in use, commercial availability and specific applications. The applications of TL phosphors in various fields are expanding and progressing rapidly and there is a constant search for suitable TL materials.

Preparation and thermoluminescence of alkaline earth sulphide phosphors has been reviewed recently in the literature[1a] and, hence, no attempt to describe them here has been made.

The thermoluminescence phenomenon has been known and observed for centuries, whenever certain fluorites and limestones were heated. Sir Robert Boyle and his colleagues reported the thermoluminescence effect as early as 1663. However, it was not until 1950 that the use of thermoluminescence as a means of radiation detection was proposed.

When a thermoluminescent phosphor is exposed to ionising radiation at a low (room) temperature, many of the freed electrons become trapped in lattice imperfections in the crystalline solid. These electrons may remain trapped for long periods when the crystals are stored at room temperature. If the temperature is raised, the probability of release is increased and electrons may be released from the traps and return to stable energy states with the emission of light.

In TL dosimetry (TLD) this property is used to measure the radiation dose to which the phosphor has been exposed. This is done by means of a TLD reader, consisting of a controlled heating element and a photomultiplier system which determines the light

* Chapter revised and updated by K. Mahesh

fluence emitted during the heating of the dosemeter material. In most TLD readers the integrated light intensity is measured as a function of the heating temperature cycle. Thus, thermoluminescence dosimetry consists of two steps: (i) radiation exposure, which leaves some excited electrons in metastable traps, and (ii) readout, which consists of controlled heating of the exposed TLD and measurement of the emitted light intensity[1b].

The basic phenomenon can be explained by using a hypothetical energy diagram of an insulating crystal exhibiting thermoluminescence due to ionising radiation. This is illustrated in Figure 4.1 to depict the fundamental processes that occur in the phosphor.

When a crystal is exposed to ionising radiation, electrons are released from the valence band to the conduction band. This leaves a hole (a postively charged site) in the valence band. The electron and the hole may move through the crystal until they recombine or until they are trapped in metastable states. These metastable states are thought to be associated with defects in the crystal such as the impurity sites. At this point, there are two possible ways in which a thermoluminescence photon may be emitted. As the crystal is heated, sufficient energy may be given to the electron so as to raise it from its trapping site to the conduction band. The electron may move around in the crystal until it recombines with a trapped hole and a thermoluminescence photon is emitted. Alternatively, the hole trap may be less stable than the electron trap and, when the crystal is heated, the hole receives sufficient energy to wander until it recombines with a trapped electron and a thermoluminescence

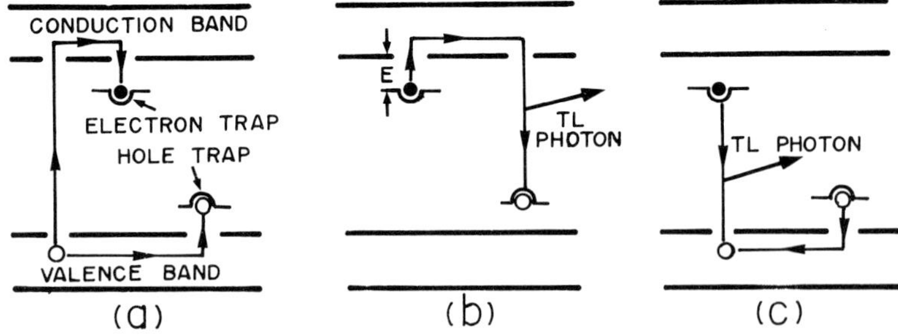

Figure 4.1. Schematic energy level diagram of a phosphor exhibiting thermoluminescence. (a) Exposure to ionising radiation. (b) Electron trap is less stable during heating, and hole trap is the emitting centre. (c) Hole trap is less stable during heating, and electron trap is the emitting centre.

photon is emitted. Since the two processes are similar, usually only the first possibility is considered in discussions of thermoluminescence.

The energy gap between the trap and conduction band determines the temperature required to release the electron and produce the thermoluminescence, and is characteristic of the particular material employed. In the practical situation, many trapped electrons and holes are produced. As the temperature of the crystal is increased, the probability of releasing any electron from a trap is increased; thus, the emitted light will be weak at low temperature, pass through one or more maxima at a higher temperature, and then decrease again to zero as no more filled traps remain.

4.2.　**Requirements of a suitable TL material**

A TL phosphor for dosimetric use should combine several properties which, as a result, limit the choice to only a few inorganic compounds. Some of the more desirable properties of a TLD phosphor are[2]:

1. A high concentration of electron (or hole) traps and a high efficiency of light emission associated with the recombination process.
2. A sufficient storage stability of the electron (or hole) traps to cause no undesirable fading even during extended storage at ambient or slightly increased temperatures (such as tropical or desert climates) or, in the case of biomedical implantation studies, the body temperature. This should also be true for the opposite extreme of climate within the Arctic Circle where temperature is low.
3. A spectrum of the emitted TL light to which the detector system (photomultiplier and filter combination) responds well, and which interferes as little as possible with the incandescence (infrared) emission of the heated phosphor and its immediate surroundings.
4. The main peak temperature (T_m) lies between 180 and 250°C. At higher temperatures the infrared emission from the hot sample and sample holder increasingly interferes with the measurement of low doses. The spectrum wavelengths between about 300 and 500 nm are most desirable unless systems specially sensitive to UV are used.

5. A trap distribution that does not complicate the evaluation procedure due to the presence of additional rapidly fading low temperature peaks or high temperature peaks (which may be difficult to anneal) and due to the retrapping processes, etc. The phosphor should preferably be completely annealed during the readout procedure itself with no changes in its sensitivity and/or background reading or linearity.

6. Resistance against potentially disturbing environmental factors, including light, humidity, organic solvent, common fumes and gases. Failing this, the detector is to be permanently sealed into a glass bulb.

7. Low photon energy dependence of response and a linear response over a wide range of dose are, for most applications, the desirable features. Depending on the particular use, a high or low thermal neutron sensitivity or a certain LET response may also be the desirable characteristics.

8. The phosphor should not be very expensive. It may be easier and cheaper to work with a 'one-way' dosemeter. It should be non-toxic and should not deteriorate during extended storage. It should be possible to use a small sample only, subject to the absence of spurious effects such as tribo- or chemiluminescence, oxygen effects, etc. It may be advantageous if the material can easily be prepared with reproducible properties in a normally equipped chemical laboratory.

4.3.   **Advantages of TL phosphors in personnel dosimetry**

For personnel dosimetry the TL systems are found to have several advantages:

  (i) TL detectors are available in powder or solid form in a wide range of size;

 (ii) TL materials are practically independent of dose rate and of the angle of radiation incidence;

(iii) TL detectors are suitable for postal service;

(iv) it is possible to have a reassessment of the absorbed dose by UV irradiation and readout of high temperature peaks (PTTL);

 (v) a lower limit of detection of about 0.2 mGy;

(vi) some TL materials have a near tissue-equivalent composition (LiF, $Li_2B_4O_7$, MgB_4O_7, BeO) which eliminates the need for energy correction;

(vii) the possibility of automated/computerised evaluation for large monitoring services. (Some automated TLD systems are now commercially available such as those manufactured by Harshaw (USA), Panasonic (Japan), Wentley (UK), and Studsvik (Sweden). With the introduction of automated TLD systems, computerised dose record keeping programs have been developed by larger personnel monitoring services.)

TL phosphors may be used to give an accurate measurement of dose from penetrating gamma radiation in situations in which a very rapid assessment of accumulated dose is required, e.g. when personnel are working in high dose rate areas. LiF has been used for this purpose because it is dose rate independent, nearly tissue-equivalent, inert to the environment and has long-term dose retention (negligible fading).

The Technical Recommendations for Monitoring Exposure of Individuals to External Radiation recommends three kinds of dosemeters: (i) a non-discriminating type basic dosemeter, (ii) a discriminating type basic dosemeter, and (iii) an extremity dosemeter. Required properties are listed in Table 4.1.

Table 4.1. Technical recommendations for monitoring exposure using TLDs.

Basic property	Non-discriminating	Discriminating
Dose range (photons) Dose range (electrons)	—	{0.2 mGy – 10 to 100 Gy}
Energy range (photons)	—	0.01–50 MeV
Energy range (electrons)	—	0.5–50 MeV
Information on photon radiation quality	Not required	Necessary in the range 10–200 keV
Overall uncertainty		For doses <0.5 Gy: –30% to +50%, or ±0.5 mGy (whichever is the greatest) >0.5 Gy: –20% to +25%
Precision		2σ <10% at 10 mGy
Photon energy dependence for body dose at 1 cm depth (over energy range given above)	–20% to +40%	±15%
Dependence on environmental conditions		Insignificant
Photon angular response		±30%
Fading		<5% over the monitoring period at 25°C

4.4. Features of TL phosphors for use in radiation dosimetry

Table 4.2 summarises the typical features of TL phosphors with regard to their use in radiation dosimetry.

4.5. Suitable TL phosphors in use

Obviously, it is difficult to meet all those requirements, as mentioned in section 4.2, in a single detector material. The various

Table 4.2. Some features of TLDs.

Property	TLD features
1. Integrating or dose-rate	Integrating
2. Normal dose range	1 mGy–100 Gy
3. Linearity of response	Supralinearity possible
4. Calibration	Secondary
5. Typical precision (coefficient of variation)	
(a) Batch calibration	$\pm 5\%$
(b) Individual calibration	$\pm 1\%$
6. Approximate tissue-equivalence	Yes
7. Dose rate dependence	No
8. Dose spectrometry	Limited
9. Measurement of dose gradients across dosemeter	Difficult
10. Relative fading of stored dose	Low
11. γ and X detected (within limits of photon energy dependence)	Yes
12. β dosimetry	Can be very good with thin dosemeters
13. Neutron dosimetry	Possible
14. Alpha dosimetry	Possible
15. Relative size	Can be very small
16. Mobility	Very good
17. Geometrical configurations available	Wide variety
18. Directional effects	Probably not
19. Sensitivity to environmental conditions	Temperature, oxygen, light, dust
20. Resistance to chemical attack	Can be very good
21. Ruggedness	Good
22. Rapid evaluation	Yes
23. Permanent dose record	Indirect
24. Re-usability	Easy
25. Measurement of cumulative doses	Not easy
26. Suitable for skin dosimetry	Yes
27. Suitable for extremity monitoring	Yes
28. Suitable for *in vivo* measurements	Yes
29. Relative cost of dosemeter	Moderate
30. Cost of reading device	High
31. Cost of reading	Low
32. Commercially available	Yes
33. Skilled personnel required for evaluation	Not for routine work

detectors listed in Table 4.3 meet these requirements to some degree[3].

Though 14 types of TLDs are listed in Table 4.3, the most important families of phosphors included are lithium fluoride (LiF), lithium borate ($Li_2B_4O_7$), calcium sulphate ($CaSO_4$), and calcium fluoride (CaF_2). Table 4.4 gives the thermal neutron dose response of different TL phosphors.

Table 4.4. Thermal neutron response of various TL phosphors.

TL Phosphor	Thermal neutron response (R per 10^{10} n.cm^{-2})
Mg_2SiO_4:Tb	0.21
$CaSO_4$:Dy (Harshaw)	0.52
$CaSO_4$:Dy (DRP)	0.38
$CaSO_4$:Tm (DRP)	0.21
$CaSO_4$:(Dy,^{6}Li) (DRP)	6.2
$CaSO_4$:(Mn,Li)	100
$CaSO_4$:(Mn,^{6}Li)	1050
$CaSO_4$:Tm	0.23
CaF_2:Mn (Harshaw)	0.6
	0.1–0.13
CaF_2:Mn (EG & G)	0.58
CaF_2:Mn (Conrad)	0.2
CaF_2 (MBLE)	3.5
CaF_2 (fluorite)	0.16
CaF_2:Mn TLD 08 (Yugoslavia)	0.77
	1.05
CaF_2:Mn (Philips)	0.07
CaF_2:Dy TLD-200 (Harshaw)	0.59
	0.5–0.65
LiF TLD-700 (Harshaw)	1.1
	0.7
	0.87–0.96
	1.0
	2.5
	1.3
LiF 7 (Conrad)	23

TL phosphor	Thermal neutron response (R per 10^{10} n.cm^{-2})
LiF TLD-100 (Harshaw)	330
	200
	200
	220
	535
	310
	65
	490
	360
LiF (Conrad)	165
LiF TLD-600 (Harshaw)	1520
	870
	2190
	1200–1700
LiF 6 (Conrad)	2650
BeO	0.45
	0.13
	0.20
	0.17–0.29
	0.2
	0.5
$Li_2B_4O_7$:Mn (Harshaw)	390
	230
$Li_2B_4O_7$:Mn (DRP)	390
$Li_2B_4O_7$:Mn (Wallace)	670
$Li_2B_4O_7$:Mn (Christensen)	300
	420
$Li_2B_4O_7$:Mn (Brunskill)	400–500

Table 4.3 General characteristics of some TLD phosphors.

TLD phosphor	Relative gamma ray sensitivity	TL emission peak (Å)	Dosimetric peak temp. (°C)	Useful dose range	Effective atomic No., Z	Photon energy dependence of response at 30 keV relative to ^{60}Co γ ray	TL fading	Significant light induced fading
1. LiF:Mg,Ti	1	4000	195	50 μGy to 10^3Gy	8.2	1.3	10%/month	No
2. MgB$_4$O$_7$:Dy	7	4800, 5700	210	—	8.4	1.5	10%/two months	—
3. Li$_2$B$_4$O$_7$:Mn	0.4	6000	210	0.1 mGy to 10^3Gy	7.4	0.98	10%/month	No
4. Li$_2$B$_4$O$_7$:Cu,Ag	1	3680	185	50 μGy to 10^4Gy	7.4	0.98	10%/month	Yes
5. Li$_2$B$_4$O$_7$:Cu	8	3680	205	—	7.4	0.98	9%/two months	No
6. CaSO$_4$:Tm	3.2	4520	210	μGy to 10^3Gy	15.0	11.5	10%/month	No
7. CaSO$_4$:Dy	38	4800, 5700	210	μGy to 10^3Gy	15.0	11.5	3%/month	No
8. CaSO$_4$:Mn	70	5000	110	μGy to 10^3Gy	15.0	11.2	60%/day	—
9. CaF$_2$:Dy	16	4800, 5700	200	μGy to 10^3Gy	16.0	15.6	12%/month	Yes
10. CaF$_2$:Mn	5	5000	260	μGy to 10^3Gy	16.0	15.4	10%/month	—
11. CaF$_2$ (natural)	23	3800	260	10 μGy to 10^3Gy	16.0	14.5	—	—
12. Mg$_2$SiO$_4$:Tb	53	5520	195	1 μGy to 10^3Gy	11.0	4.5	3%/month	Yes
13. BeO	3.1	2400–4000	180-220	0.1 mGy to 10^3Gy	7.2	0.87	5%/month	Yes
14. Al$_2$O$_3$	5	4250	250	10 μGy to 10^3Gy	10.2	4.5	5%/two weeks	Yes

4.6. Experimental methods of preparation of different TL phosphors

4.6.1. *Lithium fluoride*

LiF:Mg:Ti is currently the most widely used family of TL phosphors. Harshaw Chemicals of USA produce a commercial LiF phosphor known as TLD-100 and its isotopic variants TLD-600 and TLD-700. The isotopic constituents of Harshaw TLD-100, -600, and -700 are: ^{6}Li 7.5% and ^{7}Li 92.5% for TLD-100, ^{6}Li 95.6% and ^{7}Li 4.4% for TLD-600, and ^{6}Li 0.01% and ^{7}Li 99.99% for TLD-700.

TLD-100 (also, -600, -700) is produced by homogeneous melting of lithium fluoride, magnesium fluoride, lithium cryolite and lithium titanium fluoride, resulting in a phosphor containing 300 ppm magnesium and 10-20 ppm titanium[4]. A single crystal solidified from the melt is then pulverised and the powder grains are sieved and separated. Extruded ribbon dosemeters are formed by compressing the original mixture at an elevated temperature, causing it to extrude through an aperture. The extrusion is cut into pieces which are then polished.

By the addition of 1-2% sodium fluoride by weight, the sodium-stabilised lithium fluoride phosphor is prepared. The material can be re-used at nominal low levels of observed dose with no thermal treatment other than an initial 500°C anneal. The phosphor is available commercially as PTL 710, with isotopic variants PTL 716 and PTL 717.

The preparation of LiF is delicate. Only well equipped laboratories with trained staff can produce a material of good quality. It is generally more convenient to use commercial products.

4.6.2. *Lithium borate*

The preparation of $Li_2B_4O_7$:Mn is relatively straight-forward[5]. A mixture of 32.5 g of lithium carbonate and 108.96 g of boric acid is prepared. The mixture is added to a 5 ml aqueous solution of 75 mg $MnCl_2.4H_2O$ which is heated at 100°C for 12 h. The product is placed in a platinum crucible, heated up to 950°C and then rapidly cooled. The crystalline mass thus obtained is ground and graded between 75 and 175 μm. Silica, 0.25% by weight, can be added to reduce the influence of humidity on the lithium borate, which is very hygroscopic. The material can either be sintered at 900°C to prepare solid detectors or diffused into a plastic carrier such as Teflon or silicone rubber.

4.6.3. *Calcium sulphate*

By far the most popular $CaSO_4$ phosphors at present are those doped with dysprosium (Dy) and thulium (Tm) and were first produced by Yamashita *et al*[6]. The TLD material is a pure $CaSO_4.2H_2O$ mixed with activator material dysprosium oxide, Dy_2O_3 (or thulium oxide, Tm_2O_3) to the extent of 0.1 mol per cent with regard to $CaSO_4:2H_2O$, and dissolved in concentrated sulphuric acid to form a saturated solution of $CaSO_4$. The solution is then heated to evaporate sulphuric acid (with adequate ventilation!). Single crystals of calcium sulphate doped with dysprosium (or thulium) are observed to grow with the progress of evaporation of sulphuric acid. After cooling, the crystals are ground to a crystalline powder and are sieved to retain the 80-200 μm crystals. The powder thus obtained is readily activated by radiation. The US company, Teledyne, markets $CaSO_4:Dy$ (or Tm) embedded in Teflon.

Analysis of TL glow curves in $CaSO_4:Dy$ is reported by Azorin and Guitierrez[7]

4.6.4. *Calcium fluoride*

Manganese-activated (3%) CaF_2 is prepared by co-precipitation of CaF_2 and MnF_3 from a solution of $CaCl_2$ and $MnCl_2$ in ammonium fluoride $(NH_4)F$. The precipitate is dried and heated in a neutral atmosphere oven at 1200°C, powdered and graded. It may be pressed or sintered in the same way as described for LiF. It can also, like all TL microcrystallites, be diffused into a plastic material such as Teflon[8,9].

Harshaw markets, under the code TLD-200, a dysprosium-activated calcium fluoride $(CaF_2:Dy)$ which is considerably more sensitive than $CaF_2:Mn$, and so is thulium-activated calcium fluoride $(CaF_2:Tm)$ under the code TLD-300[10,11].

4.6.5. *Other phosphors*
(i) Beryllium oxide

Beryllium oxide (BeO) is used extensively as an electrical insulating and refractory material. It is also a neutron moderator in nuclear reactors. Several different types of BeO are commercially available and some have proved suitable for use as TL phosphors.

(ii) Aluminium oxide

Aluminium oxide (Al_2O_3), like BeO, is a refractory material and is used for many industrial applications. Its various forms contain many different impurities, including Ca, Cr, Ti, Ni, Mg, Na, Fe, etc. The TL properties of Al_2O_3 in its various forms have been widely investigated.

(iii) Magnesium orthosilicate

Magnesium orthosolicate (Mg_2SiO_4) occurs in the natural mineral forms of olivine and forsterite. Hashizume *et al* (in Reference 5) were the first to manufacture Mg_2SiO_4 TL phosphor containing the rare earth terbium (Tb) as a dopant. The characteristics of the phosphor depend on the kind of preparation used and, in particular, on the temperature of the melting process.

(iv) MgB_4O_7:Dy:Tm and $CaSO_4$:Dy:Tm

These recently produced phosphors are available in the form of physically robust sintered discs 9, 6, 5 and 3 mm in diameter.

4.7. Forms of TLD phosphors

Physically, TLD phosphors are available in two general forms: loose phosphor powder or solid dosemeters (Figure 4.2). The solid dosemeters may be composed entirely of phosphor as single crystals and polycrystalline extrusions, or as a homogeneous composite of the phosphor powder and some binding material such as polytetrafluoroethylene (PTFE, Teflon). The physical forms of some commercially available TL phosphors listed in Table 4.3 are given in Table 4.5.

Table 4.5. Physical form of some phosphors in Table 4.3

Phosphor	Available forms
LiF	Crystals, powder, bulb chips, microrods, PTFE-based discs, PTFE-based microrods, and PTFE-based tape
$Li_2B_4O_7$:Mn	Powder, bulbs, chips, PTFE-based microrods
CaF_2 (natural)	Crystals
CaF_2:Mn	Powder, bulbs, chips, microrods
CaF_2:Dy	Crystals, powder, bulbs, chips
CaF_2:Tm	Chips
$CaSO_4$:Mn	Powder
$CaSO_4$:Dy	Powder, discs, chips
BeO	Ceramic

The characteristics of single crystal phosphor dosemeters may be considerably different from those of the composites. Single crystals of TL material, while providing a very sensitive form of dosemeter, display a very non-uniform distribution of TL sensitivities, and by grinding up many such crystals and thoroughly mixing the resultant powder, a reasonably uniform TL sensitivity can be achieved. The sensitivities of powder phosphors, however, have been shown to be

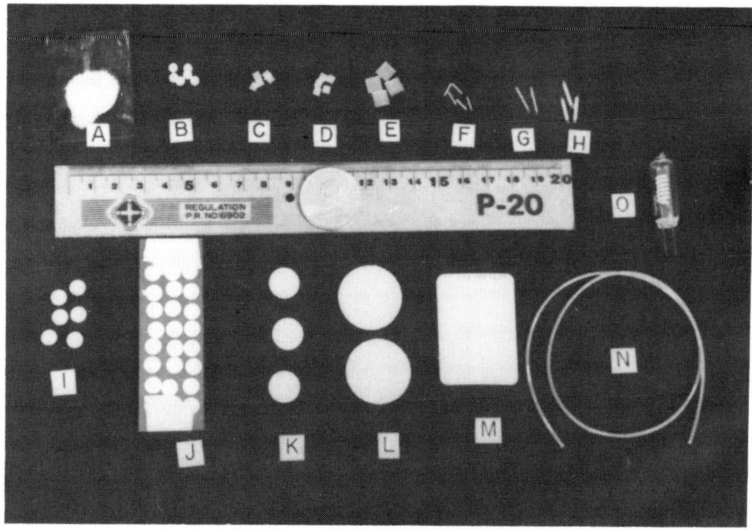

Figure 4.2. Physical forms of some TL phosphors.

A: Powder LiF, 80-200 μm
B: Ribbon CaSO$_4$:Dy, 4 mm diam.$\times$4 mm
C: Ribbon LiF, 0.125$\times$0.125$\times$0.015 inches
D: Ribbon LiF, 0.125$\times$0.125$\times$0.035 inches
E: Ribbon CaF$_2$:Dy, 6.35$\times$6.35$\times$1.00 mm
F: Rod LiF + Teflon, 1.00 mm diam.$\times$6 mm
G: Rod LiF encapsulated in glass, 1.4 mm diam.$\times$12.5 mm
H: Rod LiF encapsulated in glass, 2 mm diam.$\times$12.0 mm
I : Disc LiF + Teflon, 6 mm diam.$\times$10 mm
J : Disc LiF + Teflon, 6 mm diam.$\times$0.13 mm
K: Disc LiF + Teflon, 12.5 mm diam.$\times$0.40 mm
L : Disc LiF + Teflon, 26 mm diam.$\times$0.08 mm
M: Badge LiF + Teflon, 32$\times$45$\times$0.4 mm
N: Tubing LiF + PVC, 1.5 mm diam.$\times$7 mm (10 pieces in series)
O: Bulb CaF$_2$:Mn, 0.4 diam.$\times$1.68 inches

strongly dependent on grain size.

The extruded and hot pressed forms of dosemeter are available in two main geometries: the ribbon, commonly called the chip, and the rod, commonly called the microrod. Both are produced by compression of the normal phosphor ingredients at an elevated temperature. For extruded dosemeters the fused polycrystalline material is extruded through an appropriately shaped die, cut and polished. The hot pressed forms have a higher sensitivity than the extruded types.

By compressing and heating an homogeneous mixture of fine grain phosphor powder and PTFE powder to a temperature above the softening temperature of PTFE (327°C) in a mould, it is possible to form an intimate matrix of phosphor and PTFE. Discs and microrods are available commercially.

Dosemeters have been produced by incorporating the phosphor powder in silicone rubber, but they have not proved very popular due to the limitation that silicone rubber needs very high heating; also that it is found to give its own TL signal when exposed to UV radiation even from fluorescent room lighting.

Sintered dosemeters are formed by compression of the phosphor powder in a press mould followed by a high temperature sintering. A major limitation of some of these dosemeters is their physical fragility. They also have not found widespread use.

4.8. Glow curve and emission spectra of TL phosphors
4.8.1. *Introduction*

The return (stimulated by heating) of the electrons (or trapped holes) to the stable state is associated with a release of energy. This small fraction of the energy may be released as visible or UV light; this is thermoluminescence. If several traps of different depths are involved in this process, the recording of the light emission as a function of heating time (or temperature) results in a glow curve consisting of several glow peaks.

There are numerous parameters that affect the shape of the glow curve, such as LET and dose level of the radiation, thermal treatment (annealing) of the phosphor, pre-irradiation to high doses, pre-treatment with certain chemicals, etc.

The spectral energy distribution of the TL light that is emitted and forms the main peaks is the emission spectrum. Variations in the spectral response of the photomultiplier and/or the optical filters in different TL readers may explain, at least in part, the difference in the reported glow curves of the same phosphors by different authors. For the same reason, it is impossible to compare absolute sensitivities of phosphors having different emission spectra without normalising the response characteristics of the readout systems.

A typical glow curve would show one or more peaks (maxima) corresponding to the ionisation of traps at various energy levels. The relative amplitudes of the peaks indicate approximately the relative populations of the electrons in the various traps. The area under the glow curve or the height of the peak may be used as a measure of the absorbed dose in the phosphor or the exposure in air. When the peak height is used to measure absorbed dose or exposure, the heating cycle must be closely reproducible to avoid peak-height fluctuations[12].

4.8.2. *Glow curves*

The following glow curves refer mainly to the major families of phosphors, such as LiF, $Li_2B_4O_7$, $CaSO_4$, and CaF_2, and are taken from Proceedings of International Conferences on Luminescence Dosimetry[13-16].

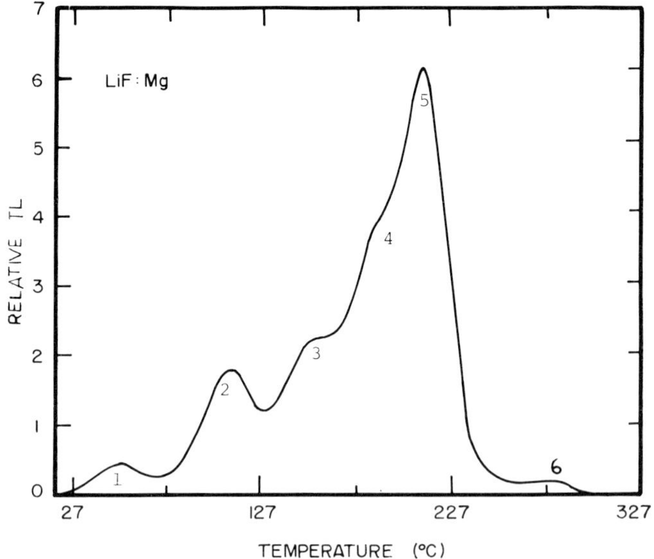

Figure 4.3. Glow curve of LiF family.

(i) LiF

Figure 4.3 shows the glow curve recorded 5 min after irradiation of LiF:Mg of the LiF family. Dose was approximately 5 Gy. Peak positions indicated are those found to give the best theoretical fit for this sample using computer assisted curve fitting techniques.

A recent study of TL in TLD-100 over the low temperature range 90-300 K has been made by Jain[17]. A total of nine glow peaks are observed in this temperature range and the most intense peak occurs at 137 K. At temperatures above 200 K, the emission spectrum contained only the 420 nm band. The sample had been irradiated by an X ray machine operating at 40 kV_p and 25 mA. The sample was exposed to a dose rate of 1.6×10^3 $R.min^{-1}$ through a Be window in the cryostat.

(ii) $Li_2B_4O_7$

Figure 4.4 shows the glow curve of $Li_2B_4O_7$:Mn. There are two peaks: one a rapidly decaying peak at about 55°C, the other a stable peak at 200°C.

(iii) $CaSO_4$

Figure 4.5 shows the glow curve of $CaSO_4$:Dy or $CaSO_4$:Tm.

(iv) CaF_2

The family of CaF_2 has different glow curves. Figure 4.6 shows the glow curves of this family.

(v) Others

Figure 4.7 shows the glow curve of nuclear reactor grade BeO phosphor. Figure 4.8 shows the glow curve of Mg_2SiO_4:Tb phosphor irradiated with an exposure dose of 0.0258 $C.kg^{-1}$. Figure 4.9 shows the glow curve of Y_2O_3–0.5 Al_2O_3:Tb (2×10^{-4} g atom mol. Y_2O_3) irradiated with 70 kVp X rays at $6.45 \times 10^{-4} C.kg^{-1}$. Figure 4.10 shows the glow curves of typical samples of various sulphates activated with rare earth ions.

4.8.3. *Emission spectra*

The emission spectrum can be measured using a wide spectrum photomultiplier, e.g. RCA 31034 with an AsGa

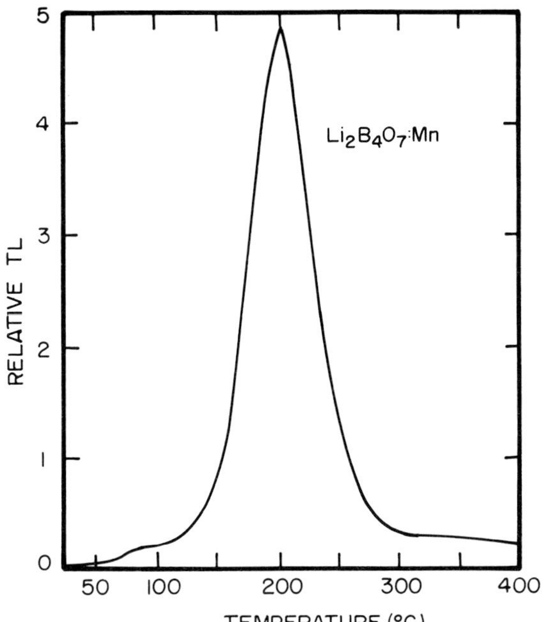

Figure 4.4. Glow curve of $Li_2B_4O_7$:Mn.

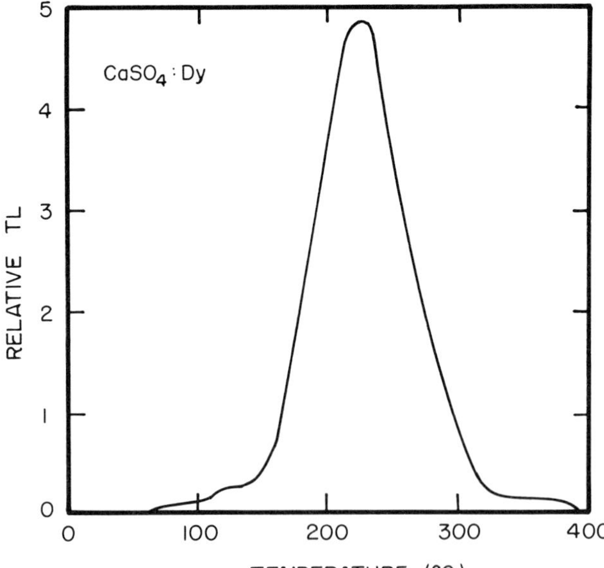

Figure 4.5. Glow curve of $CaSO_4$:Dy or $CaSO_4$:Tm.

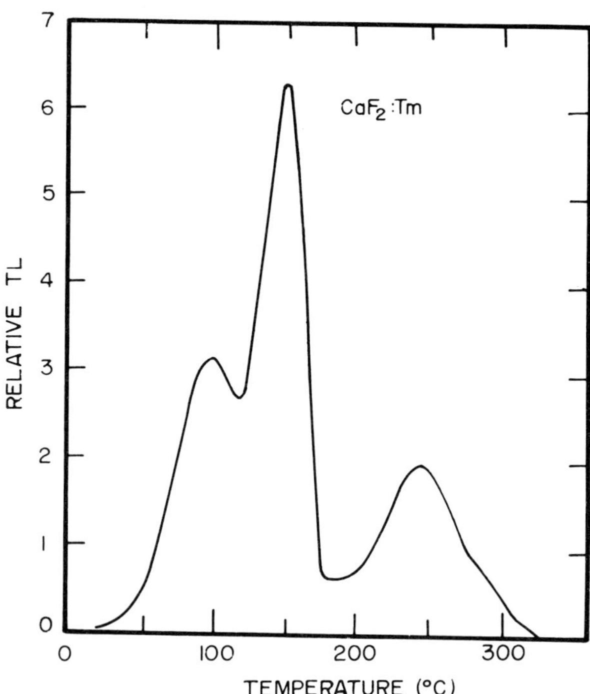

Figure 4.6. Glow curves of CaF$_2$ family.

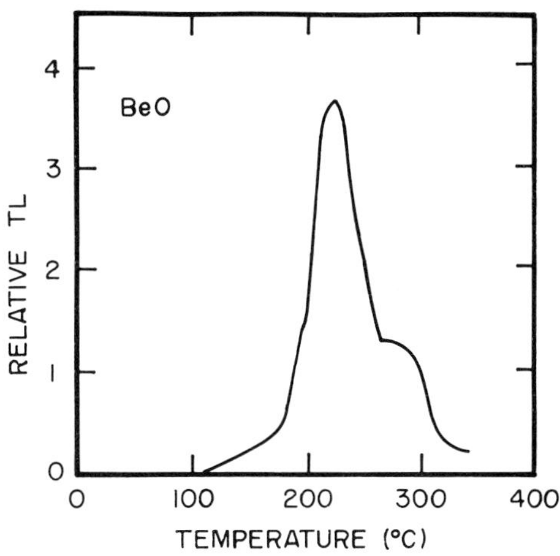

Figure 4.7. Glow curve of BeO.

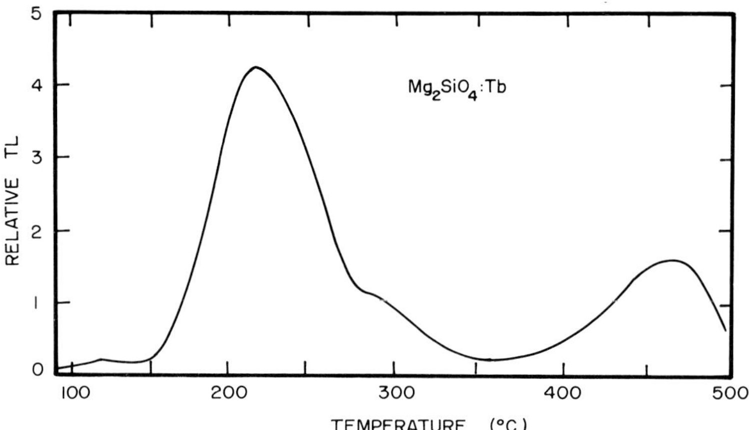

Figure 4.8. Glow curve of Mg_2SiO_4:Tb.

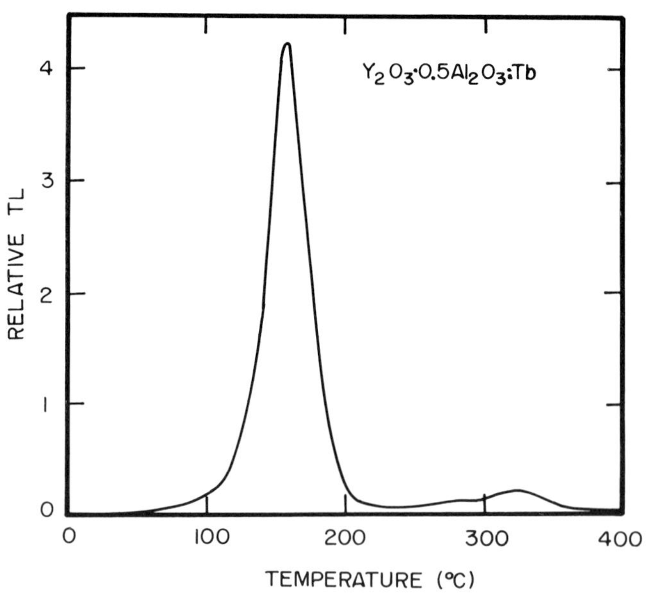

Figure 4.9. Glow curve of Y_2O_3–0.5 Al_2O_3:Tb.

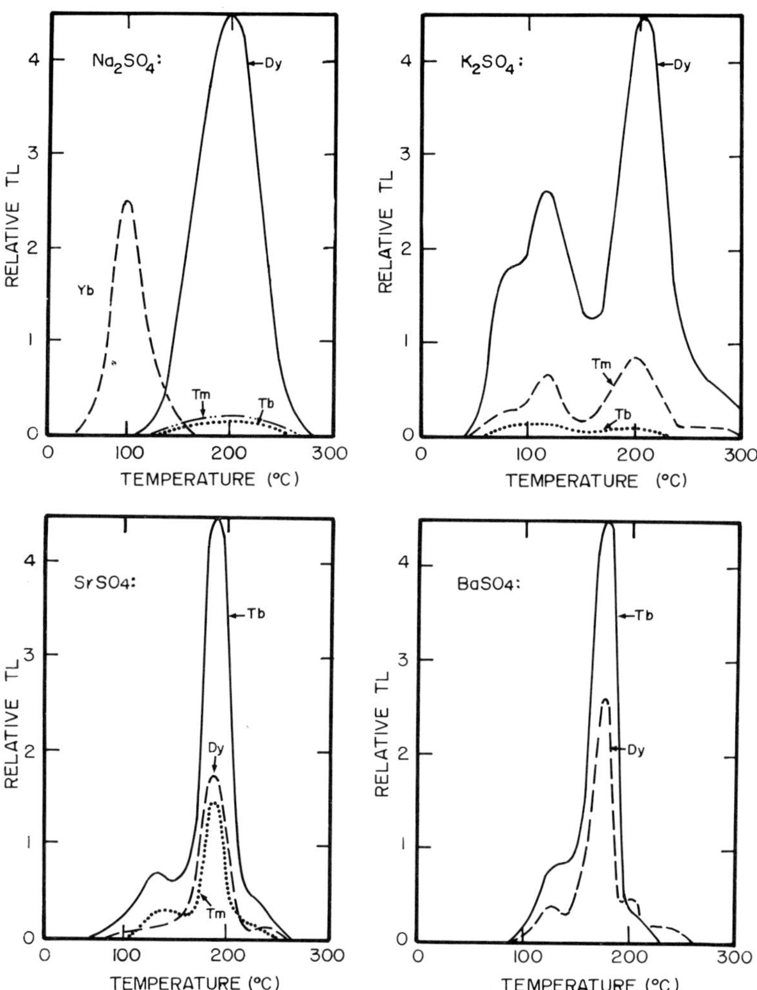

Figure 4.10. Glow curves of typical samples of various sulphates activated with rare earth ions.

photocathode, and sets of low-pass filters with progressive cut-off wavelengths. The following emission spectra refer mainly to the major families of phosphors such as LiF, $Li_2B_4O_7$, $CaSO_4$, and CaF_2.

(i) LiF:Mg,Ti

The spectral emission peaks of both LiF:Mg,Ti and PTL phosphors are at 400 nm in the blue region with full width at half maximum (FWHM) of approximately 115 nm. This emission corresponds with the peak maximum efficiency of S-11 and bi-alkali photomultiplier (PM) tubes[18].

(ii) $Li_2B_4O_7$:Mn

The spectral emission from the 200°C glow peak of $Li_2B_4O_7$:Mn phosphor is yellow-orange in colour, and the maximum appears at 600 nm, while that for the $Li_2B_4O_7$:Cu phosphor it is at about 360 nm[19].

(iii) $CaSO_4$:Tm and $CaSO_4$:Dy

The spectral emission peak of $CaSO_4$:Tm phosphor is at 452 nm, and at 480 and 570 nm for the Dy dopant. It has been shown that Dy gives essentially the same emission spectrum regardless of the host lattice.

(iv) CaF_2:Dy

The spectral emission peaks are at 460 and 480 nm in the blue and 570 nm in the yellow region.

(v) Others
(a) BeO. The spectral emission of BeO phosphor extends from the blue region of the visible spectrum into the UV, down to a 200 nm wavelength. A higher detection sensitivity can be achieved using a PM tube fitted with a quartz window and UV transmitting optics.
(b) Al_2O_3. The spectral emission varies considerably with the type of Al_2O_3. The spectral emission for ruby (Al_2O_3:Cr) is at 700 nm in the red region, while an emission peak at 425 nm in the blue is associated with those phosphors containing titanium.
(c) Mg_2SiO_4:Tb. The spectral emission peak of this phosphor is at 552 nm.
(d) MgB_4O_7:Dy,Tm. The spectral emission peaks of this phosphor

are at 480 and 570 nm.

A summary of all spectral emission maxima mentioned above is given in Table 4.3, and the spectra of some phosphors are illustrated in Figure 4.11.

4.9. Characteristics and dosimetric features of TL phosphors

The characteristics and dosimetric features of TL phosphors include dosimetric peaks, sensitivity, energy dependence, fading, annealing procedures, and dose ranges. While a summary of most items mentioned above is given in Tables 4.1-4.5, the annealing procedures are described below.

To prepare the dosemeter material for re-use, it must be heated again or annealed. After the traps have been emptied by heating at a high temperature for a sufficient length of time, and the phosphor has been allowed to cool, it is ready to be re-used. The exact prescription for the heat treatment of the crystal depends on the material itself and on the intended use of the crystal, e.g. the exposure level range. Thus, while the re-usability of TLD is one major advantage over other dosemeters, the annealing process causes a loss of earlier information about dose stored in the dosemeter.

There are a large number of thermoluminescent materials; in

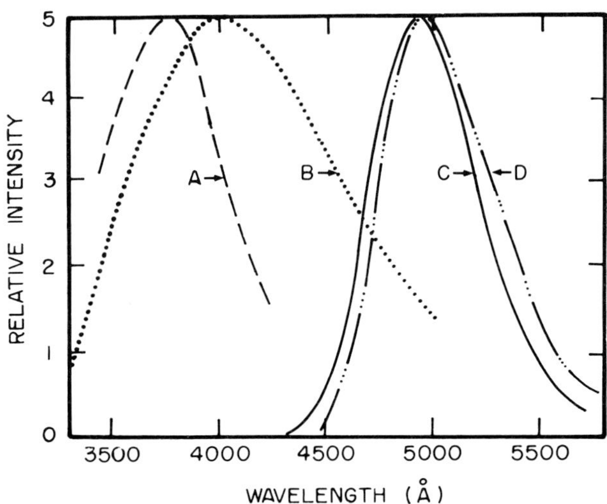

Figure 4.11. Emission spectra of TL from different phosphors. Curves: A, CaF_2:natural; B, LiF (TLD-100); C, CaF_2:Mn; and D, $CaSO_4$:Mn.

fact, as stated earlier, most materials thermoluminesce to some extent. To be more useful for a normal dosimetric application, a thermoluminescent phosphor must have a relatively strong light output and be able to retain trapped electrons for reasonable periods of time at temperatures encountered during the proposed application. This requirement limits useful phosphors to those with glow peaks at or above 80°C.

The annealing procedure is similar for every TL material and, in some cases such as LiF, it is very critical because if the procedure is not strictly the same, one can obtain significantly different results from repeated irradiations of phosphors to the same exposure.

The annealing procedures for several TL materials used in practice are summarised in Table 4.6. As can be seen, the heat treatment needed to anneal LiF (TLD-100) and to stabilise its response is somewhat complex and it must be preformed in a very rigorous and reproducible way.

The different annealing procedures given in Table 4.6 are necessary in order to empty all shallow and deep traps. Many TL materials also have some peaks at low temperatures. These peaks are likely to be emptied even at room temperature, particularly those which are below 100°C. Thus, in many cases it is necessary to perform a partial annealing before the reading in order to avoid a significant loss of information. This procedure is known as pre-readout annealing, and is mentioned in Table 4.6.

References

1. a. Rao, R. P. *The preparation and thermoluminescence of alkaline earth sulphide phosphors.* J. Mater. Sci. **21**, 3357 (1986).

Table 4.6. Annealing procedures.

TL phosphor	Annealing procedures	Pre-readout annealing
LiF (TLD-100)	1 h at 400°C + 24 h at 80°C*	10 min at 100°C
$Li_2B_4O_7$:Mn	30 min at 300°C	10 min at 100°C
$CaSO_4$:Dy	1 h at 400°C	
$CaSO_4$:Tm	1h at 400°C	
CaF_2:Dy (TLD-200)	1 h at 400°C	
CaF_2:Mn (TLD-200)	1 h at 400°C	
CaF_2:Dy (TLD-300)	1.5 h at 400°C	
CaF_2:Tm (TLD-300)	1.5 h at 400°C	
Mg_2SiO_4:Tb	1 h at 300°C	

*Alternatively, 5 min, at 400°C followed by 15 min at 100°C[20].

1. b. Eichholz, G. G. and Poston, J. W. *Principles of Nuclear Radiation Detection* (Ann Arbor Science Publishers) pp. 195-196 (1982).

2. Becker, K. *Solid State Dosimetry* (Cleveland, OH: CRC Press) pp. 34-36 (1973).

3. Pradhan, A. S. *Thermoluminescence Dosimetry and Its Applications.* Radiat. Prot. Dosim. **1**, 153-167 (1981).

4. McKinlay, A. F. *Thermoluminescent Dosimetry* (Bristol: Adam Hilger) pp. 32-33 (1981).

5. Oberhofer, M. and Scharmann, A. (eds) *Applied Thermoluminescence Dosimetry* (Bristol: Adam Hilger) (1981).

6. Yamashita, T., Nada, N., Onishi, H. and Kitamura, S. *Calcium Sulfate Activated by Thulium or Dysprosium for Thermoluminescence Dosimetry.* Health Phys. **21**, 295-300 (1971).

7. Azorin, J. and Gutierrez, A. *Determination of thermoluminescence parametres from glow curves in CaSO$_4$:Dy.* Nucl. Tracks Radiat. Meas. **11**, 167 (1986).

8. Weng, P. S. and Huang, C. Y. Health Phys. **26**, 104 (1974).

9. Oberhofer, M. and Scharman, A. (eds) *Applied Thermoluminescence Dosimetry* (Bristol: Adam Hilger) p. 113 (1981).

10. Furetta, C. and Li, Y. K. *Annealing and Fading Properties of CaF$_2$:Tm (TLD-300),* Radiat. Prot. Dosim. **5**, 57-63 (1983).

11. Jao, J. C. *Study on the Characteristics of TLD-300 (CaF$_3$:Tm).* MS Thesis, Institute of Nuclear Science, National Tsing Hua University, Hsinchu, Taiwan, ROC (1984).

12. Hsu, P. C. and Weng, P. S. *Application of Total TL Output and Peak Height in Dose Measurement.* Institute of Nuclear Science, National Tsing Hua University, Hsinchu, Taiwan, ROC (unpublished, 1984).

13. Schulman, J. H., Kirk, R. D. and West, E. J. *Use of Lithium Borate for Thermoluminescent Dosimetry.* IN Proc. 1st Int. Conf. on Luminescence Dosimetry, Stanford. US AEC CONF 650637 (Springfield, VA: NTIS) pp. 113-117 (1965).

14. Sakamoto, H., Hitomi, T. and Kotera, N. *Thermoluminescence of Y$_2$O$_3$-Al$_2$O$_3$:Tb Phosphors.* IN Proc. 2nd Int. Conf. on Luminescence Dosimetry, Gatlinburg. US AEC CONF 680920 (Springfield, VA: NTIS) pp. 27-42 (1968).

15. Nakajima, T. *On the Sensitivity Factor Mechanism of some Thermoluminescence.* IN 3rd Int. Conf. on Luminescence Dosimetry, Riso Report 249, pp. 466-479 (Danish AEC Research Establishment, Riso) (1971).

16. Yamashita, T. *A Series of Sulfate Phosphors Having Different Effective Atomic Numbers for Evaluating Radiation Energy.* IN Proc. 4th Int. Conf. on Luminescence Dosimetry, Krakow, Poland (Krakow: Inst. of Nucl. Phys) pp. 467-479 (1974).

17. Jain, V. K. *Thermoluminescence mechanism in LiF (TLD-100) from 90 to 300K.* J. Phys. D: Appl. Phys. **19**, 1791 (1986).

18. Knoll, G. F. *Radiation Detection and Measurement* (New York: John Wiley) p. 277 (1979).

19. Visocekas, R., Lorrain, S. and Marinello, G. *Evaluation of a Preparation of Li$_2$B$_4$O$_7$:Cu for Thermoluminescence Dosimetry.* IN Joint Symposium on Radiation Safety and Dosimetry, Rome University, Italy, 5-8 June 1984. Nucl. Sci. J. **22**, 61 (1985).

20. Regulla, D. F. IN *Instrumentation in Radiotherapy.* Report GSF-S 471 (Gesellschaft für Strahlen-und Umweltforschung mbH, Munchen, FRG) (1978).

CHAPTER 5

Thermoluminescence Instrumentation

P. S. Weng

5.1.   Introduction

Back in the years 1930-1950, the technique of measuring TL by glow curves was developed[1]. Since then this technique has been simplified so that many TL materials can be studied easily. TL can be observed visually, for example, by pouring an irradiated phosphor onto a heated electric iron in the dark. Instrumentation is therefore not very complicated: it consists of a heat source and a quantitative light detector. During the early days, most researchers in the field built their own readout equipment, usually based on electrical resistance heaters, photomultipliers (PM), and filter systems for discrimination against spurious signals and the infrared emission from heated components.

Instruments for the study and quantification of the TL from a phosphor are described in many reports; some are available commercially, and the commercial readers are of most interest to the average TLD user. They fall into two types: those specific to a particular design of dosemeter, and those which can accept a variety of dosemeter forms or phosphor types. They may be manual, semi-automated, or fully automated systems. The essential features common to every TLD reader are[2-4]: (1) a phosphor heating system, (2) a light collection and detection system, (3) a signal-measuring system, and (4) a display and recording system.

5.2.   Basic TLD reader

A simple TLD reader has been described by Cameron *et al*[5,6]. In 1967 its cost was less than US$200 for parts, it required about 20 h of construction time, had a variable heating rate and time, and accepted a large variety of sample sizes. Shutter action was accomplished by moving the PM tube mounted on a metal sheet, and the planchet was heated by a standard projection light bulb.

A few simple, general rules for the design of a TLD reader can be given here. For the heating of the phosphor, electric resistance heating and hot gas heating have been widely used. The reader should be designed in such a way that the PM tube sees as much of the phosphor, and as little of the other heated areas as possible. The spectral response of the PM tube should cover spectrum of the TL to be measured. The complex electronic circuitry of modern TLD readers is beyond the scope of this book. Most TL readers nowadays employ an integrator because the integrated TL light is more closely related to the dose and easier to measure reproducibly than the TL peak height. The instruments usually provide a digital data display with automatic switch of range. Additional provisions for an *X-Y* glow curve recorder, data print-out system, and even a microprocessor are also available. A schematic diagram illustrating features common to all TLD readers is shown in Figure 5.1.(a), while a typical Harshaw model is shown in Figure 5.1.(b).

5.3. Automatic and portable TLD reader systems

The existing commercial reader system can be roughly categorised as follows:

1. Moderately priced, rugged, simple-to-operate readers for routine use by technicians with little training, mostly in medical physics departments, but also for training or routine personnel dosimetry in small installations.

2. In the second category are the more sophisticated, versatile, and expensive laboratory type instruments which can be used for a wide variety of dosemeter types, and which normally offer adjustable heating programmes, cooled PM tubes, precise temperature control, and higher stability. Their main use is in

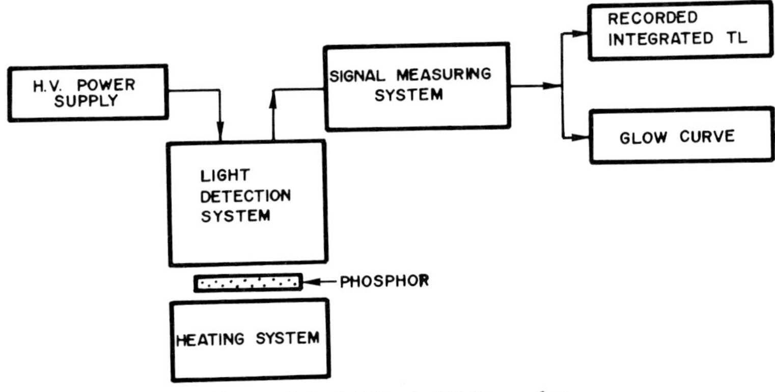

Figure 5.1. (a) Basic TLD reader.

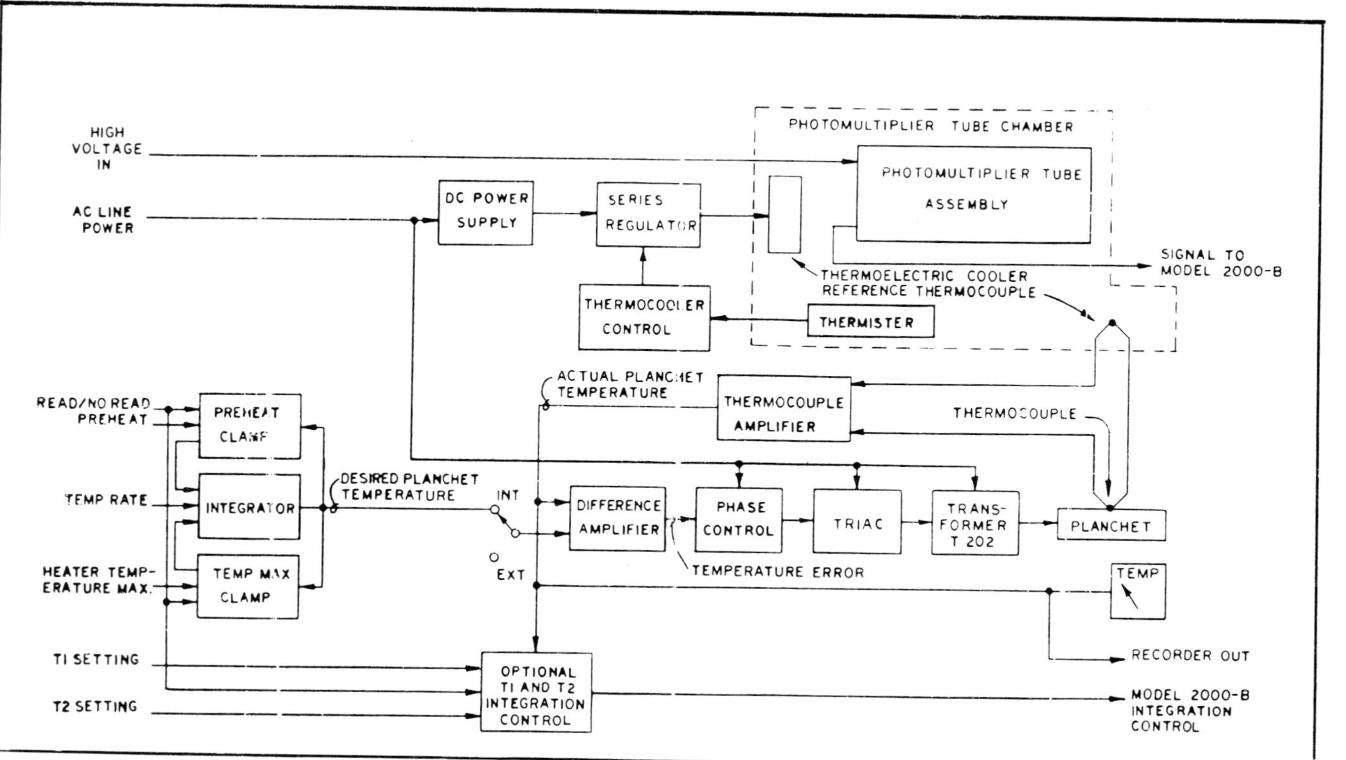

Figure 5.1. (b) Block diagram of Harshaw Model 2000-A and 2000-C TL detectors. (Courtesy Harshaw–Filtrol, U.S.A.)

research, and for special problems such as accurate measurement of low doses.

3. The final category is that of the automatic systems for large scale TLD personnel dosimetry for evaluation of a special type of badge. Such readers, whose use may become more widespread in the future in countries with high labour costs, usually also provide for automatic reading and print-out of the number of the individual dosemeter. They may require a back-up system or a dual system in case of malfunction.

Several larger installations built their own automatic readers. One at the Chalk River Nuclear Laboratories in Canada, for example, employs a semi-automatic reader for fingertip dosemeters and a fully automatic reader for the normal personnel dosemeters. The diagram is given in Figure 5.2.

In the nuclear power stations of Taiwan, ROC, the Harshaw automated TL analyser system (ATLAS) is in use. The ATLAS consists of a Model 2000 D TL detector, a Model 2000 B automatic integrating picoammeter, a vacuum pump and a printer. A heated laminar flow of nitrogen gas is used to heat the TLD chips. A turntable disc, which can accommodate 50 chips, is inserted into the Model 2000 D unit and the cover closed. The TLDs are brought into the read-out position by a vacuum probe which protrudes through holes in the turntable and lifts the detectors into the reading chamber. One 50-chip sequence is completed in approximately 12

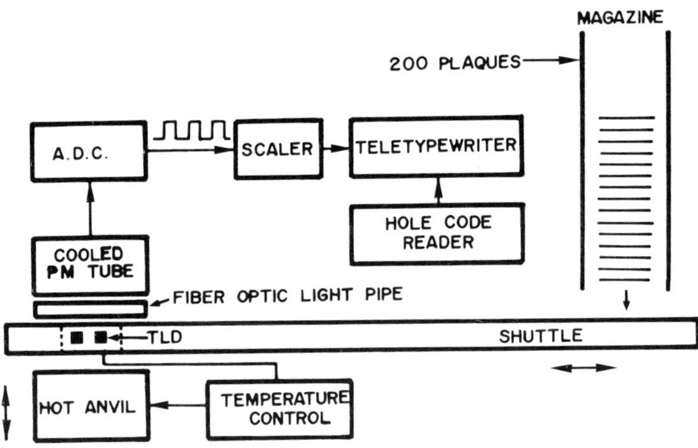

Figure 5.2. Schematic diagram of an automatic TLD personnel dosemeter reader.

min. Key parameters are monitored continuously and will stop the system and prevent further operation if a fault is detected. The system sensitivity may be checked whenever the system is not processing dosemeters by using an internal reference light source.

The portable TLD system consists of a small battery operated TLD reader and a special TLD bulb containing $CaSO_4$:Tm or $CaSO_4$:Dy TLD powder[7,8]. The $CaSO_4$:Tm or $CaSO_4$:Dy TLD powder is held to the heating current by a special glue. The heating current reaches the tray via gold coated contacts. The glass envelope is made from a low potassium content glass. This bulb dosemeter is inserted into a pen-like holder which enables astronauts, for example, to handle it easily in space. The bulbs are put into shielding capsules so that they can be worn in the spacecraft and to flatten their energy response. The bulb dosemeters were subjected to extremely severe vibration (20 *g*) and shock (300 *g*) tests without spoiling their technical parameters.

The block diagram of the TLD reader is shown in Figure 5.3. The bulb dosemeter is put into a light-proof compartment at the right hand side of the TLD reader. On turning the bulb in a clockwise direction a signal is generated for the electronic circuits by an optical switch and at the same time the bulb opens a light-proof window in front of the PM tube. The bulb is automatically evaluated according to the time diagram generated by the control unit. After the start signal, the high voltage unit only operates for a given time to stabilise the sensitivity of the PM tube, then the heating current switches on, but still the TL light from the bulb is not detected (preheating) in order to eliminate the short term fading originating from the low temperature glow peak of the $CaSO_4$:Tm or

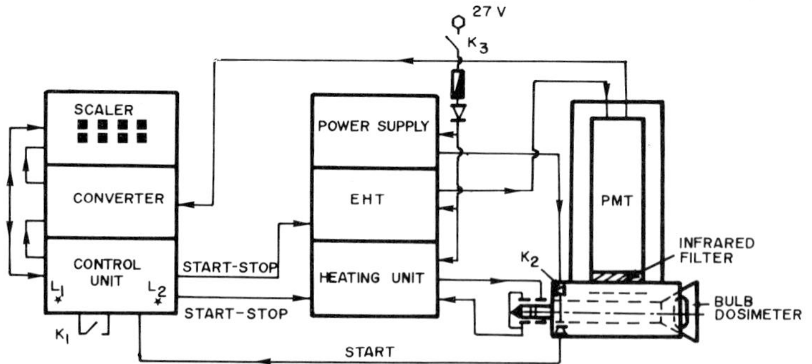

Figure 5.3. Block diagram of a portable TLD reader.

$CaSO_4$:Dy. After this preheating period the converter and the scaler begin operating to detect the light from the main glow peak of the bulb (evaluation). After this, the bulb is still heated but the light is not detected (post-annealing); this is done to minimise the residual dose of the bulb. In this way the bulb can be used to measure the dose from the ionising radiation without any additional annealing. After the evaluation the measured TL signal (in mrad) is displayed for about 5 s only by the 4-digit LED to minimise the power consumption of the reader. By pressing a button, the content of the scaler can be displayed any time until the next evaluation.

The whole reading process of one TLD bulb takes only one minute. The main technical details of the TLD reader are as follows: the volume is about 1 dm^3; the mass is about 1 kg; its electrical power consumption is especially low, being about 5 W. The measuring dose range is from 10 μGy to 100 mGy. In this range the linearity of the reader is better than 5%. The accuracy of the dose measurements above 0.1 mGy is better than 5%. The individual sensitivity of the bulbs did not differ from one irradiation to the other by more than 2-3%. The sensitivity of the reader was tested against the ambient temperature (the gain of the PM tube varies according to ambient temperature). A special PM tube, tested against vibration and selected for low dark current, is used in the reader (EMI 9824 NA), with a magnetic shield. It was found that the reader sensitivity increased by 8% when the ambient temperature decreased by 10°C. The sensitivity of the reader can be corrected for changing ambient temperature. In order to protect the PM tube against excessive shock and vibration a special container, a 3-dimensional isolator based on the frictional principle, was designed. This container was tested for shock and continuous sinusoidal vibration excitation.

Portable TLD systems can also be used in environmental monitoring. The evaluation of a bulb dosemeter needs only a few minutes, and the measured dose value can be reported back by radio. This is an important aspect during an emergency situation[9].

5.4.   Heating systems and DC power supplies

In the TLD system, various types of heating are used. They may be classified as planchet heating, gas heating, microwave heating, and laser heating.

5.4.1. *Planchet heating*

Planchet heating, or ohmic heating, is simply the heating of a read-out pan or tray on which the TLD is placed, either directly by the passage of a current through the pan or tray, or indirectly by bringing it into contact with an electrically heated element or block. The heating element referred to has low thermal capacity while the heating block provides a high thermal capacity heating. In both methods, temperature regulation is achieved by means of a thermocouple. Planchet heating is by far the most widespread form of heating.

5.4.2. *Gas heating*

Some TLD readers have been built utilising heated nitrogen as the heat transfer medium. This is only suitable for solid forms of TLD and not for loose powder. The heating is very rapid, and is particularly suited to automated TLD readers. The TLD, generally brought into the heating cavity by means of a vacuum needle, is heated by one or more hot nitrogen jets. The advantages of gas heating are a short heating cycle of about 10 s, low background signal, good reproducibility, easy loading procedures, and high versatility.

5.4.3. *Microwave heating*

Microwave heating, or RF heating, is the heating of graphite or other suitable material in which the TLD is placed. The graphite is heated by the induced current produced by a microwave induction heating coil. In one design the TLD is sealed onto a graphite tray using silicone resin adhesive. The TLD is positioned by a drawer system above a flat, water-cooled microwave coil (1 MHz), where it reaches its maximum temperature of 300°C in about 10 s. Disadvantages of microwave heating are the large amount of microwave power needed, and the difficulties of controlling the heating cycle. Direct temperature measurement is possible.

5.4.4. *Laser heating*

A laser is an active electron device that converts input power into a very narrow, intense beam of coherent visible or infrared light. The input power excites the atoms of an optical resonator to a higher energy level, and the resonator forces the excited atoms to radiate in phase.

Laser heating is one type of optical heating. The feasibility of non-contact laser heating of TL materials was studied by Braunlich *et*

al[10]. They showed that the laser heating technique yields a dramatic increase of the heating rate up to 2.0×10^{4}°C.s^{-1}, and thus increases the sensitivity of TLDs. The experiments performed by Braunlich *et al* were with continuous wave (CW) CO_2 lasers emitting up to 8 W of power. The use of infrared (IR) lasers has a number of advantages over visible or ultraviolet (UV) lasers. They are: (1) no inter-band transitions occur, and defect level-band transitions are negligible at the power densities employed, (2) IR photons produce no new defects that may take place as a result of excitation recombination, and (3) optical discrimination between the IR photons and the TL emission is readily achieved with proper choice of filters, e.g. a thin sapphire or quartz window between the PM tube and the emitting phosphor.

In the following, only $CaSO_4$:Dy will be used to illustrate laser heating. In order to minimise thermal losses, thin layers of $CaSO_4$:Dy powder were precipitated onto thin glass microscope cover slides instead of thick optical surface plates. Prior to exposure of the samples to the focussed beam of a CW CO_2 laser with a focal

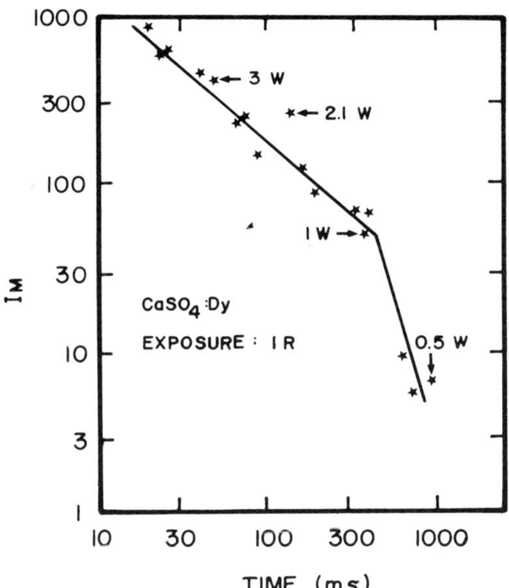

Figure 5.4. Intensity of the nominal 210°C TL glow peak of $CaSO_4$:Dy plotted against time of peak occurrence after onset of laser heating pulse.[10]

spot of 0.6 mm diameter, the TL material was exposed to 2.5×10^{-6} to 2.58×10^{-4} C.kg^{-1} from a ^{137}Cs source. Up to 25 different sites could then be evaluated on one sample, which could be moved with the aid of an *X-Y* positioner. The glow curve of CaSO$_4$:Dy (Figure 4.5 of Chapter 4) is known to consist of a prominent peak at 210°C preceded by a weaker one at 110°C which could be removed readily by slowly preheating the entire sample to about 110°C on a hot plate. The maximum intensity I_m of the 210°C peak as a function of time of its occurrence after onset of the laser exposure is shown in Figure 5.4 for various laser powers.

Of interest in connection with microdosimetry is the small size and weight of the TL sample heated during laser exposure of a single site on the microscope cover slide. The data displayed in Figure 5.4 were taken with a 45 μm thick layer. The exposed volume (<1 mm diam.) had a weight of only 40 μg.

Appendix IV gives further details on laser heating of TL phosphors.

5.5.   High voltage power supply

The DC power supply in a typical commercial TL reader, with a negative voltage output used for the PM tube in the TL set-up, is illustrated in Figure 5.5. Circuitry for the adjustable –500 V to

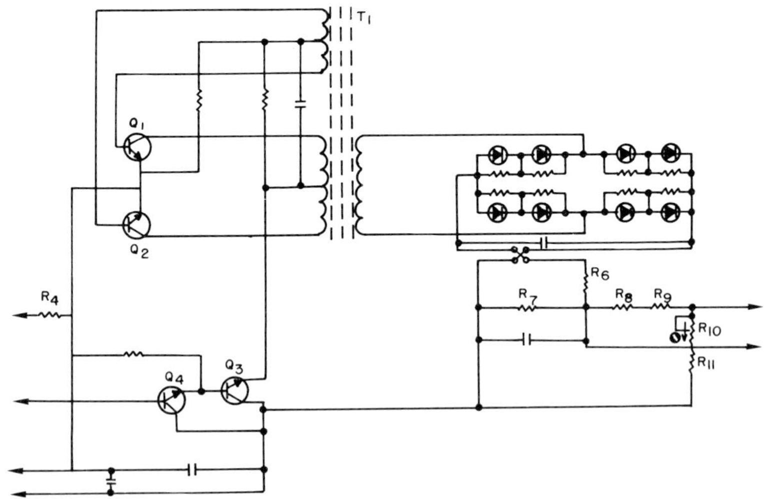

Figure 5.5. High voltage power supply.

1500 V DC power supply is located on the high voltage board.

The primary of the power supply consists of transistors Q1 and Q2 operated in an astable multivibrator circuit. The feedback necessary to sustain oscillation is obtained from a separate winding on transformer T1. The exact frequency of oscillation is not critical and may vary slightly with output voltage.

Output voltage is adjusted by selecting the operating voltage applied to the collectors of Q1 and Q2. This is accomplished by a resistor divider consisting of R4 and a potentiometer on the panel meter. The resistance of the potentiometer determines the base voltage of Q4. Q3 is an emitter follower which sets the operating voltage applied to the collectors of Q1 and Q2.

The transformer secondary winding is connected to a full wave bridge rectifier circuit. Each of the diodes in the bridge is shunted by a resistor which equalises the current through each diode. The positive leg of the bridge is connected to the amplifier ground. The negative leg is connected to the high voltage output through a divider formed by R6 and R7.

Resistors R8, R9, R11 and potentiometer R10 form another voltage divider network that permits the high voltage to be displayed on the panel meter. R10 is a meter scale calibration adjustment.

5.6. Temperature controllers and programmers

Different forms of TLDs vary considerably in their thermal properties. The purpose of the heating cycle is to heat the phosphor until the electrons are liberated from the dosimetry traps. An additional function is to remove rapidly any low temperature traps by means of a low temperature hold (pre-read). At high exposure levels it is often necessary to use a high temperature hold (anneal) to ensure complete erasure of all electrons from dosimetry traps and this is generally performed in separate ovens. A heating cycle used to read out LiF Teflon disc dosemeters is illustrated in Figure 5.6. The recorded integrated TL signal is denoted by the shaded area. Research readers should be capable of reading out many different phosphor materials with a wide range of glow peak temperatures. They should be capable of providing a wide range of linearly controlled heating rates. On the other hand, high-throughput personnel dosemeter reader systems require fast and reproducible heating that is not necessarily linear. Most commercial readers offer a compromise between the heating requirement for a research reader system and a personnel dosemeter reader system. Some of them offer flexible research heating modules as plug-in accessories.

5.6.1. *Linear heating rate*

In the early days a linear temperature rise was applied to most TLD readers. This method is still in use. The glow curve analysis can be performed using a linear heating rate. In most cases the temperature rise is followed by a plateau as shown in the temperature curve of Figure 5.6. Response measurements can often be done by choosing a distinct temperature or time interval for the integration of the TL signal. It is a programmable heating cycle.

5.6.2. *Hyperbolic heating rate*

For large scale routine measurements, the linear heating rate tends to be too time consuming. A rapid rise in temperature such as that achieved by a hyperbolic heating rate is then preferred. The only requirement is that the temperature, as a function of time, should be highly reproducible. This requirement is also applied to the linear heating rate.

Generally, it is not easy to select a special integration interval, as the TL begins too quickly. The pre-read heat treatment and the post-

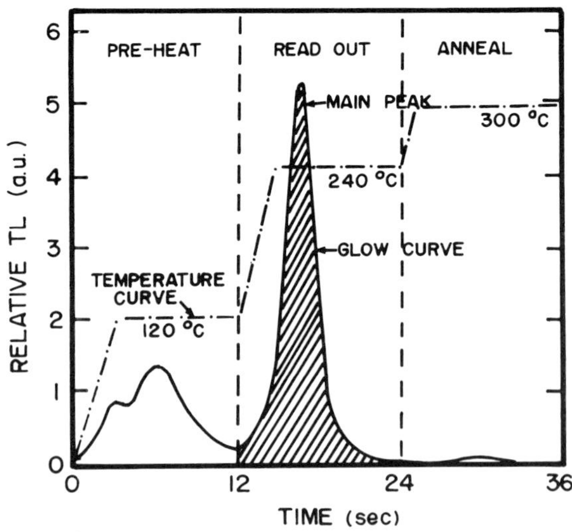

Figure 5.6. Heating cycle and glow curve for readout of LiF Teflon disc.

read anneal can be performed in an oven. Pure nitrogen gas is most commonly used in suppressing spurious signals when the TLD is heated.

5.7. Furnaces for phosphor preparation and annealing

It has been stated that ovens are needed for TLD pre-read heat treatment (up to 150°C) and post-read anneal (up to 400°C). Good temperature control and constancy of the pre-selected temperatures by means of a precision temperature regulator over extended periods of time are required. Forced convection is desirable since there are then no temperature gradients within the oven. It is convenient to have CO_2 cooling for controlled cooling down of the TLDs. The oven should be brought up to the pre-selected temperature and allowed to stabilise thoroughly before TLD insertion, and it should be reserved exclusively for TLD annealing in order to avoid contamination.

5.8. Thermocouples and calibration

In planchet heating (section 5.4.1), a heating element with low thermal capacity is mentioned. Because of the low thermal capacity of the heater planchet, it is fairly simple to control its temperature versus time profile, just by controlling the generating voltage. A thermocouple is used to give a negative feedback signal to the voltage control unit. The thermocouple is welded to the heater element, which can be brought into close contact with the planchet.

It is not easy to establish the precise temperature of the phosphor during readout, even when using a heating element. The temperature indication actually relates to the location where the thermocouple is welded. There are several methods for temperature calibration. They are, for example, infrared (IR) measurement, observing the boiling of alcohol, water, or glycerine on the planchet, and test runs with TLD phosphors that have well established peaks.

5.9. Infrared filters

Commercial readers are generally equipped with filters that reflect and absorb photons in the IR wavelength region but have high transmission in the whole visible region. The optical filters are to isolate the housing of the light detector thermally from the readout chamber. The isolation is obtained by preventing heat convection to the light detector and by reflection and absorption of IR radiation. If only one TL material is used the signal-to-noise ratio (S/N) can often be increased significantly by an optical filter with a narrower band pass. The increase is due to reduction of the noise caused by loss in the detected incandescent light and the non-radiation-induced signals from the TLDs.

5.10.   Photomultipliers and current amplifier

The light emitted by a TLD after being exposed to 2.58×10^{-7} C.kg^{-1} is of the order of 10^{-3} lumen (lm). Only PM tubes are considered to be applicable in TLD to measure such low level radiation. They are highly sensitive and have a large dynamic range, 10^{-13}–10^{-6} lm.

The use of the photometric term lumen in this context is not entirely appropriate as it refers to the photopic response of the human eye (peak 550 nm). However, it is used by PM tube manufacturers.

A PM tube is chosen for optimal cathode sensitivity, matching the wavelength typical of the TL material under consideration. A low response to other wavelengths such as IR is preferred.

Sensitivity and S/N are affected by a number of parameters as described below.

When a PM tube is operated in complete darkness some electrons are still emitted from the photocathode mainly by thermionic emission. The current due to these electrons is called the dark current. It may be reduced by cooling the tube below normal ambient temperature. The passage of electrons along the length of the PM tube is influenced by the presence of external electric and magnetic fields. The effect can be minimised by the use of a mumetal shield surrounding the PM tube and maintained at cathode potential. Fatigue may result if a PM tube is subjected to prolonged or repeated exposure to a source of light equivalent to a tube current of $>10^{-6}$A, reducing the S/N of the system. In short, dark current, ambient temperature[11], fatigue, ageing, magnetic fields, discharges due to high tube voltage, high tension variations, and current leakage are the parameters to be considered.

The PM signal is typically at the 10^{-12} A level. Generally it needs further amplification. Two methods are usually applied: peak height measurement and peak area measurement.

The peak height with maximum amplitude in the glow curve is taken as the relevant signal. It is measured by converting the PM current into a voltage by which a capacitor is charged. The maximum charge of the capaictor is then measured.

For the peak area the digitised PM signal obtained by a charge-to-pulse converter is fed into a counter. The number of pulses at the end of the readout time is proportional to the integral light output of the TLD. A ratemeter, connected to the converter output, provides the signal used for plotting glow curves. This is the method most often applied in TLD.

The method of photon counting is still used[12]. The PM anode charge pulses started by single photons interacting with the photocathode are counted. When the photon counting technique is used, glow curve recording has to be done with a ratemeter or a multichannel counter such as a pulse height analyser operated in the multiscale counting mode.

5.11.   Readout and display devices

Modern TLD readers are usually equipped with additional devices which extend their capabilities such as glow curve plotting, output writing and response handling, and automation.

Several types of TLD reader are provided with a special output of the ratemeter signal to which an X-Y plotter or X-t plotter can be connected. It is sometimes convenient to use a pulse height analyser to record the TL signal against time. The glow curve signal can also be fed directly into a computer or microprocessor for pattern recognition.

Almost all TLD readers can be equipped with an output recorder, such as a simple printer, teletype or similar device. More sophisticated systems sometimes have the capability of setting a lower and an upper level, subtracting a predetermined background, or multiplying the response by a calibration factor. They are generally computer compatible, and all data handling and instrument control is done by a computer.

Automatic TLD readers are designed either for personnel monitoring purposes or for automatic readout of loose solid TLDs. Automation usually improves accuracy considerably. The greatest reproducibility can also be obtained by performing no annealing and using a single optimum heat treatment in an automatic reader.

5.12.   **Simultaneous recording of glow curves and emission spectra**

The emission spectrum can be measured using a wide spectrum photomulitiplier (type RCA 31034, with an AsGa photocathode) and sets of low pass filters with progressive cut-off wavelengths[13]. These filters are mounted four at a time on a rotating support. Thus, glow curves of $Li_2B_4O_7$:Cu for four domains are measured at a time as shown in Figure 5.7. As compared with the use of a spectrometer on-line, this method gives a poorer spectral

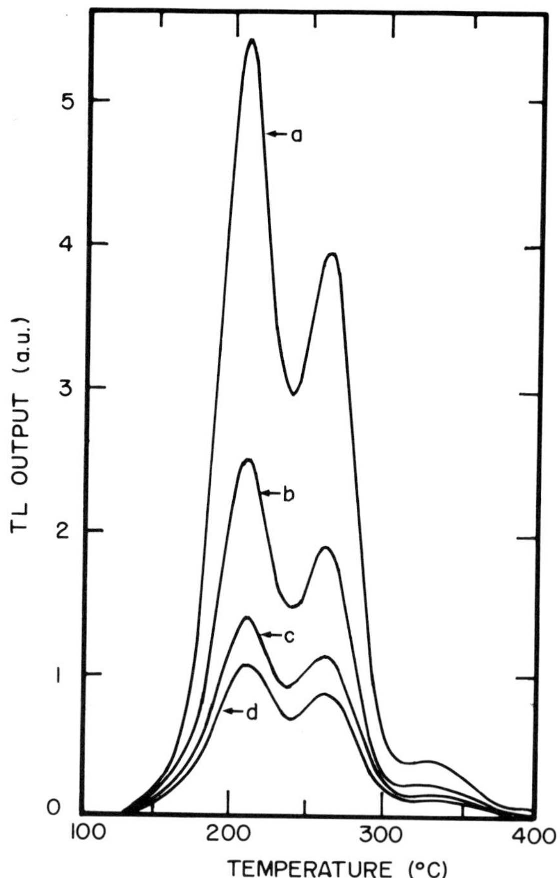

Figure 5.7. Glow curves of $Li_2B_4O_7$:Cu. Curve a, cut off at $\lambda < 345$ nm, curve b, <400 nm, curve c, <420 nm, and curve d, <435 nm.

definition but a better sensitivity. Spectra are obtainable at any temperature where a TL emission can be detected.

The glow curves shown in Figure 5.7 were obtained after 90 Gy irradiation and 120 days storage with 5 min holding at 127°C. Four filters on a rotating support were used together during the same heating period yielding four simultaneous glow curves. Curve a was cut off at $\lambda <345$ nm, curve b <400 nm, curve c <420 nm, and curve d<435 nm.

Thus, the emission spectrum of $Li_2B_4O_7$:Cu shown in Figure 5.8 can be obtained. It confirms emission spectra already published. In addition, it appears to be invariable all along the TL process and for all the TL peaks including the low temperature unstable one.

This spectrum is found identical for all batches of the phosphor and for various doses considered, up to 100 Gy. It is efficiently covered by the spectral response of the PM, comparable to the spectrum of LiF but in contrast to that of $Li_2B_4O_7$:Mn. Effective shielding against thermal radiation, having its wavelength lying outside the spectral response of PM, is also automatically provided.

Only 12% of the energy is emitted for λ greater than 410 nm, while 37% is emitted in the UV domain with λ shorter than 345 nm. A good transparency of the optics used in the reader down to 300 nm is strongly advisable.

This spectrum is very similar to the emission spectrum of

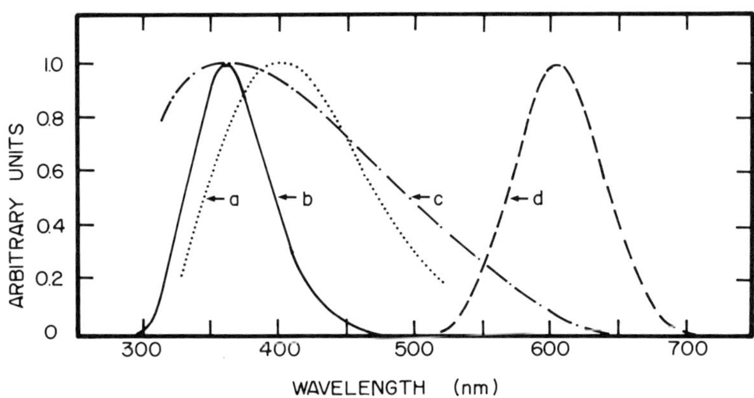

Figure 5.8. Emission spectra of $Li_2B_4O_7$:Cu and other TLDs. Curve a, LiF; curve b, $Li_2B_4O_7$:Cu, curve c, spectral sensitivity for bialkali photocathodes; curve d, $Li_2B_4O_7$:Mn.

monovalent copper used as a dopant in different lattices such as alkali halides, which therefore confirms its attribution to the copper activator.

5.13.   Monochromators for recording emission spectra

A monochromator is an instrument used to supply a beam of light having some desired narrow range of wavelengths. Although sometimes used in photochemistry experiments as a single instrument, it is usually part of a spectrophotometer. In that case, it has the following component parts: (1) an entrance slit producing a well defined beam of heterochromatic radiation, (2) a prism or diffraction grating dispersing the incident radiation into a continuous spectrum, (3) some device to rotate the prism or grating so that the desired wavelengths of exit radiation are obtained, and (4) an exit slit allowing a narrow band of wavelengths. A monochromator becomes a spectrophotometer if it is preceded by a source of continuous radiation and followed by a detector, an amplifier, and a device for measuring the amplified output signal.

The common form resembles a prism spectroscope. White light from a heated TLD, entering the fixed collimator as usual through a narrow slit, is dispersed by a special 4-sided prism after one internal reflection. The image of the resulting spectrum falls on a metal plate, and a second, fixed, narrow slit in this plate allows light of approximately only a single wavelength to emerge. The purity of this beam depends on the narrowness of the two slits, also upon diffraction at the slits and scattering by the prism. The prism is mounted so that it can be rotated by a tangent screw, thus bringing different parts of the spectrum to the second slit as desired. The tangent screw may be provided with a graduated head reading directly in wavelengths; but too much reliance must not be placed on the indicator, since the adjustment is affected by mechanical factors.

Instead of a prism, a diffraction grating may be used to disperse the light. Such instruments are called grating monochromators.

Measurements of TL spectra are of value since the emission spectrum may give information about the recombination process. Knowledge of the emission spectrum is also required in order to optimise the function of the light detector and the optical filter.

There are two main problems connected with emission spectrum measurements: (1) the low intensity light level at the monochromator input requires heavily irradiated TL samples, and (2) the light signal varies with time. The low light levels at the monochromator input cause worry partly due to the differential nature of the measurements. Often, a compromise is necessary between wavelength resolution and sensitivity. The sensitivity problem is considerably reduced when the spectra are integrated over the entire readout cycle, however, the emission spectrum may change with the readout temperature and if this is the case it is necessary to use a spectrometer with a rapidly scanning monochromator.

Three types of rapidly scanning monochromators have been applied to TL measurements. Harris and Jackson[14,15] used a highly sensitive grating spectrometer which allowed emission spectra measurements at low dose in the linear portion of the TL dose-response curve. This is an important achievement in the testing of models relating to supralinearity, e.g. in determining whether the creation of luminescence recombination centres (new optical absorption bands) is contributing to the TL dose response. The Harris-Jackson monochromator consisted of a prism and a rotating mirror. With this type of monochromator the design of the wavelength control system is critical because of the small dispersion of the prism for long wavelengths. Spectrometers with rapidly scanning grating monochromators have also been used[16-19] but these spectrometers are usually less light sensitive than prism spectrometers. A third type of rapidly scanning spectrometer has been described by Bailiff *et al*[20]. The monochromator, which consisted of 16 interference filters mounted on the periphery of an aluminium disc rotating at 60 rpm had a relatively low wavelength resolution of approximately 10 to 20 nm, determined by the bandwidth of the filters. An advantage of this type of spectrometer is that large area TL samples can be employed.

Huntley *et al*[21] describe a TL spectrometer which employs a custom built concave holographic grating and microchannel plate photomultiplier (2.5 cm diam. photocathode). It can record spectrum from 300 - 750 nm while the sample is being heated. The spectrometer is sufficiently sensitive to allow the measurement of spectra from naturally occuring minerals found in sediments.

None of the above instruments are sensitive enough to measure emission spectra for non-radiation-induced signals and therefore, in these measurements, considerably cruder methods (e.g. optical filters) must be employed. Since it is often difficult to reproduce measurements of non-radiation-induced TL, two PM tubes must be

Table 5.1. A few commercial TLD readers.

	Teledyne Isotopes Model 7300C	Harshaw Model 2000	Pitman Model 654	Victoreen Model 2800	Eberline Model TLR-5	Panasonic Model 505	Harshaw Model 3000*
N_2 flowmeter and front panel valve, standard	yes	no	yes	no	no	uses hot air	no
Linear non-range changing integrator	yes	no	yes	no	yes	no	no
Glow curve voltage output	yes	yes	no	yes	yes	no	yes
Temperature voltage output	yes	yes	yes	no	yes	no	yes
Fixed preheat, readout and anneal cycle	yes	no	yes	yes	no	is not required	no
Interchangeable heating elements	yes	with difficulty none for Teflon	yes	no	no	yes	with difficulty non for Teflon
Ultra-stable H.V. supply	yes	no	no	no	no	no	no
Fixed preset sensitivity range	yes	no	no	no	no	yes	no
Versatile ramp heating	yes	yes	yes	no	no	no	yes
Low background phototube, standard	yes	no	yes	yes	no	yes	no

Designed specifically for Teflon, chips, and powder	yes	no	yes	no	no	no – requires Panasonic dosemeters	no
Thermoelectric photomultiplier tube cooler	no	yes	no	no	no	no	no
Reads to less than 1 mR by experience	yes	yes	yes	no	no	yes	no
Built-in light source	yes	yes	yes	yes	yes	yes	yes
Temperature meter	yes	yes	yes	yes	yes	no	no
High voltage meter	yes	yes	no	no	no	no	no
Number of count decades	6	4	6	3½	5	3	3½
Zero supression	no	yes	yes	yes	no	no	yes
Front panel control of all parameters	yes	no	yes	no	no	no	no
Instant instrument calibration method	yes	no	yes	no	no	no	no
Plug-in I.C.'s for easy servicing	yes	no	no	no	no	no	no
PRICE. USA	$5,190	~$6,000	~$8,500	~$3,500	~$2,500	~$5,500	~$4,000

used simultaneously, one measuring filtered TL and the other measuring unfiltered TL for reference purposes.

5.14.   Instrumentation for TL dosimetry

There are several commercial TLD readers. A few of them are included in Table 5.1. The prices shown are simply an indication since they are subject to change. Further information should be obtained from the manufacturer.

5.15.   Instrumentation for TL dating

In archaeological dating by TL techniques the dominant instrumental problem is to achieve a sensitivity high enough for detection of the weak TL from archaeological samples. All the procedures for increasing the S/N ratio must therefore be considered when designing an instrument for TL dating. Special attention must be paid to the problems imposed by the relatively high readout temperatures ($\sim500°C$). Reduction of incandescent light emission and PM tube stabilisation are, therefore, especially important. The pre-dose TL dating technique for example, uses the $110°C$ glow peak of quartz and involves heating the quartz samples to $500°C$. This annealing can be carried out either in the TL reader or in an external oven. If performed in the TL reader, it may be necessary to use a cooling system for the readout chamber to avoid excessive heating of the sample changer and the readout chamber due to repeated heatings. A systematic error may otherwise be introduced by the increasing temperature since the $110°C$ glow peak in quartz has a relatively short half-life even at room temperature, about 3.5 h at $110°C$, and the fading rate increases with increasing temperature.

Samples used in TL dating often consist of small grains of quartz or feldspar. Since the grains have a relatively large surface-to-volume ratio, non-radiation-induced signals may also constitute a significant problem. The main methods for reducing these signals are readout in an oxygen-free atmosphere, e.g. nitrogen, and the use of optical filters to suppress selectively the non-radiation-induced TL. The purity of the inert gas is critical; the oxygen content must not be greater than a few parts per million. A nitrogen flow of several litres per minute is more efficient in reducing the non-radiation-induced signals than a static nitrogen atmosphere, probably because of desorption, in the former case, of oxygen from the sample during heating.

The TL dating of an archaeological object requires many TL measurements on different samples so that manual sample changing is overly time consuming. An automated TL readout instrument especially designed to circumvent this problem is necessary. The instrument contains a ^{90}Sr-^{90}Y β source for irradiation of the samples. Two irradiation series, each followed by readout of the samples, are performed for the determination of the archaeological dose. The individual calibration allows sample mass corrections in a relatively easy way. The whole irradiation and readout procedure is controlled by a microprocessor and takes about two hours[2].

References

1. Daniel, F., Boyd, C. A. and Saunders, D. F. *Thermoluminescence as a Research Tool.* Science **117**, 343 (1953).

2. Spanne, P. *TL Readout Instrumentation.* IN Thermoluminescence and Thermoluminescent Dosimetry, Vol. III, ed. Y. S. Horowitz (Boca Raton, Florida: CRC Press) (1984).

3. McKinlay, A. F. *Thermoluminescence Dosimetry* (Bristol: Adam Hilger) p. 118 (1981).

4. Julius, H. W. *Instrumentation in Thermoluminescence Dosimetry.* Radiat. Prot. Dosim. **17**, 267 (1986).

5. Becker, K. *Solid State Dosimetry* (Cleveland, OH: CRC Press) p. 48 (1973).

6. Cameron, J. R., Suntharalingam, N. and Kenney, G. N. *Thermoluminescent Dosimetry* (University of Wisonsin Press, Madison, Wisconsin, USA) p. 75 (1968).

7. Fehér, I., Deme, S., Szabó, B., Vágvölgyi, J., Szabó, P. P., Csöke, A., Ránky, M. and Akatov, Yu. A. *A New TLD System for Space Research.* KFKI-1980-33 (Hungarian Academy of Sciences, Central Research Institute for Physics, Budapest) (1980).

8. Fehér, I., Szabó, B., Vágvölgyi, J., Deme, S., Szabó, P. O. and Csöke, A. *New Advanced TLD System for Space Dosimetry.* KFKI-1983-99 (Hungarian Academy of Sciences, Budapest) (1983).

9. Szabó, P. O., Fehér, I., Deme, S., Szabó, B., Vágvölgyi, J. and German, E. *Environmental Monitoring with a Portable TLD System.* Radiat. Prot. Dosim. **6**, 100-102 (1984).

10. Braunlich, P., Gasiot, J., Fillard, J. P. and Castagné, M. *Laser Heating of Thermoluminescent Dielectric Layers.* Appl. Phys. Lett. **39**, 769 (1981).

11. Lan, C. Y. and Hsu, P. C. *Noise Study of Photomultiplier Tube.* Nucl. Sci. J. **19**, 232 (1982).

12. Hsu, P. C., Lan, C. Y. and Weng, P. S. *Photon Counting in Thermoluminescent Measurement for the Determination of Low Exposure.* Nucl. Sci. J. **14**, 13 (1977).

13. Visocekas, R., Lorrain, S. and Marinello, G. *Evaluation of a Preparation of $Li_2B_4O_7$:Cu for Thermoluminescence Dosimetry.* Nucl. Sci. J. **22**, 61 (1985).

14. Harris, A. M. and Jackson, J. H. *A Rapid Scanning Spectrometer for the Region 200-850 nm: Application to Thermoluminescent Emission Spectra.* J. Phys. E: Sci. Instrum. **3**, 374 (1970).

15. Harris, A. M. and Jackson, J. H. *The Emission of Thermoluminescent Dosimetry Grade Lithium Fluoride.* J. Phys. D: Appl. Phys. **3**, 624 (1970).

16. Strash, A. M. and Madey, R. *The Thermoluminescence Spectrum and the Energy Conversion Efficiency of Lithium Fluoride.* IN Proc. 2nd Int. Conf. Luminescence Dosimetry, US AEC CONF-680920 (Springfield, VA: NTIS) p. 607 (1968).

17. Oltman, B. G., Kastner, J. and Paden, C. *Spectral Analysis of Thermoluminescence Curves.* IN Proc. 2nd Int. Conf. on Luminescence Dosimetry, p. 623 (1968).

18. Fairchild, R. G., Mattern, P. L., Lengweiler, K. and Levy, P. W. *Thermoluminescence of LiF TLD 100 Dosimeter Crystals.* IEEE Trans. Nucl. Sci. **NS-21**, 366 (1974).

19. Fairchild, R. G., Mattern, P. L., Lengweiler, K. and Levy, P. W. *Thermoluminescence of LiF TLD 100: Emission Spectra Measurements.* J. Appl. Phys. **49**, 4512 (1978).

20. Bailiff, I. K., Morris, D. A. and Aitken, M. J. *A Rapid Scanning Interference Spectrometer: Application to Low-level Thermoluminescence Emission.* J. Phys. E: Sci. Instrum. **10**, 1156 (1977).

21. Huntley, D. J., Godfrey-Smith, D. I., Thewalt, M. L. W. and Berger, G. W. *Thermoluminescence spectra of some mineral samples relevant to thermoluminescence dating.* J. Lumin. **39**, 123 (1988).

CHAPTER 6

Models and Theories of Thermoluminescence

K. Mahesh and C. Furetta

6.1.   Introduction

Despite the complicated nature of the TL phenomenon, a considerable amount of theoretical work has been done, making at least the more important features of TL understandable. Principles and methods featuring the theoretical aspects of TL were described in Chapter 3; they are, at best, approximate and analytical only, and do not always agree with the experimental observations. For this reason a more general approach, or one that is specific to a given material, is attempted. These attempts not withstanding, Niewiadomskii[1] has analysed in detail the experimental results for TL in LiF by taking into account various possible models of luminescent centres and concludes that none of the models that have been developed explain clearly the observed TL behaviour of LiF phosphors with a variety of possible dopings. Braunlich[2] has described many theoretical developments prior to 1968. A tabulation of the expected variation of the TL intensity with temperature has been given by Fleming[3]. These computations have been made for both first and second order kinetics over a trap depth range from 0.1 to 1.5 eV and for T_m ranging from 100 to 700 K. The temperatures at which the intensity equals 1/2, 2/3 and 4/5 of peak value I_m have been considered. The tabulation is of direct help in the analysis of recorded glow curves for determination of parameters such as trap depth, frequency factor (or pre-exponential factor), re-trapping and recombination cross sections. Fleming's work[3] is based on the exact analysis of the recombination processes carried out by Kelly and Laubitz[4].

6.2.   Two-trap model

Sweet and Urquhart[5] have explained the low temperature TL results of ZnS single crystals by assuming two 'overlapping' trap states (thermally connected) that have a close energy difference and produce overlapping glow peaks. This is an extention of the 'single centre' model in which only one trapping state is assumed to participate in the TL process (see Figure 6.1) and is, for instance, found to be inadequate in the analysis of the observed glow peaks exhibiting low values of T_m such as those shown in Figure 6.2. The following definitions are used to describe the model.

$$
\begin{aligned}
E_1 &= \text{trap depth (eV)} \\
h_1 &= \text{concentration of trapping centres (m}^{-3}) \\
h_{\bar{1}} &= \text{concentration of trapped electrons (m}^{-3}) \\
a,a^+ &= \text{concentration of recombination centres and trapped} \\
&\qquad \text{holes (m}^{-3}) \\
n,p &= \text{concentration of free electrons (in CB) and free holes (in} \\
&\qquad \text{VB, assumed negligible) (m}^{-3}) \\
\beta_1 &= \text{trapping rate constant (m}^{3}.\text{s}^{-1}) \\
\alpha &= \text{recombination rate constant (m}^{3}.\text{s}^{-1}) \\
\gamma_1 &= \text{transition probability for electrons from trap to} \\
&\qquad \text{conduction band}
\end{aligned}
$$

The second trap is considered to have an energy E_2 ($>E_1$). The symbols for this trap are similarly defined using a subscript 2. The electron-hole recombination process can be described as

$$\mathrm{d}h_{\bar{1}}/\mathrm{d}t = -\gamma_1 h_{\bar{1}} + \beta_1 n(h_1 - h_{\bar{1}}) \tag{6.1}$$

$$\mathrm{d}h_{\bar{2}}/\mathrm{d}t = -\gamma_2 h_{\bar{2}} + \beta_2 n(h_2 - h_{\bar{2}}) \tag{6.2}$$

$$(\mathrm{d}n/\mathrm{d}t) + (\mathrm{d}h_{\bar{1}}/\mathrm{d}t) + (\mathrm{d}h_{\bar{2}}/\mathrm{d}t) = -\alpha\, n\, a^+ \tag{6.3}$$

$$\gamma_1 = s_1 \exp(-E_1/k_\text{B}T) \tag{6.4}$$

$$\gamma_2 = s_2 \exp(-E_2/k_\text{B}T) \tag{6.5}$$

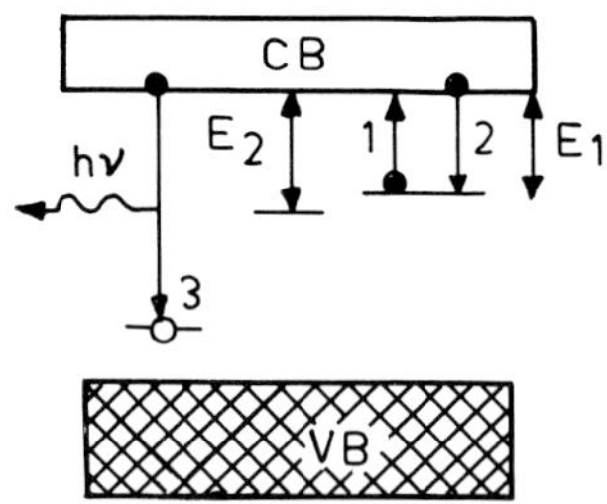

Figure 6.1. Single centre model of TL in which only one trap participates. 1, thermal stimulations; 2, re-trapping; 3, recombination centre. ●, electron; ○, hole.

The condition for charge neutrality under the assumption of zero free hole concentration ($p=0$) is given by

$$a^+ = n + h_1^- + h_2^-$$

(6.6)

Further, a linear rate of heating is assumed, i.e. $T = T_i + qt$ where

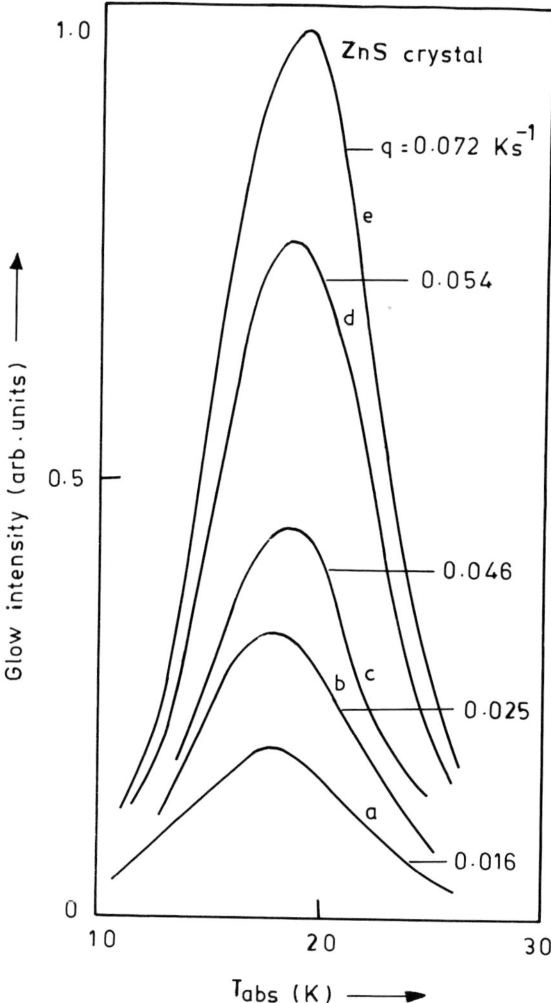

Figure 6.2. Experimental glow curves of ZnS for different heating rates, q[5].

T_i is the initial temperature; or

$$q = \mathrm{d}T/\mathrm{d}t \tag{6.7}$$

In order to solve these equations numerically, the following simplifying approximations are used[6,7] in the two-centre model:

$$n \ll h_1^- \text{ and } (\mathrm{d}n/\mathrm{d}t) \ll (\mathrm{d}h_1^-/\mathrm{d}t) \tag{6.8}$$

$$n \ll h_2^- \text{ and } (\mathrm{d}n/\mathrm{d}t) \ll (\mathrm{d}h_2^-/\mathrm{d}t) \tag{6.9}$$

We process these equations in a manner similar to the one-centre model case described by Sweet and Urquhart[8] for determining $I(T)$. It is as follows:

Using Equations 6.8 and 6.9 in Equation 6.6

$$a^+ = n + h_1^- + h_2^- \approx h_1^- + h_2^- \tag{6.10}$$

Also, using Equations 6.6 and 6.3

$$-(\mathrm{d}a^+/\mathrm{d}t) = -(\mathrm{d}n/\mathrm{d}t) - (\mathrm{d}h_1^-/\mathrm{d}t) - (\mathrm{d}h_2^-/\mathrm{d}t) = \alpha\, na^+ \tag{6.11}$$

Further, from Equation 6.10

$$(\mathrm{d}a^+/\mathrm{d}t) \approx (\mathrm{d}h_1^-/\mathrm{d}t) + (\mathrm{d}h_2^-/\mathrm{d}t)$$

Using this in Equation 6.11 we find

$$\mathrm{d}n/\mathrm{d}t \approx 0 \tag{6.12}$$

Thus Equation 6.3, in which Equation 6.12 has been used, gives

$$\mathrm{d}n/\mathrm{d}t = 0 = -\alpha na^+ - (\mathrm{d}h_1^-/\mathrm{d}t + \mathrm{d}h_2^-/\mathrm{d}t)$$

Using Equations 6.1, 6.2 and 6.10

$$0 = -\alpha\, n(h_1^-+h_2^-) - [-\gamma_1 h_1^- + \beta_1 n(h_1-h_1^-) - \gamma_2 h_2^- + \beta_2 n\,(h_2-h_2^-)]$$

or

$$-n\left[\alpha(h_1^-+h_2^-) + \beta_1(h_1-h_1^-) + \beta_2(h_2-h_2^-)\right] = -\gamma_1 h_1^- - \gamma_2 h_2^-$$

or

$$n = \frac{\gamma_1 h_1^- + \gamma_2 h_2^-}{[\alpha(h_1^-+h_2^-) + \beta_1(h_1-h_1^-) + \beta_2(h_2-h_2^-)]} \tag{6.13}$$

The glow curve intensity is given by the rate of decrease of trapped holes due to recombination. Hence, using Equations 6.11 and 6.10 and taking $c = 1$,

$$I(T) = -c\,(da^+/dt) = (-da^+/dt) = \alpha\,na^+ = \alpha\,n(h_1^- + h_2^-) \quad (6.14)$$

Using Equation 6.13,

$$I(T) = \frac{\alpha(\gamma_1 h_1^- + \gamma_2 h_2^-)\,(h_1^- + h_2^-)}{\alpha(h_1^- + h_2^-) + \beta_1(h_1 - h_1^-) + \beta_2(h_2 - h_2^-)} \quad (6.15)$$

Equation 6.15 depicts the two-centre model (the one-centre model can be obtained from Equation 6.15 by setting parameters with subscript 2 as zero). It has been computer-calculated with the parameters s (s values for the two traps are, for simplicity, taken as equal; also, $\beta_1/\alpha = \beta_2/\alpha \equiv \beta/\alpha$), δ ($= \beta/\alpha$), E_1, E_2 and the relative concentration (h_1/h_1^-) and (h_2/h_2^-) all adjusted in order to simulate an $(I(T)\text{–}T)$ glow curve that is identical , as far as possible, to that observed at the given experimental value of q. This simulation is shown in Figure 6.3. The parameters for the best fit are:

$$s_1 = s_2 = 3 \times 10^5\ \text{s}^{-1}$$

$$(h_1/h_1^-) = 64;\ (h_2/h_2^-) = 100$$

$$E_1 = 0.0180\ \text{eV},\ E_2 = 0.0225\ \text{eV}$$

$$\delta = 90$$

6.3. Single trap model

The single trap and single luminescent centre model is actually the well known Randall-Wilkins model in which the re-trapping and recombination mechanisms are described by first order kinetics (no re-trapping).

In terms of trap parameters, the general expression for the glow peak intensity $I(T)$ can be given by Equation 6.15 in which second trap parameters are ignored and thus $h_1 \equiv h$, $h_1^- \equiv h^-$, $\gamma_1 \equiv \gamma$ are used for simplicity,

$$I(T) = \frac{\alpha\,\gamma\,(h^-)^2}{\alpha h^- + \beta(h - h^-)} = \frac{s\alpha\exp(-E/k_B T)\,(h^-)^2}{\alpha h^- + \beta(h - h^-)} \quad (6.16)$$

In Equation 6.16 the presence of 'thermally disconnected' traps (deep electron-occupied traps) has been ignored. Otherwise, αh^- will be replaced by $\alpha(h^- + H^-)$ where H^- is the concentration of electrons in such deep traps. Since

$$I(T) = -da^+/dt = -dh^-/dt = -q\,(dh^-/dT)$$

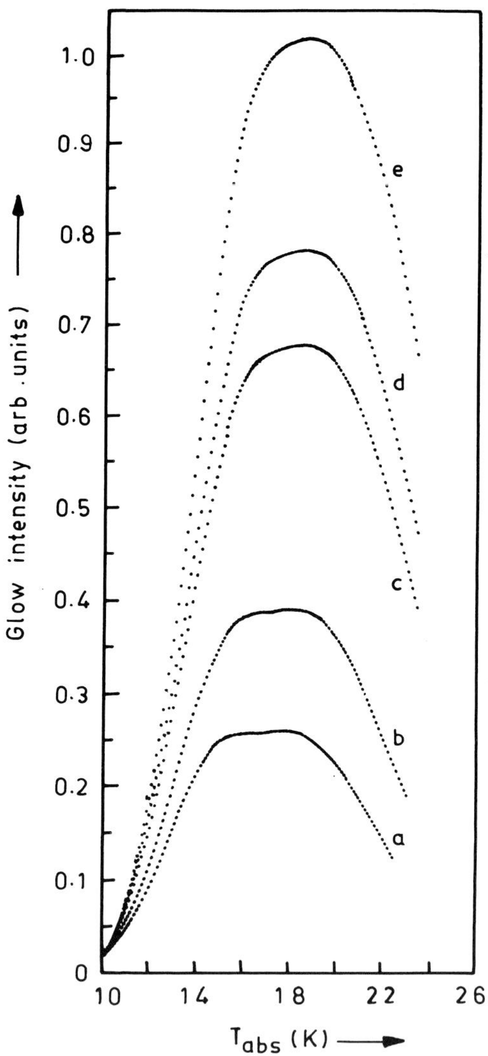

Figure 6.3. Glow curves generated by the computer simultation for comparison with those in Figure 6.2[5]. q values, key as Figure 6.2.

integration of the above gives

$$h^-(T) = h^-(T_i) - \frac{1}{q} \int_{T_i}^{T} I(T') \, dT' \tag{6.17}$$

In the Randall-Wilkins model, as the re-trapping is negligible ($\beta/\alpha=0$) Equation 6.16 returns to the familiar form for first order kinetics:

$$I(T) = s\, h^- \exp(-E/k_B T) \tag{6.18}$$

For second order kinetics, $\beta/\alpha = 1$ and we get

$$I(T) = s'\, (h^-)^2 \exp(-E/k_B T) \tag{6.19}$$

where $s' = (s/h)$ is the pre-exponential factor.

Cooke and Gavathas[9] have shown conspicuous success of the Randall-Wilkins model for TL emission in Al_2O_3 and 5' deoxycytidine monophosphate H_2O (5' dCMP). The computer-simulated glow curve using the Randall-Wilkins formula, i.e.

$$I(T) = cn_o\, s \exp(-E/k_B T) \exp\left[(-s/q) \int_{T_i}^{T} \exp(-E/k_B T') \, dT'\right]$$

fitted the experimental curve well. This is shown in Figure 6.4. They further showed that glow peaks occuring at low temperatures ($T_m = 74$ K and $E = 0.095$ eV for 5' dCMP) are characterised by small frequency factors ($s<10^5$ s^{-1}). Hence small values of E point to low values of s, in contrast to the higher values of 10^8–10^{14} s^{-1} often quoted in the literature.

6.4.   Associated pair model

The traps and recombination centres show the possibility of forming associated pairs. Thermal stimulation raises the trapped electron to an excited level of recombination centre from which it may either be re-trapped or, alternatively, emit luminescence by returning to the ground state of the recombination centre. Let n_e be the concentration of such thermally excited electrons. If r (s^{-1}) is the probability of retrapping, then the number of electrons re-trapped is $r\, n_e$. The number of electrons excited thermally is given by $-dh^-/dt$ which is

$$-dh^-/dt = s\, h^- \exp(-E/k_B T) - rn_e \tag{6.20}$$

and

$$I = - \mathrm{d}a^+/\mathrm{d}t = bn_e \tag{6.21}$$

where b (s^{-1}) is the probability of recombination of the thermally excited electron.

The charge neutrality condition is

$$a^+ = h^- + n_e \tag{6.22}$$

It is assumed that $n_e \ll h^-$, so that Equation 6.22 becomes

$$\mathrm{d}a^+/\mathrm{d}t \approx \mathrm{d}h^-/\mathrm{d}t \tag{6.23}$$

Hence from Equations 6.20 and 6.21

$$I = - \mathrm{d}a^+/\mathrm{d}t = - \mathrm{d}h^-/\mathrm{d}t$$

$$= (s\,h^-)\,\exp(-E/k_{\mathrm B}T) - rn_e = bn_e$$

or

$$n_e\,(b+r) = (sh^-)\,\exp(-E/k_{\mathrm B}T)$$

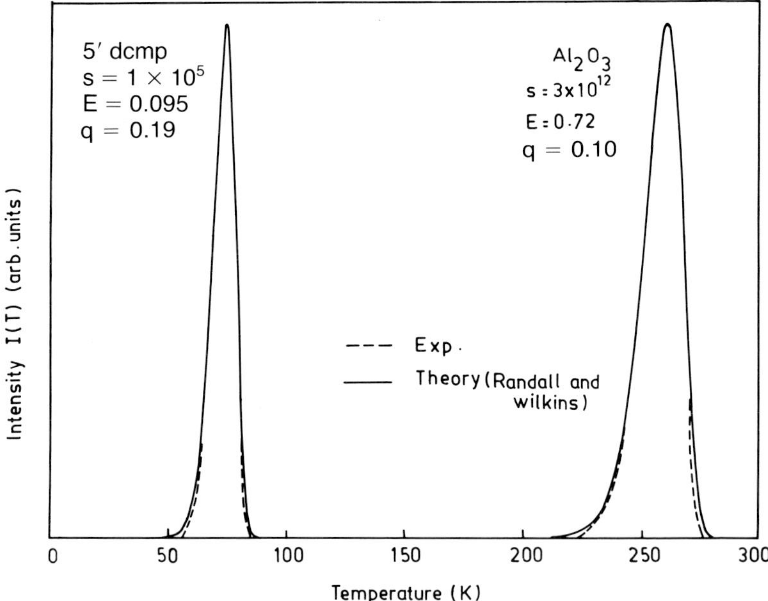

Figure 6.4. Comparison of experiment with Randall-Wilkins theory[9]. $s(\mathrm{s}^{-1})$, $E(\mathrm{eV})$ and q $(\mathrm{K.s}^{-1})$.

Hence $I(T) = bn_e$

$$= [b/(b+r)]\,(sh^-)\,\exp(-E/k_B T) \qquad (6.24)$$

For a very high recombination probability, $b \gg r$, Equation 6.24 reduces to the well known Randall-Wilkins equation.

6.5.   Dipole model

This model has been postulated by Grant and Cameron[10] to explain some of the glow peaks in LiF:Mg,Ti (TLD-100, a typical commercial product containing 100-200 ppm of Mg and 10 to 20 ppm of Ti, manufactured by Harshaw). A typical TL glow curve structure contains six peaks when TLD-100 is irradiated at room temperature and subsequently warmed to 300°C. This is shown in Figure 6.5. The impurity vacancy dipole constitutes an electron trap. The dipole trap model is based on the fact that magnesium goes into the crystal lattice as a divalent (2+) ion and is attracted to the monovalent lithium ion vacancy. Thus the Mg^{2+} ion of the

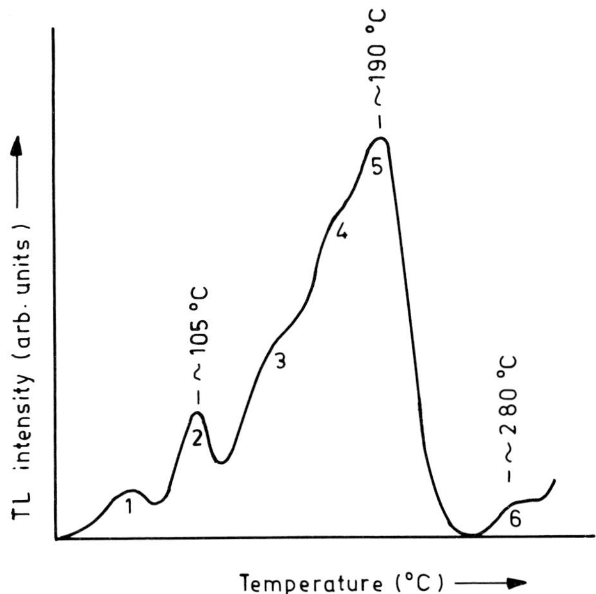

Figure 6.5. Thermoluminescence glow curve for LiF (TLD-100) with glow peaks numbered in conventional fashion. Each glow peak results from a trapping level.

impurity and the Li$^+$ ion vacancy of the host lattice form a dipole which acts as an electron trap responsible for peaks 2 and 3. These dipoles can further aggregate and form higher order complexes, e.g. two dipoles join together to make a dimer, three dipoles aggregate to form a trimer which gives rise to peaks 4 and 5. Peaks 1 and 6 are believed to be due to intrinsic crystal defects, not related to the dipoles.

The dipoles and aggregates of dipoles are thermally unstable in the sense that dipoles can aggregate and then, later, may disintegrate under specified temperature conditions. This view is evidenced by the changes in glow curve after thermal treatment. Peaks 2 and 3 are considerably reduced due to cluster formation when TLD-100 is annealed at temperatures <100°C. When the phosphor is heated to temperatures between 120 and 300°C, the clusters disintegrate and peaks 2 and 3 are restored.

6.6. Track-interaction model

Claffy *et al*[11] introduced this model to explain supralinearity and TL sensitisation observed in TLD-100. Attix[12] has stated that it is sufficiently general and can be applied to account for these effects in other phosphors as well. The basic feature of the model is that electron-hole pairs are produced by the ionising radiation along the tracks of secondary charged particles such as Compton electrons and photoelectrons and are trapped near the track. Some of the centres so produced are luminescent centres. When the phosphor is heated, one kind of charge carrier is released (considerable evidence[13] indicates that electrons are the charge carriers concerned) which migrates to the oppositely charged trapped carriers. They recombine and in some cases, when the centre appears to be a luminescent centre, TL is emitted.

For low doses the secondary electron tracks are sufficiently separated, as shown schematically in Figure 6.6. The released charge carriers in this case tend to recombine with the luminescence centres of the same track. The TL response as a function of dose is linear in this dose range. When the average separation of adjacent electron tracks is comparable with the average separation of luminescent centres along each track, as the dose is increased more luminescent centres are available within the migrating range of the released charge carriers and the TL output rises more rapidly than the corresponding proportion of the dose. This is the supralinear

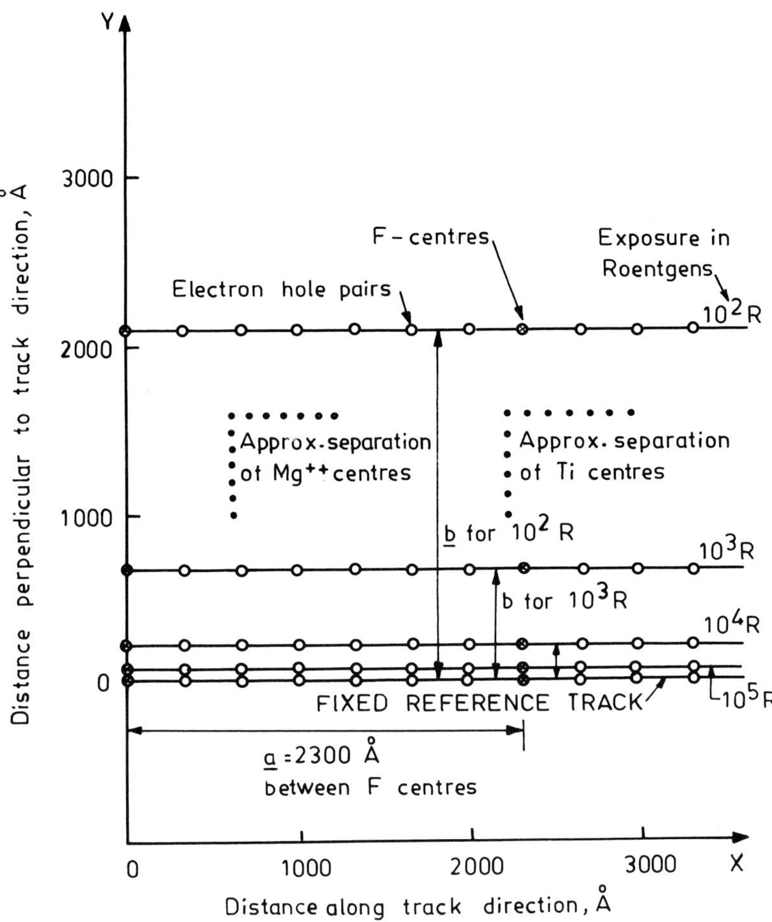

Figure 6.6. Schematic diagram of the separation of Compton electron tracks at various levels of ^{60}Co γ ray exposure in LiF (TLD-100). The tracks are 10^7 Å long. The electron-hole pairs are indicated by $\bigcirc$, and the F centres by $\otimes$. The spacing of the Mg^{2+} centres (with associated Li^+ vacancy) and Ti centre are also shown. The lattice constant in LiF is about 4 Å. The space between adjacent Li^+ and F^- ions is therefore 2 Å. a is the average spacing of luminescent centres along the track and b is average spacing between the centres along the Y-direction[12].

behaviour of the TL phosphor.

To account for the phenomena of TL sensitisation, which corresponds to the enhanced TL efficiency of the phosphor subjected to radiation exposure and heating after it has already gone once through the TL readout cycle, it is postulated that most luminescence centres are not reached by a released charge carrier during the readout cycle and thus continue to exist. The later radiation exposure assists the population of luminescence centres surviving from the previous exposure.

6.7.   Competing trap model

The model explains the mechanism for the radiation induced increase in sensitivity in LiF by postulating an increase in the probability of TL emission during readout. The removal of competitors for the existing luminescent recombination centres due to irradiation, as proposed by Mayhugh *et al*[14], is assumed to be the cause of the increase in TL emission probability. Zimmerman[15] has used TL and TSEE studies in support of the relationship between supralinearity and recombination and excludes the possibility of an increase in the number of luminescent centres due to the increase in radiation dose. The presence of Mg-OH vacancy complexes in LiF crystals is found to decrease the TL sensitivity and supralinearity. These complexes may, therefore, be regarded as playing the role of the competitors mentioned above. These competing traps (CT) can initially trap electrons during irradiation. During readout, the electrons released from TL traps could be re-trapped at a CT or could go to the normal recombination centre giving a TL output proportional to the radiation dose. At higher dose rates the CT begin to saturate. At this stage the thermally released electrons have an enhanced chance of being involved in the recombination process.

6.8.   Mayhugh's model

In Mayhugh's model[16] the mechanism of the TL phenomenon is studied in terms of transitions and consequent interactions of charge carriers moving between metastable energy levels in the forbidden gap. The energy levels of hole traps are assessed by means of light absorption measurements from which a relation between energy levels and positions of peaks of the glow curve can be obtained. In fact, if the absorption measurements and the heating process are carried out together, the simultaneous bleaching of some absorption bands and the appearance of certain TL peaks may be observed.

As already described, the irradiation of a TL crystal produces a certain number of free electrons and holes that are soon trapped by their respective centres. The size of the population of trapped carriers depends on the physical properties of the crystal lattice. A centre consisting of an electron trapped at a halide ion vacancy is called a colour centre or F centre and a centre consisting of a hole trapped at a positive ion vacancy is called a V centre or recombination centre. In the subsequent readout process the crystal is supplied with an amount of thermal energy sufficient for the progressive liberation of trapped charge carriers.

As far as holes are concerned, it must be noted that the excited states are located 'below' the ground level, i.e. between the ground level and VB. The holes liberated during the reading cycle may fall either directly into the valence band or back into some excited states of V centres from which they originated. If a V centre, or also an F centre, produces absorption band fading at room temperature, it means that its excited states are close to VB, or CB respectively. At the same time a certain number of electrons are drawn out from F centres into some excited state from which they may fall into those centres close to VB that contain holes, i.e. V centres. These V centres play the role of activators, as they may contain, at the same time, electrons and holes, which recombine giving out the recombination energy in the form of a light quantum. Its frequency depends on the energy levels through which the final configuration has been reached. This, in turn, depends on the lattice structure of the TL material. A direct recombination process of V and F centres without any intermediate activation, even if possible in theory, has not been observed experimentally.

According to the model, electrons are supplied by F centres and holes are supplied by V centres; activators are due to metallic impurities of valency different from that of the halide metal. For instance, in the case of LiF, in addition to Mg and Ti (responsible for F and V centres), there are also Al impurities, responsible for activators. In short, the TL features of a crystal bear a close relationship to the quantity and the quality of doping elements, while the sign and charge of the resulting ions decide whether they behave as activators, V centres, or F centres.

The probability for transition of electrons from F centres into activators is strongly influenced by their mutual position. As a consequence, TL emission varies in the presence of two different conditions: (i) random production of F centres, and (ii) production of most of the F centres in the proximity of activators. In the former case, only a small number of F centres are close to activators, just about enough to undergo TL transition, so that the overall efficiency of the material is poor; in the latter case, on the contrary, most of the F centres lead to recombination and the resulting efficiency is high. Since experimental results evidence a relatively high efficiency, it is concluded that the latter assumption is more probable.

We now consider in detail one of the most frequently employed TL materials, namely LiF:Mg,Ti. The glow curve features five peaks below 250°C; each of them is associated with an observed absorption band and, therefore, a specific trap level. In particular, both peaks 4 and 5 are associated with the central absorption band at ~3100 Å and can be distinguished because of the different bleaching rates when irradiated by light of the same wavelength. Peak 5 is associated with a fast component and peak 4 with a slow one. A secondary light emission which clears the centres associated with peaks 4 and 5, proves that there exists a second excited state, close to CB, from which the electrons can escape by means of energy from light at room temperature. In addition, the clearance of some special V centres, called V_3 centres, is simultaneously

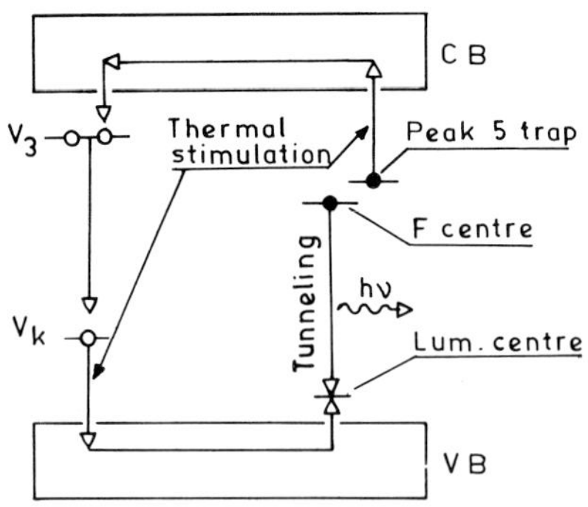

Figure 6.7. Sketch showing the occurence of peak 5 in LiF:Mg,Ti phosphor glow curve.

observed. Altogether, the behaviour is similar to that of F centres. This suggests that the centres related to peaks 4 and 5 are electronic traps close to F centres, but somehow perturbed by interaction with neighbouring ion impurities.

A V_3 centre is a negative ion vacancy that can trap up to two holes. When an electron emerges from a trap into CB by the action of thermal stimulation, it may fall into a V_3 centre, recombining with one hole out of two. As a consequence, the V_3 centre is transformed into a so called V_k centre, i.e. a negative ion vacancy located at a deeper level, that can trap only one hole. Figure 6.7 sketches the descripton of this model.

The TL efficiency of the crystal can be improved by the process of pre-irradiation at a very high dose followed by annealing at 300°C. Due to this process, new electron trapping centres are probably created, depending on some interaction between F centres and existing impurities. Because of this improvement effect, it must be concluded that these centres play the role of activators as well. Alternatively, the improvement effect can be explained as a decreased probability for non-radiative transitions due to annealing.

Light absorption measurements prove that the centres responsible for peaks 4 and 5 are analogous as they can be converted into one another by light exposure at 3100 Å. Nevertheless, this is not so clear when we look for interpretation of

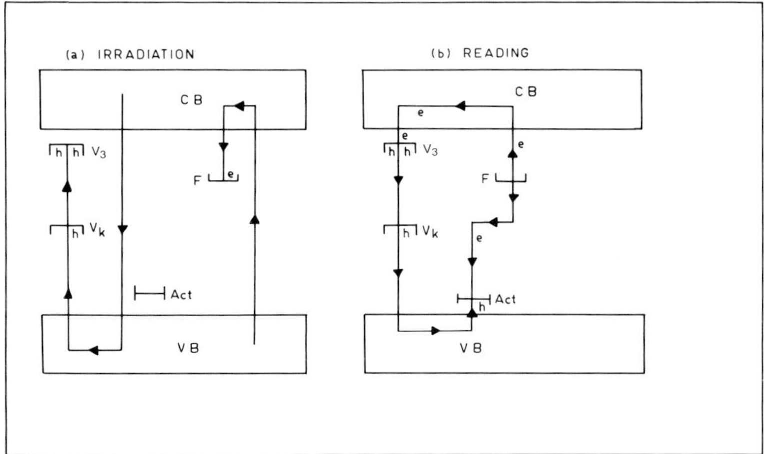

Figure 6.8. Mayhugh's model[16]. Key: Act, activation/recombination centre, e, electron; h, hole.

the corresponding TL output. However, it may be assumed that, as far as peak 5 is concerned, the probability for TL emission is higher than the probability for a conversion process. This means that the centres responsible for peak 5 are far from activator centres so that their presence cannot be detected by TL measurements. Mayhugh's model[16] is illustrated in Figure 6.8. Cooke[17] has extended this model with a view to explaining the UV emission observed at low temperatures. This model also removes the objections made by Podgorsak *et al*[18] to Mayhugh's model. Cooke has postulated that almost simultaneous release of shallow trapped electrons and V_k centres accounts for the UV emission at low temperature. At temperatures >170 K, the original mechanism proposed in the Mayhugh model is operative.

6.9.   Nink and Kos' model (Z centre model)

In Nink and Kos' model[19] it is assumed that the doping of lithium fluoride by magnesium and titanium produces different effects in lattice structure; namely, Mg ions produce electron trapping, and Ti ions produce the hole trapping centres. The transitions which are at the origin of TL emission are then explained according to this assumption.

When LiF:Mg,Ti is irradiated, we observe two absorption bands at 3800 and 3100 Å, corresponding to the photon energies of 3.3 and 4.0 eV respectively and are, therefore, located beneath the basic F level. In particular, the 4.0 eV band corresponds to peak 5, which is of major interest for our purposes, since it features an outstanding proportionality between TL peak height and irradiation dose, and implies a very constant sensitivity over a wide range.

In addition, a third band is detected at 2250 Å, corresponding to 5.5 eV, above the basic F level. This band is most probably produced by interaction between an F centre and an Mg ion. Finally, with an even deeper level at about 9 eV, there is associated another band, related to the previous one, and therefore due to Mg impurities.

On the other hand, Ti ions produce a band directly below CB at 6.2 eV, corresponding to 2000 Å. Moreover, in the proximity of VB at 11 eV, an absorption band is observed which is probably of intrinsic origin and may be related to V centres.

This is, in short, the topography of the forbidden gap. The problem now consists in describing the detailed structure of the centres responsible for the observed bands, and discovering a suitable mechanism that would explain the possibility of conversion of the centres into one another and the transitions corresponding to the observed TL emission.

In the Nink-Kos model, the electron trapping centres, called Z centres (defined in section 2.3 of Chapter 2) are F centres locally perturbed by neighbour Mg^{2+} ions. There is a family of Z centres, Z_2 and Z_3 consist of a Mg ion interacting with an F' centre and an F centre respectively, whereas Z_1 and Z_4 are not exactly defined. Probably, Z_1 is produced by a single electron perturbing a Mg ion, and Z_4 is produced by an electron trapped by a system consisting of a Mg^{2+} ion and a positive vacancy.

Let us now consider what happens when the crystal is subjected to external excitation. First of all, the basic F band, located at about 5 eV below CB, slowly fades away by thermal action; nevertheless, a temperature higher than 300°C is necessary to bleach it out. Then, the band at 4 eV, due to Z_2 centres, bleaches out at temperatures between 100 and 200°C. Finally, the behaviour of the band located at 5.5 eV, due to Z_3 centres, is quite different. If we compare it to that of the Z_2 band, we can understand how the different centres are related to different thermal excitations. In fact, the two bands are characterised by constant intensity for temperatures up to 100°C. Between 100 and 200°C, an increase in the Z_3 band and a decrease in the Z_2 band are observed; at a higher temperature, the Z_3 band has practically disappeared, while Z_2 is constant. Only at still higher temperatures, does the Z_2 band start decreasing slowly, approaching a zero intensity.

It must be noted that all these bands are similarly observed in LiF crystals doped with either Mg or Mg and Ti although the deep band at 6.6 eV is not verified in crystals doped with Mg only. It is concluded that the former bands are essentially due to Mg, and the latter to Ti impurities.

As stated above, the fact that band Z_3, when increasing, is 'fed' by band Z_2, indicates a conversion of centres; further, it has been experimentally observed that one electron is liberated from Z_2 centres. Not only is this in agreement with the proposed model, but it also leads to a basis for a study of quantitative relations between TL emission, treatment parameters (irradiation and annealing), and physical constants of the crystal (efficiency and concentration of centres).

The conversion $Z_2 \rightarrow Z_3$ is verified according to the process:

$Z_2 \rightarrow Z_3 + e$ (in the annealing process).

Thus, the maximum concentration of Z_3 centres should be equal to the initial concentration of Z_2. However, this is in conflict with experimental evidence, which suggests that it can actually be higher, i.e. the number of Z_3 centres thermally produced is eventually higher than the number of Z_2 centres thermally destroyed. A rough assessment of the concentrations of different centres can be made by measuring the area of the respective bands. This means that there must exist some kind of deeper centres that play a storage role for Z_2 and Z_3. They are referred to as Z_0 centres, and are most probably due to positive vacancies interacting with Mg impurities. At high temperatures the conversion $Z_3 \rightarrow Z_0$ is the prevailing process, according to

$Z_3 \rightarrow Z_0 + e$ (annealing).

As a final result, the thermal process leads the crystal into an initial condition where all Z bands are erased, the corresponding centres are emptied, and only Z_0 centres, located at about 9 eV below CB, remain. Z_0 centres, acting as storage during a subsequent irradiation release a certain number of active Z centres. Altogether, the irradiation process can be described by the following relations

$$Z_0 + e \rightarrow Z_3$$

$$Z_3 + e \rightarrow Z_2$$

resulting in an equilibrium condition in which the size of the Z centre population is quantitatively related to the radiation dose and the physical parameters of the crystal.

Let us assume that n_i is the concentration of centres Z_i, released by irradiating the crystal with ionising radiation. Then, heating the crystal to about 250°C, a conversion process of the type $Z_2 \rightarrow Z_3$ is produced, with free electron emission:

$$n_2.Z_2 \xrightarrow{T} n_2.Z_3 + n_2.e$$

The initial concentration of Z_3 centres is not modified, i.e. $n_3.Z_3 \xrightarrow{T} n_3.Z_3$ and a fraction $f (f \leq 1)$ of released free electrons gives rise to luminescence phenomena (L)

$$n_2.e \rightarrow f.n_2.L + (1-f).n_2.e$$

The remaining free electrons, interacting with Z_0 centres, produce new Z_3 centres,

$$(1\text{–}f)\, n_2.\text{e} + n_0.Z_0 \rightarrow (1\text{–}f)\, n_2.Z_3 + [n_0\text{–}(1\text{–}f)n_2]\,.\, Z_0$$

In conclusion, in the Z centre model the TL process is analytically described by:

$$n_0.Z_0 + n_2.Z_2 + n_3.Z_3 \rightarrow [n_0\text{–}(1\text{–}f).n_2].\, Z_0 + [n_3 + (1\text{–}f).n_2]\,.\, Z_3 + fn_2.L$$

This relation shows that the number of TL events is proportional to the initial concentration of Z_2 centres in the crystal and the fraction f of released free electrons; thus f is a measure of the TL efficiency of the crystal.

The heating process reduces the initial concentration of Z_0, and increases the concentration of Z_3 centres. The increase depends on the efficiency. In the two limiting cases, this is

$$\triangle n_3 = 2n_2 \;(\text{for } f = 0); \quad \triangle n_3 = n_2 \;(\text{for } f = 1)$$

Additional physical properties of Z_0 centres are deduced from experiments, e.g. they are located in high energy levels (9.5 eV) and evidenced only in Mg doped crystals; their concentration is not affected by thermal treatments up to 100°C, but decreases because of the production of new 'shallow' Z centres if the crystal is irradiated. If the same material is then immediately irradiated a second time, the increase in Z_2 concentration is higher as a consequence of the fact that this time, not only the Z_0 centres but the Z_3 centres contribute to the process as well. In fact, they are responsible for the first step. Therefore, this type of treatment offers a real improvement in TL sensitivity (measured TL output per unit dose). This, however, is verified only in the low dose range. If the administered dose is very high, the overall effect is permanent damage to the TL centres in the lattice structure and, consequently, a reduced sensitivity.

The two-step mechanism of Z_2 production is evidenced by a further observation. If the TL material, previously irradiated at room temperature and then heated at 250°C, is immediately submitted to a second irradiation, the increase in Z_2 concentration is again higher and for the same reasons.

The activation effect of Ti impurities on both Z_3 and F bands is particularly evident when the crystal is heated from 200 to 350°C. In LiF:Mg,Ti these bands fade continuously, until they completely disappear, whereas in LiF:Mg a plateau is observed over the whole temperature range. This can be explained by assuming that Ti ions act as activators for thermally liberated electrons. The destruction process of Z_3 centres is then gradual, becoming total at a specific temperature. If, however, the amount of these activators is negligible, the electrons liberated from Z_3 centres remain free for a considerable time and the probability for reconstruction of Z_3 centres is high. The result is a dynamic equilibrium situation in which a stable value of Z_3 concentration is maintained.

In conclusion, the Nink and Kos model is efficient in explaining the mechanism of TL phenomena of a suitably pretreated crystal, particularly LiF:Mg,Ti. It displays a definite relation between TL peaks and Z centres. Peak 1, as previously described, is bleached out by a special annealing procedure; peaks 2 and 3 are related to the presence of Z_1 centres characterised by low activation energy; peaks 4 and 5 are related to the presence of Z_2 centres. Finally, peak 6, which is peculiar for high temperatures, is related to Z_3 centres.

According to the model, centre F is not directly related to the TL emission process; it is a basic centre which, interacting with the various impurities, is responsible for the production of Z centres. Since it features a very low decay process, a possible TL peak would be detectable with great difficulty in the form of a weak, continuously distributed background signal.

One problem is not solved by the Nink and Kos model; that is the kinetics that rule the conversion process between the different centres, with particular reference to the role of Z_0 centres. Even if several absorption bands due to alpha centres, from which Z_0 centres originate, have been observed, further studies are necessary.

The Nink and Kos model may be summarised as follows[20-22]. During irradiation, the 3.3 eV centres (which are Z_2 centres with an additional electron), Z_2 centres and Z_3 centres are created. The dipoles formed by Mg^{2+} cation vacancies which act as trapping centres are converted during irradiation into Mg^{2+} electron occupied anion vacancies. The latter form the family of Z centres. In the course of the annealing process (a 300°C anneal after high X ray irradiation increases the height of dosimetric peak 5 of LiF:Mg,Ti by a factor of up to 4) firstly the 3.3 eV centres transform to Z_2 centres liberating electrons which cause peak 2. As the temperature increases, Z_2 centres are transformed to Z_3 centres releasing electrons which cause peak 5. At temperatures above 200°C, Z_3 centres give up their electrons, which causes TL peak 6[*]. The varying TL peak heights can be interpreted in terms of thermal stability of Z centres. The Z_2 centre is neutral because the Mg^{2+} ion is compensated by two electrons in the anion vacancy. The weakly bound additional electron with the Z_2 centre makes it a 3.3 eV band centre which is thermally quite unstable and gives the TL peak 2 at 120°C. The Z_3 centre is charged positively because the adjacent anion vacancy has only one electron. This makes Z_3 much more stable against thermal release of electrons which, therefore, cause only a feeble peak 6 at $\approx$ 280°C.

An illustration of the Nink and Kos model[19] is given in Figure 6.9. Figure 6.10 explains the general mechanism of TL emission in a phosphor following irradiation.

6.10.   Phototransferred thermoluminescence (PTTL)

The PTTL technique has been used over the years for UV dosimetry and, more importantly, for the dating of geological and

[*] A. R. Lakshmanan, in Radiat. Prot. Dosim. **6**(1-4), 52 (1984), has commented on the Nink and Kos model. On the basis of the argument that the 4 eV absorption band is in fact at 4.4 eV, he has ascribed Z_2 and Z_3 centres to the TL peaks 7 (260°C) and 10 (400°C) respectively.

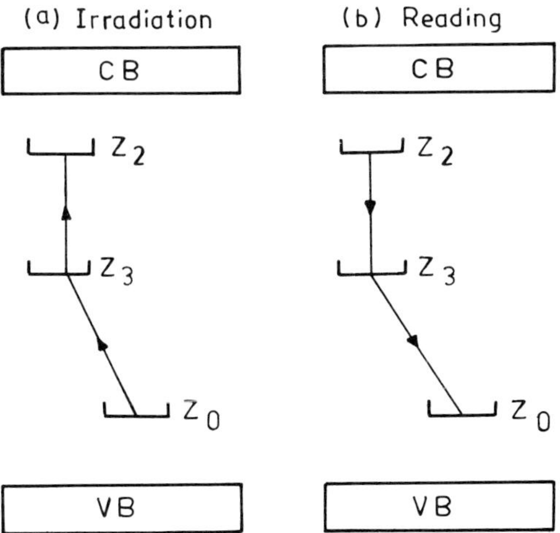

Figure 6.9. Nink and Kos model of TL in LiF:Mg,Ti phosphor[19].

archaeological samples[23-25]. In the phototransfer process the electrons from deep traps are transferred into shallow traps by illumination of the sample with ultraviolet light. The filled shallow traps then give the TL glow peak which had been thermally bleached previous to any UV irradiation.

The deep traps in the phototransfer process act as donors and the shallow traps as acceptors of carriers.

PTTL has been investigated in various materials. In LiF (TLD-100) phosphor, which is widely used in radiation dosimetry, the phenomenon has been studied by, among others, Sunta and Watanabe[26], in Mg_2SiO_4:Tb phosphor by Lakshmanan and Vohra[27]; in $CaSO_4$:Dy by Nagpal *et al*[28] and Pradhan *et al*[29]; in clay minerals (zircon and fluorapatite) by Bailiff[30]; and in CaF_2 by Sunta[31]. The use of the phototransfer technique in thermoluminescence dating has been reported by Bailiff *et al*[32] on the quartz extracted from pottery and on other minerals, e.g. zircon and fluorapatite.

A desirable feature in the PTTL studies is that the sample has donor states belonging to high temperature ($\approx$300–500°C) TL peaks which are highly stable and for which the fading is therefore negligible.

Considering the well known example of LiF, if it is given a gamma pre-exposure and then annealed at a temperature below 400°C, some of the high temperature traps (>400°C) (residual traps) still

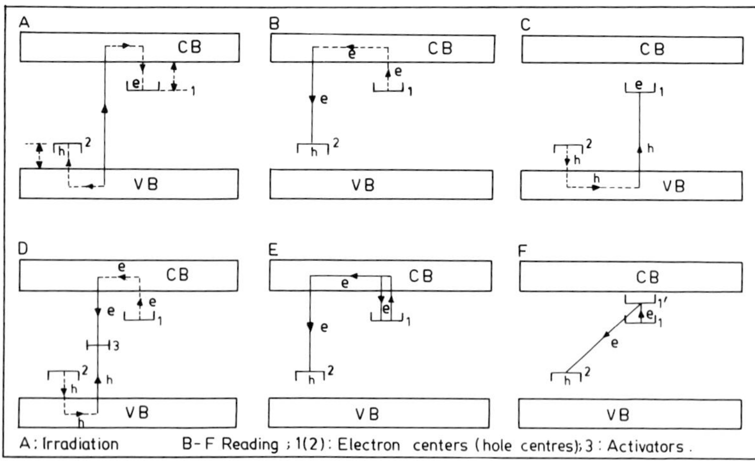

Figure 6.10. General mechanisms of TL emission.

remain filled. Subsequent UV exposure depletes these traps and regenerates the lower temperature traps. Hence one observes a reduction in residual TL (RTL) and regeneration of the already thermally bleached TL glow peak. It is usual to assume the UV induced TL glow peak intensity to be directly proportional to de-trapping of the donor traps.

If we plot the UV induced glow peak intensity as a function of UV energy, the maxima in the plot occuring at an energy E_{opt} is known as the optical trap depth or optical activation energy of the donor traps. This is shown in Figure 6.11. For archaeological studies the donor level population is a measure of the amount of dose received by the sample (such as quartz) extracted from pottery etc. If the dose rate in the vicinity of the buried sample is estimated by other techniques, then the age of the sample can be determined, as the PTTL output is proportional to the donor level population. In the case of quartz, the donor traps are those associated with 325°C and 375°C TL peaks and the acceptor traps correspond to 110°C. The donor levels in quartz are populated by electrons. The PTTL

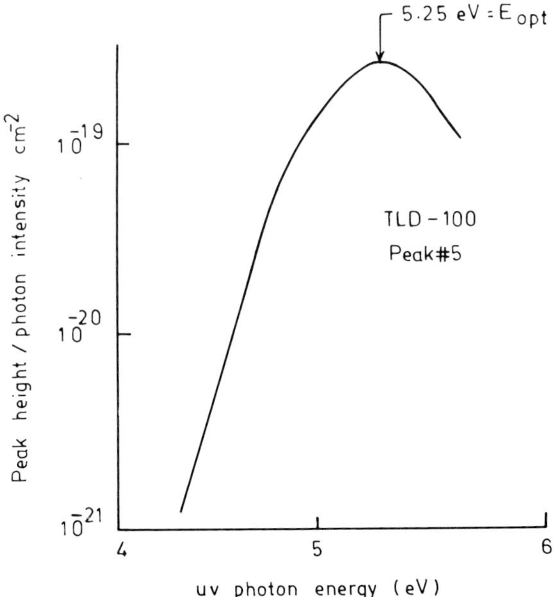

Figure 6.11. UV induced glow peak height plotted against the UV photon energy[26].

technique is of particular significance in those materials in which the dosimetric peaks exhibit excessive fading (anomalous fading) and, as such, can not be used for TL readout. Hence the stable deeper levels are sought for the purpose.

The strength of PTTL (110°C glow peak height) is related to the proportion of the donor population transferred (i.e. carriers) to the glow peak. It is expressed as

equivalent dose transferred (EDT)

$$= (1/\chi) \text{ (height of re-excited 110°C peak)}$$

where χ is the beta sensitivity per rad of the 110°C TL peak.
The procedure for the measurement of PTTL for the 110°C trap is as follows:

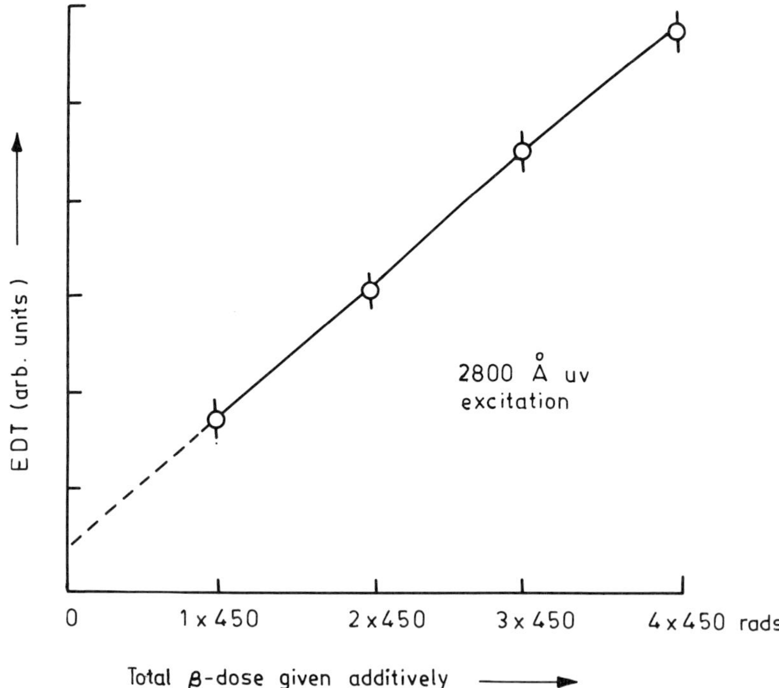

Figure 6.12. Dose calibration of PTTL in quartz using 110°C TL glow peak[23].

 (i) Drain archaeological sample by heating at 500°C, so that donor traps are evacuated.
 (ii) Give beta dose of several hundred rads.
(iii) Heat sample for one or two minutes at 210°C to give 'thermal wash' to the 110°C.
(iv) Give a test dose of beta particles (from ^{90}Sr/^{90}Y beta source) equal to 1 rad and measure TL of 110°C peak. This gives χ.
 (v) Expose the sample to monochromatic UV light (260 nm) for about one minute and measure the height of the 110°C TL glow peak.
(vi) Administer beta dose, as in step (ii), to another sample (of the same archaeological find) to a value such that EDT comes out to be the same as from the artificially unexposed archaeological sample which has been exposed to natural radiation only over the period of antiquity. This procedure gives the measure of the accumulated dose of the archaeological sample. Dose calibration in the quartz sample is shown in Figure 6.12. The additive β dose signifies that for each measurement of EDT at successively increasing dose values, a new portion (each having equal weight) of the sample is used. This is to take care of the changed TL sensitivity of the minerals due to heating during the first readout. PTTL methodology, in addition to its use in UV dose determination (regenerated TL output in a phosphor irradiated by a given β or γ radiation is proportional to the UV dose), offers certain advantages over conventional TL in age determination as observation of the high temperature glow peak in the latter suffers from the limitations of thermal quenching of TL and masking of TL signal by the thermal radiation from the heater plate and sample as well as the spurious TL. The PTTL dating of fossil bones and teeth is particularly attractive because of avoidance of decomposition of the sample at higher temperatures.

6.11.   **Band model for PTTL and photoluminescence**

Referring to Figure 6.13, if M_1 is the concentration of thermally stable trap levels (these are essentially the activator levels occupied by electrons as well as holes) and M_2 is the concentration of shallower trap levels, then carrier transitions from deeper to shallower traps under the influence of photo-excitation will be governed by the following equations.

$$dn/dt = k_1 (M_1 - P) + k_2 N - \beta n(M_2 - N) - \alpha n P \qquad (6.25a)$$

$$dN/dT = -k_2 N + \beta n(M_2 - N) \qquad (6.25b)$$

$$dP/dt = k_1 (M_1 - P) - \alpha n P \qquad (6.25c)$$

$$P = N + n \text{ (charge neutrality)} \qquad (6.25d)$$

where

n = concentration of electrons in the CB,
N = concentration of electrons in the trapping levels,
P = concentation of holes at the activator levels,
α = probability of recombination ($\alpha = \sigma_1 u$, where σ_1 is the effective capture cross section of an electron by a hole at the activator level and u is the mean velocity of a conduction electron),
β = probability of trapping ($\beta = \sigma_2 u$, where σ_2 is the effective capture cross section of an electron by a hole at the trapping level),
k_2, k_1 = probability of stimulation of an electron into the conduction band from the trapping level, or activator level.

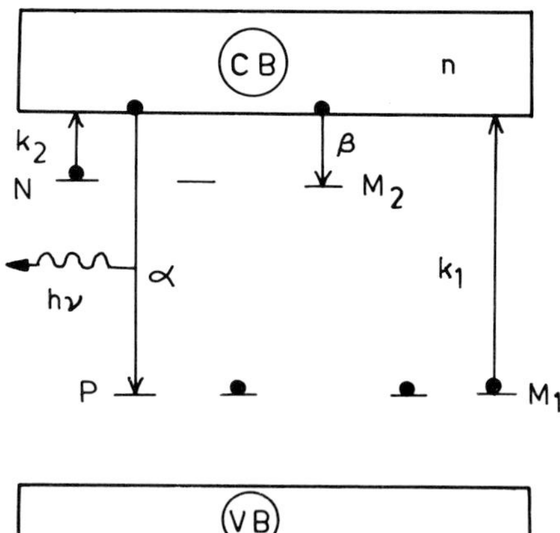

Figure 6.13. Energy band model of a crystal phosphor showing transfer of electron from deeper M_1 to shallower M_2 trap; as well as the photoluminescence. (E. I. Adirovich, quoted in *The Physics of the Electrophotographic Process.* by V. M. Fridkin (London: Focal Press) (1973).

If s_1 and s_2 are the cross sections of the interaction of the activator and trapping levels with a photon and I is the luminous intensity; I (photoluminescence) is given by

$$k_1 = s_1 I \text{ and } k_2 = s_2 I$$

or

$$I = (k_1 + k_2)/(s_1 + s_2) \tag{6.26}$$

The intensity of the PTTL is proportional to the electron filling of the M_2 levels. This is given by the term $\beta n(M_2{-}N)$ where n is given by Equation 6.25a.

If the absorption of light does not take place at the activator levels but in the valence band, then the kinetics of trap filling will be described by the following set of equations:

$$dn/dt = k_1 M_1 + k_2 N - \beta n (M_2{-}N) - \alpha nP \tag{6.27a}$$

$$dN/dt = -k_2 N + \beta n(M_2{-}N) \tag{6.27b}$$

$$dP/dt = k_1 M_1 - \alpha nP \tag{6.27c}$$

$$P = N + n \text{ (charge neutrality)} \tag{6.27d}$$

Equations 6.27 result from Equations 6.25 under the assumption that electron transition takes place from VB for which $P{\approx}0$.

References

1. Niewiadomski, T. *Confrontation of TL Models in LiF with Experimental Data.* Report No. 936/D (Institute of Nuclear Physics, Krakow, Poland) (1976).
2. Braunlich, P. IN Thermoluminescence of Geological Materials, ed D. J. McDougall (New York: Academic Press) pp. 61-88 (1968).
3. Fleming, R. J. Can. J. Phys. **46**, 1569 (1968).
4. Kelly, P. J. and Laubitz, M. J. Can. J. Phys. **45**, 311 (1967).
5. Sweet, M. A. S. and Urquhart, D. Phys. Status Solidi a **59**, 223 (1980).
6. Chen, R. J. Mater. Sci. **11**, 1521 (1976).
7. Kivits, P. J. Lumin. **16**, 119 (1978).
8. Sweet, M. A. S. and Urquhart, D. J. Phys. C: Solid State Phys. **14**, 773 (1981).
9. Cooke, D. W. and Gavathas, E. P. Preprint, 1982; Also see Cooke, D. W. and Close, D. M. J. Chem. Phys. **76**, 2174 (1982).
10. Grant, R. M. and Cameron, J. R. J. Appl. Phys. **37**, 3791 (1966).
11. Claffy, E. W., Klick, C. C. and Attix, F. H. IN Proc. Second Int. Conf. on Luminescence Dosimetry, Gatlinburg. USAEC Conf. 680920 (Springfield, VA: NTIS) p. 302 (1968).
12. Attix, F. H. J. Appl. Phys. **46**, 81 (1975).

13. Christy, R. W. and Mayhugh, M. R. J. Appl. Phys. **43**, 3216 (1972).

14. Mayhugh, M. R., Chirsty, R. W. and Johnson, N. M. IN Proc. 2nd Int. Conf. on Luminescence Dosimetry, Gatlinburg. USAEC Conf. 680920, p. 294 (1968).

15. Zimmerman, J. J. Phys. C: Solid State Phys. **4**, 3277 (1971).

16. Mayhugh, M. R. J. Appl. Phys. **41**, 4776 (1970).

17. Cooke, D. W. J. Appl. Phys. **49**, 4206 (1978).

18. Podgorsak, E. B., Moran, P. R. and Cameron, J. R. J. Appl. Phys. **42**, 2761 (1971).

19. Nink, R. and Kos, H. J. Phys. Status Solidi a **35**, 121 (1976).

20. Nink, R. and Kos, H. J. Nucl. Instrum. Meth. **175**, 15, 24 (1980).

21. Kos, H. J. and Takeuchi, N. Phys. Status Solidi a **57**, K171 (1980).

22. Kos, H. J. and Nink, R. Phys. Status Solidi a **57**, 203 (1980).

23. Bowman, S. G. E. PACT **3**, 381 (1979).

24. Sasidharan, P., Sunta, C. M. and Nambi, K. S. V. PACT **3**, 401 (1979).

25. Mobbs, S. F. PACT **3**, 407 (1979).

26. Sunta, C. M. and Watanabe, S. J. Phys. D: Appl. Phys. **9**, 1271 (1976).

27. Lakshmanan, A. R. and Vohra, K. G. Nucl. Instrum. Meth. **159**, 585 (1979).

28. Nagpal, J. S., Kathuria, V. K. and Gangadharan, P. Phys. Med. Biol. **25**, 549 (1980).

29. Pradhan, A. S., Bhatt, R. C. and Supe, S. J. J. Phys. D: Appl. Phys. **17**, 1699 (1984).

30. Bailiff, I. K. Nature **264**, 531 (1976).

31. Sunta, C. M. Phys. Status Solidi a **53**, 127 (1979).

32. Bailiff, I. K., Bowman, S. G. E., Mobbs, S. F. and Aitken, M. J. J. Electrostat. **3**, 269 (1977).

CHAPTER 7

Thermoluminescence-related Phenomena

K. Mahesh

7.1.   Introduction

Thermoluminescence is one among several thermally stimulated processes in solids[1] which are either strongly correlated with TL (such as thermally stimulated conductivity, TSC) or are of complimentary (ITC/DC) and supplementary (ESR) value to TL studies. Electron spin resonance (ESR) investigation of TL samples can, in some cases, help identification of an ongoing mechanism of luminescent centres[2]. In polymers the observed TL and TSC peaks can correspond to the onset of molecular motion which seems to affect strongly the de-trapping of carriers[3]. TL correlation measurements thus provide support to the study of defects in many solids.

7.2.   TL-ESR correlation

Electron spin resonance is a well documented technique applied to the study of defects related to paramagnetic impurities in insulators. ESR data can provide definite information on the structure of the dopant surroundings, such as the distribution of charge over the surrounding nuclei, charge and aggregation state of impurity, and characterisation of new paramagnetic centres if produced during the irradiation. Despite the fact that there is a considerable difference in the relative sensitivity of the two measurements, i.e. the ESR measurements are sensitive to relatively high concentrations of imperfections (10^{-10}–10^{-7}) and the TL process is sensitive to much smaller concentrations (up to about 10^{-15}), strong correlations, as will be evident in subsequent pages, between the two studies do exist.

7.3.   Principles of ESR

We observe the splittings between electron energy levels of the defect/impurity due to the interaction of the uncompensated (paramagnetic) electronic spin with crystal fields, nuclear spins and the applied magnetic field. Consider, for the purpose of illustration, a free ion having a resultant electron spin ($\Sigma s = 1/2$). The applied magnetic field (H_A) removes spin degeneracy and splits the energy level into two spin non-degenerate levels characterised by $m_s = 1/2$ and $-1/2$ as shown in Figure 7.1. The extent of splitting is $\triangle E = g\,\mu_B H_A$ where $\mu_B = e\,\hbar/2\,m_e c$ is the Bohr magneton and g (Landé g factor) $= 2$, for a free electron. If electromagnetic radiation (microwave) is impressed upon the material then it will be strongly absorbed (resonance absorption) if its frequency ν is

$$\nu = g\,\mu_B H_A/h \tag{7.1}$$

or

$$g\,H_A = (h/\mu_B)\nu = 7.14 \times 10^{-7}\,\nu\,\text{(cgs)}$$

As for a free electron $g = 2$, and if $\nu = 10^{10}$Hz, then $H_A = 3570$ gauss. It is convenient to use a fixed frequency and vary the magnetic field H_A to observe spin resonance. The ESR mechanism is illustrated in Figure 7.2.

7.4. TL-ESR studies
7.4.1. *NaCl:Mn^{2+} phosphor*
TL and ESR techniques in parallel have been employed[4] to investigate various types of impurity centres produced by X irradiation at room temperature and to determine the processes leading to their annihilation. When a solid sample is irradiated by X rays or nuclear radiation, defects, other than the existing traps due to impurities and crystalline imperfections, are produced. Irradiation can also fill the already existing traps. During the

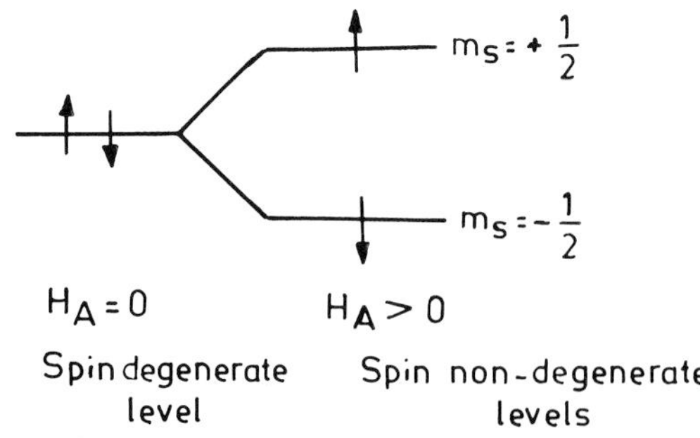

Figure 7.1. Lifting of spin degeneracy.

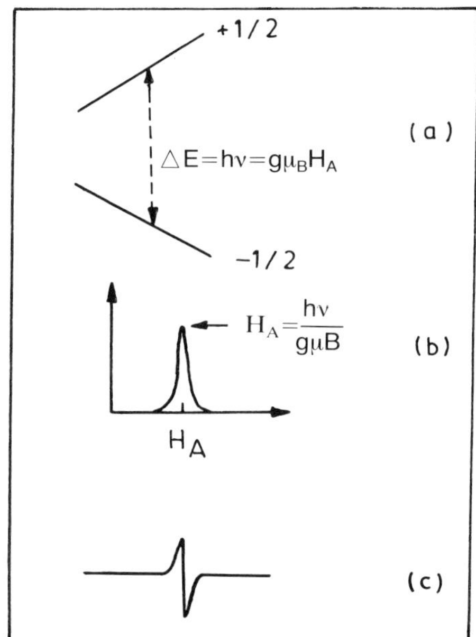

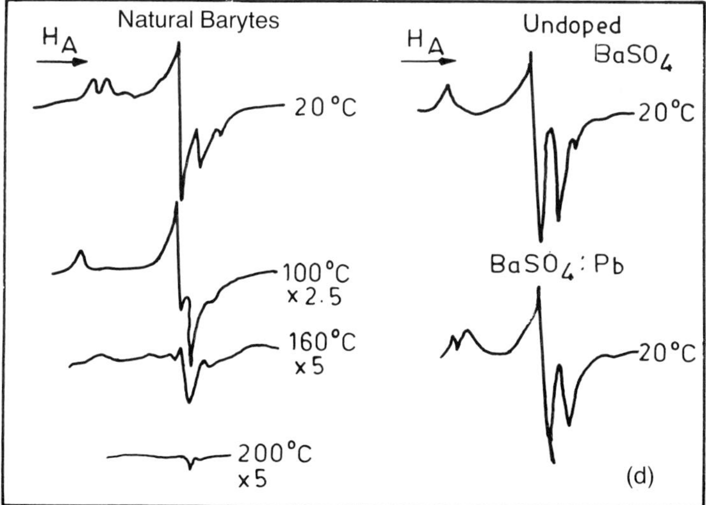

Figure 7.2. ESR signal mechanism. (a) Electron energy-level splitting. (b) Resonance absorption signal. (c) Derivative of absorption signal. Peak to peak separation is the linewidth. (d) ESR spectra of natural undoped barytes, and Pb-doped barium sulphate samples after gamma irradiation at room temperature: typical thermal annealing characteristics of ESR signals of irradiated natural barytes samples[12].

cycle and from the observed ESR signal characteristics and their comparison with the corresponding TL glow curve features, defects responsible for TL output are studied. Quenched samples of NaCl:Mn^{2+} show marked enhancement in TL light output. The quenching is done by heating the samples to 350°C for 30 min and then promptly transferring them onto a copper block at room temperature. During irradiation with X rays, the sample holder is cooled by a water flow to prevent heating due to irradiation. ESR studies in this sample show the presence of free Mn^{2+} and Mn^{2+} - vacancy dipoles as nearest and next nearest neighbours before irradiation. On irradiating these samples, growth of new ESR lines corresponds to substitutional Mn^{+}, interstitial Mn$^{\circ}$–C (C: cation) and Mn$^{\circ}$–D (D: dipole). Many of the observed glow peaks in the 5950 Å band correspond to the annihilation of the above mentioned Mn centres.

7.4.2. *NaCl:Cu^{+}*

A more evident correspondence between TL glow and ESR is provided in X irradiated NaCl:Cu^{+} single crystals[2]. The sample, containing Cu^{+} ions to about 2×10^{18} cm^{-3} concentration, is heated to 500°C followed by quenching to room temperature and is then X

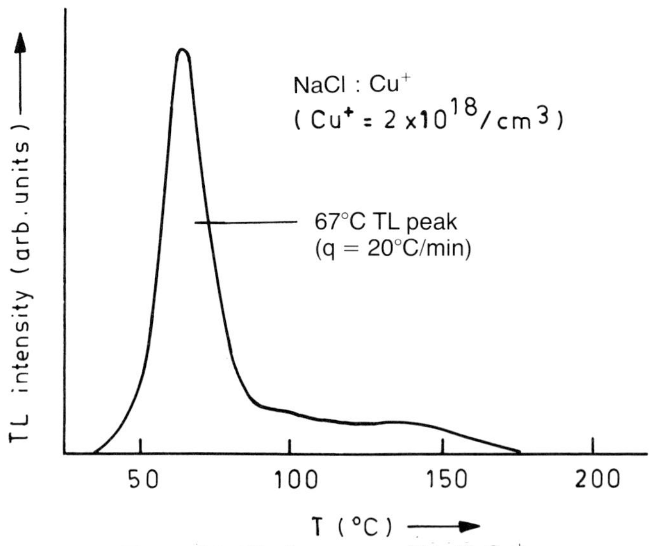

Figure 7.3. TL glow curve of NaCl:Cu^{+}.

irradiated to 5×10^3 R, and shows a prominent TL peak at 67°C ($q = 20°C.min^{-1}$) (shown in Figure 7.3) in addition to three other glow peaks observed at 105, 135 and 195°C. The feeble peaks become relatively more pronounced for Cu^+ concentration equal to 1×10^{17} cm^{-3} but disappear again when the concentration reduces further to 1×10^{16} cm^{-3}. The emission band for glow peak 1 shows a maximum at 3.55 eV, which is characteristic of Cu^+ emission[5]. ESR measurements on the sample are made at −140°C and it is found that no ESR signal is observed before X irradiation. The irradiation causes the creation of Cu^{++} centres in the material and an intense ESR spectrum is observed. When the sample is heated in steps to appropriate temperatures so that TL peaks at 67°C, 105°C etc. are thermally bleached, the ESR spectrum of the sample heated to 67°C shows a clear absence of the signal. This indicates the presence of a 67°C peak due to the Cu^{++} recombination centre which captures a thermally excited electron from the F centre. Such a correlation is not observed in respect of other TL peaks. Hence, it is difficult to identify the origin of these peaks from ESR analysis. Correlation of the 67°C peak with the corresponding ESR signal is clearly identifiable in Figure 7.4.

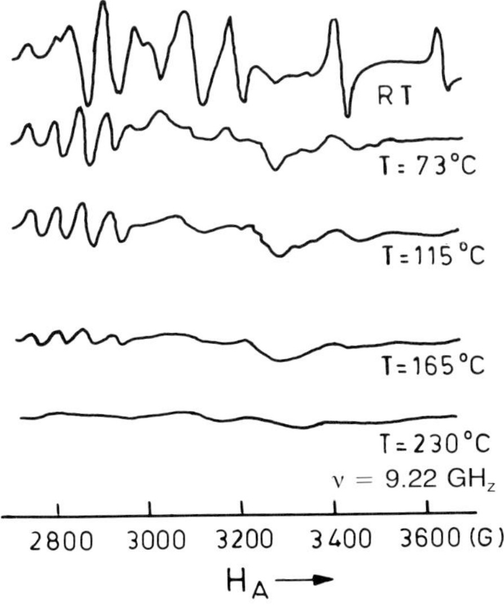

Figure 7.4. ESR signals measured at −140°C after heating stepwise to higher temperatures[2].

7.4.3. **TLD-100**

Thermal treatment of irradiated (10^6R) TLD-100 powder and the consequent shape and size of the ESR signal have been utilised to study the defect structures responsible for TL emission[6]. In Figure 7.5, g values for different observed signals are shown. These are determined using the known signal position of DPPH (diphenyl picryl hydrazyl; $g = 2.0036$) as the calibration point (g marker). This is measured experimentally by placing a quartz tube containing DPPH in the second cavity while the irradiated sample powder, in a clean and dry matching quartz tube, is placed in the first cavity of the dual cavity ESR spectrometer. The difference of the g factor as compared to its free-electron value ($g = 2$), i.e. $\triangle g$ ($= g_{bound} - g_{free}$) is a measure of the strength of binding of the electron within the defect. The sign of $\triangle g$ gives the charge state of the

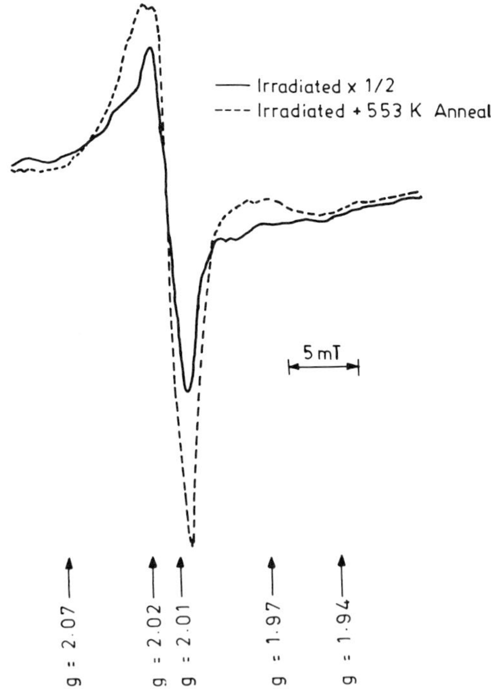

Figure 7.5. Central ESR signals from LiF (TLD-100) powder after irradiation to 10^6R (shown at half intensity) and after annealing at 280°C for 10 min[6].

defect. The electron gives a negative, and a hole, a positive, value of Δg, respectively.

The observation of a broad ESR band is usually related to the presence of centres in a precipitate or colloid form. The precipitates form during the irradiation, a fact which can be verified as the ESR band grows in intensity during the first few irradiation cycles, after which the saturation is exhibited.

7.4.4. *CaSO₄:RE*

The use of $CaSO_4$:RE (RE, rare earths) TL phosphors in gamma radiation dosimetry has been attempted. These phosphors are thought to show better sensitivities in addition to the ease and economy with which they can be prepared in the laboratory[7]. The similarity of the ESR spectra due to the unpaired trapped charge carriers in the material of irradiated powder samples with different RE dopants allows one to deduce that while the traps are the basic feature of the host material, the added dopants only influence the relative population of traps. The paramagnetic host lattice radicals like SO_4^-, SO_3^-, O_3^- etc. produced by the gamma irradiation constitute the different host lattice traps. The ESR spectrum is shown in Figure 7.6. Another example of the effect of sample anneal (which erases the given TL glow peak) on the ESR spectrum is provided in Figure 7.7 As previously stated, in a sample preheated to a temperature corresponding to a given glow peak, the glow peak is erased (thermal bleaching) and the ESR signal for this peak will also be absent. In this manner various TL glow peaks are identified in the ESR spectrum and one can, therefore, easily assign g values to various TL glow peaks. The possibility of a different mechanism in which RE^{3+} ions may be reduced to RE^{2+} by capturing an electron during gamma irradiation also exists – particularly in case of TL of CaF_2(RE) phosphors[8]. On heating, if holes are released, then these can recombine with electrons trapped at the RE^{2+} site giving rise to RE^{3+} in the excited state on account of absorption of recombination radiation by the RE^{3+} ion. The observed TL in this case will be the fluorescence emission from the excited RE^{3+} ion.

7.4.5. *SrCl₂(Eu)*

Another significant example of the TL-ESR correlation studies is that of $SrCl_2$ doped with europium. Eu, when used as

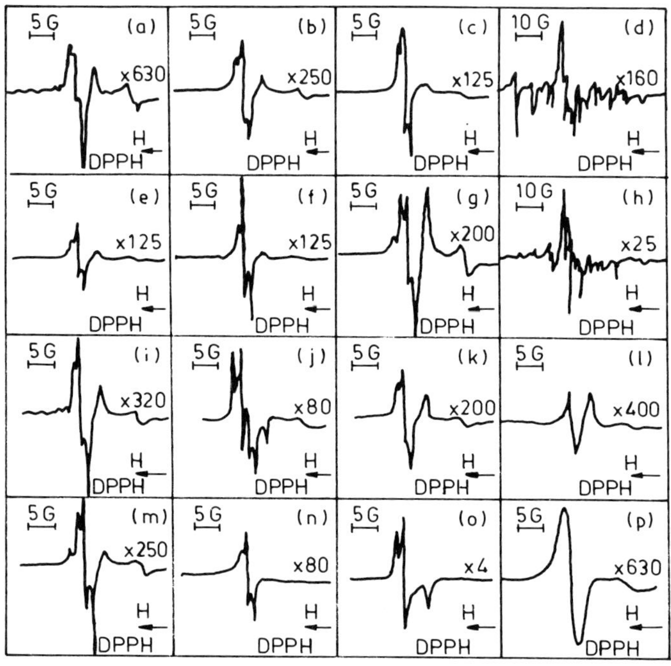

Figure 7.6. ESR spectra of rare earth doped $CaSO_4$ phosphors after gamma irradiation at room temperature: (a) undoped; (b) Sm; (c) Er; (d) La; (e) Ce; (f) Tb; (g) Tm; (h) Lu; (i) Pr; (j) Dy; (k) Yb; (l) Eu; (m) Nd; (n) Ho; (o) Y; (p) Gd[7].

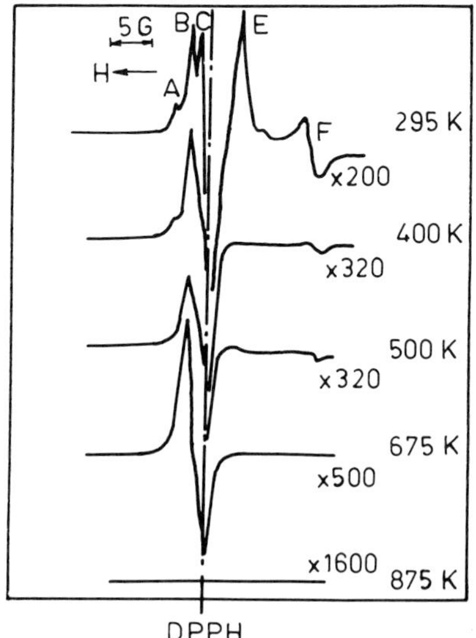

Figure 7.7. Typical thermal annealing characteristics of ESR signals of gamma irradiated powders. Sample: $CaSO_4(Tm)$[7].

dopant in $SrCl_2$, goes at least partly as divalent Eu^{2+} ions. This is confirmed by a strong ESR spectrum characteristic of Eu^{2+} as shown by the unradiated sample. The Eu^{2+} spectrum is found to be dependent on the gamma dose and also the thermal and atmosphere-exposure history[9]. The Eu^{2+} signal is considerably diminished when the sample is exposed to about 3×10^6 R of gammas. This observation is attributed to the formation of diamagnetic Eu^{3+} ions because a hole is trapped by Eu^{2+}. The temperature dependence of the ESR signal of the irradiated sample shows the strong emergence of a Eu^{2+} signal at about 500 K. At this temperature there is also a glow peak (3, see Figure 7.8) which means that it is connected with the reduction of Eu^{3+} to Eu^{2+} by the trapping of a thermally released electron. If, however, the sample is exposed to the atmosphere, the sample, being probably hygroscopic, absorbs water. As a result, OH centres are formed which are revealed in the ESR spectrum. In the $SrCl_2(Eu)$ sample exposed to 3×10^6 R gammas in the open atmosphere for 12 h, peak 3 becomes nearly extinct. A doublet with a separation of about 13 gauss at $g = 2.02$ is observed in the ESR spectrum, which is

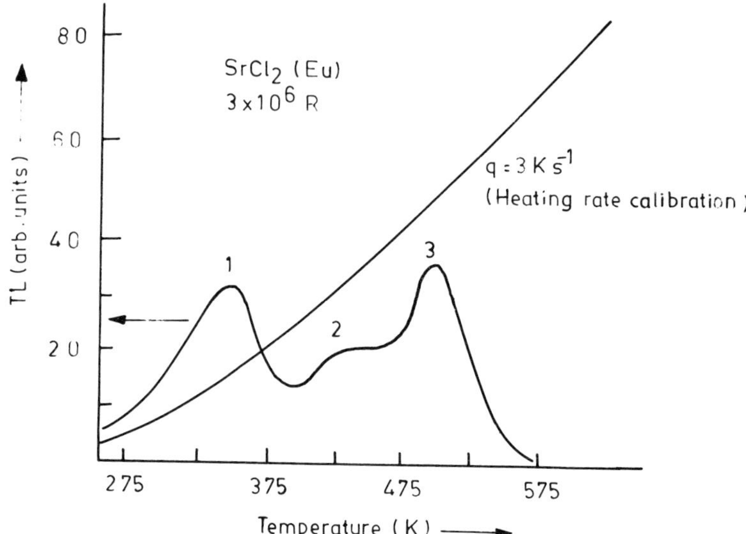

Figure 7.8. TL glow pattern of $SrCl_2(Eu)$ gamma irradiated to 3×10^6R and the sample sealed from the atmosphere[9].

attributed to a neutral OH radical. In the temperature dependence of the ESR spectrum the doublet disappears at around 420 K. As this temperature corresponds to glow peak 2, this peak is, therefore, associated with the release of a hole by OH so as to become an OH ion. The hole recombines at an electron trap site giving rise to TL glow peak 2. The temperature dependence of the ESR spectrum of $SrCl_2(Eu)$ is shown in Figure 7.9.

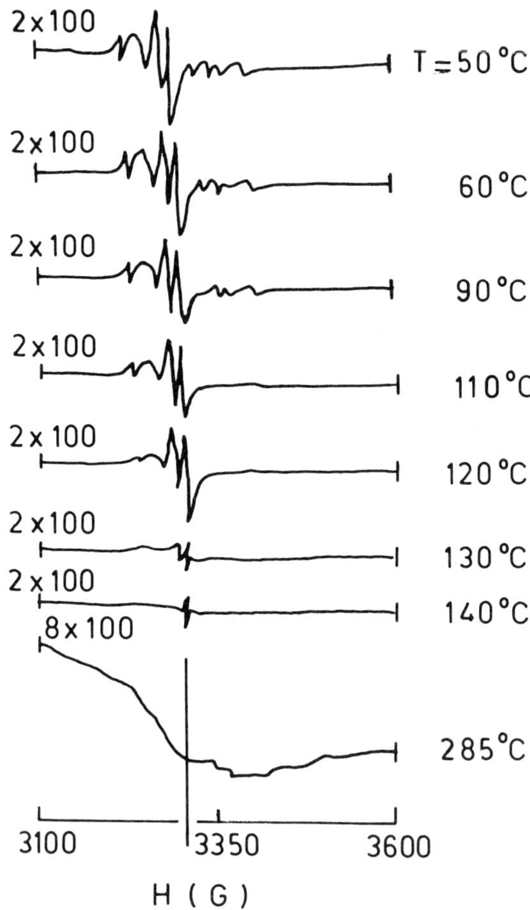

Figure 7.9. The temperature dependence of the ESR spectrum of $SrCL_2$:Eu, gamma irradiated to 3 MR. The sample was exposed to the atmosphere for 12 h during irradiation[9].

7.4.6. *BaSO₄, BaSO₄:Pb and natural barytes*

TL and the role of different dopants has been studied by a number of workers[10-12]; studies which help to bring out the complimentary features of TL and ESR techniques. The comparison of natural barytes, synthetic barytes (undoped $BaSO_4$) and doped $BaSO_4$ can be made from their respective ESR spectra. The ESR signals are ascribed to anion radicals, SO_4^-, SO_3^-, SO_2^- and O_3^-, resulting from gamma irradiation ($\approx 10^6$ rad). These anions act as hole traps. The ESR spectra of γ irradiated barite samples is more or less identical with that of γ irradiated $BaSO_4$ and shows similarity with the spectra of samples doped with different impurities (such as Al, Pb, Ag, Mn).

The effect of thermal annealing on TL glow curves and ESR spectra is similar. It is found that natural barytes, barium sulphate and Pb^{2+}-doped barium sulphate samples have a similar decay of the ESR signals at the temperatures corresponding to the peaks of the TL glow curves. The decay of ESR spectra of γ irradiated natural barytes along with the room temperature spectra of $BaSO_4$ and Pb^{2+}-doped $BaSO_4$ samples has been shown in Figure 7.2(d).

7.4.7. *Fading features of TL and ESR signals*

Striking correspondence in the time decay features of many samples are also sometimes observed. An example that shows this feature clearly is the decay with time in $CaSO_4$:Tm of the TL and ESR signal intensity at room temperature (a similar pattern is also found at higher temperatures). This is shown in Figure 7.10.

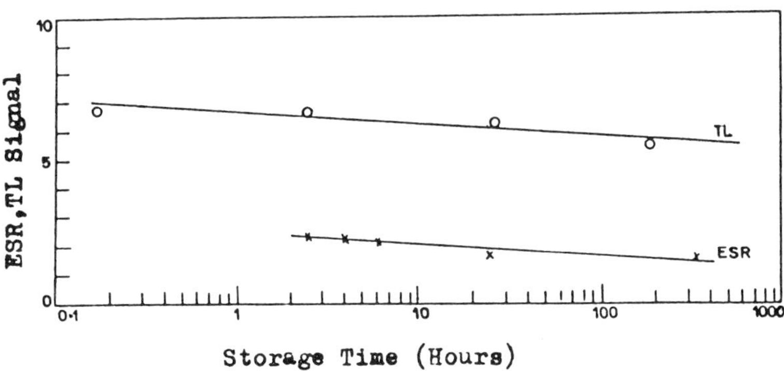

Figure 7.10. Decay of TL and ESR signal of gamma irradiated $CaSO_4$:Tm as a function of storage time of the sample kept at room temperature[11].

7.5.   Thermally stimulated exoelectron emission (TSEE)[13-20]
7.5.1.   *Introduction*

Another important physical phenomenon that aids the interpretation of effects underlying TL and other information concerning the distribution and nature of traps is the release of trapped electrons (called the exoelectrons) upon heating an insulator or a semiconductor that has been pre-exposed to X, γ, or β rays. The emitted exoelectrons leave the solid and can be detected by an electrode placed above the sample and kept at a positive potential. Just as in case of TL and TSC, the intensity of the effect (i.e. number of emitted exoelectrons) is directly related to the radiation dose received by the sample. For this reason TSEE has been exploited as a radiation dosimetric technique.

The special feature of TSEE is the fact that since only the electrons (and not holes) can be released from the sample when heated, the observed TL peaks can be identified as arising from

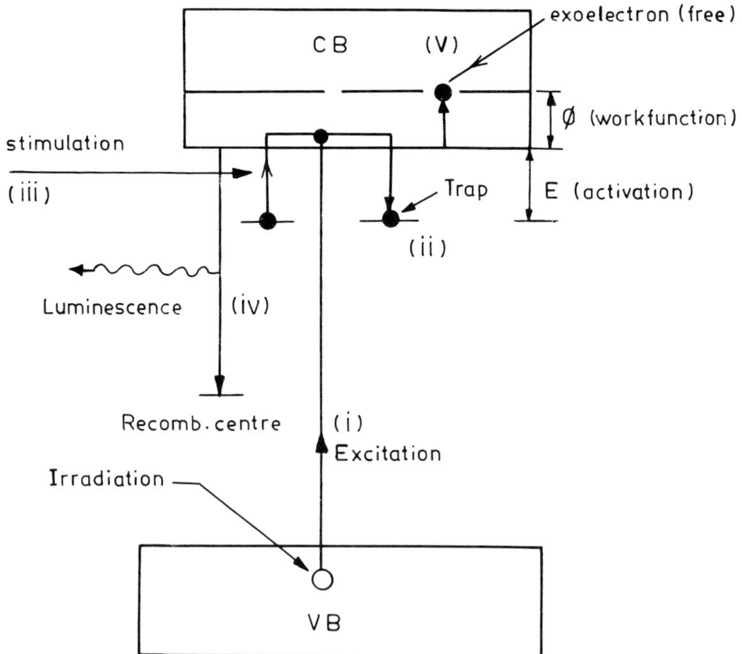

Figure 7.11. Schematic diagram to show: (i) excitation; (ii) trapping; (iii) de-trapping (stimulation); (iv) recombination and (v) exoelectron emission.

either the electron or hole traps. TSEE has application particularly in low level radiation dosimetry.

7.5.2.   *TSEE phenomenon and technique*

As stated, the exoelectron is the electron thermally released from the trap; it leaves the surface of the insulator to be detected by the electron detector (point counter, GM counter or a proportional counter). The same electron could, under different circumstances, eventually recombine at a recombination centre and give rise to luminescence. Figure 7.11 depicts the essential features of the exoelectron emission process. (Fintlemann[21] has pointed out the possible inadequacy of thermal stimulation as a mechanism of exoelectron emission. Exoelectrons have energies of a few eV. The equivalent thermal energy will correspond to a temperature of several tens of thousands of Kelvin; thermal stimulation as the expected cause of exoelectron emission is therefore doubtful. On the other hand, the optically stimulated exoelectrons (OSEE; or PSEE-photostimulated exoelectrons) have practically the same energies. It is postulated that exoelectrons are emitted by the photoelectric effect or through the Auger conversion process. For the surface exoelectrons, probably some kind of chemical reaction on the surface due to the changes in adsorbed layers of atoms or molecules, e.g. the reaction $H_{ads} + OH_{ads} \rightarrow H_2O + energy$, may be responsible. The chemical energy of the reaction is thought to be taken up by the electron held in the nearby trap). TSEE is essentially a two way process involving the 'surface traps' and the 'volume traps' within the crystal. Surface traps can arise from adsorbed species which may be atmospheric constituents and other dirt as well as surface lattice imperfections. These traps do not contribute to TL emission. The volume centres are those that also participate in thermoluminescence (F and M centres in LiF, for instance). Hence it is these volume centre electrons that show a close relationship between TSEE and TL. In some samples the TSEE and TL peak maxima, as well as the fading rates of the two effects, are reasonably identical. However, in the case of those centres (dilute rare earth luminescent centres) in which the conduction band is not involved in the TL emission, one may observe only the latter without any corresponding exoelectron peaks. Also, internally converted by the 'Auger process', the exoelectron may be emitted and the TL light may stimulate another electron and make it overcome the work function Φ. Despite these limitations to the TSEE and TL parallelism, there are many ionic crystals with enough band gap for TSEE and TL to correlate. Experimentally, it is advisable to measure TSEE and TL simultaneously in a sample during its heating cycle[22]. A schematic diagram of the experimental set-up for recording the exoelectron glow curve is shown in Figure 7.12. The electron detector may be a gas flow counter or an electron multiplier with associated electronics. The samples are powders compressed into a Cu mesh and the compressed layer is thick enough to cover the wires of the mesh completely. One can use the Randall-Wilkins model to analyse the glow curve and determine the activation energy E. A simple estimate is possible using the Bohun-Nassenstein relation[23,24]. This is an identical relation (see Equation 3.18 of Chapter 3) given by Urbach for TL process in a KCl:Tl phosphor (E in eV and T_m in K)

$$E = T_m/500$$

The most widely used method is however that of Balarin and

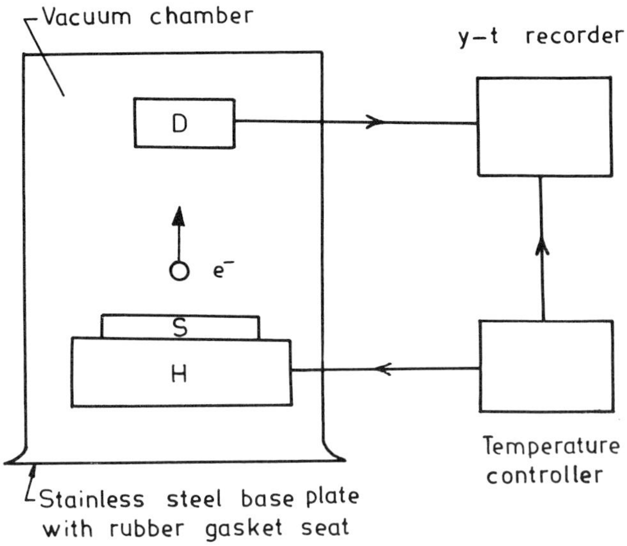

Figure 7.12. Schematic diagram of an experimental set-up for recording an exoelectron glow curve. H, heating block. S, sample. D, electron detector.

Zetzsche[25]. In this method the whole glow peak is used for the determination of E. At the start of the heating of the sample, there is a certain concentration, C, of occupied traps due to prior excitation. During heating, the concentration of occupied electron traps decreases as a function of time (or temperature). The number of electrons emitted per unit time is proportional to $-dC/dt$. Since $q = dT/dt$ (heating rate); we find

$$-dC/dt = -q(dC/dT)$$

In the experiment we record the number of electrons $N(T)$ emitted per second as a function of temperature using a constant heating rate. We plot $\ln [\ln 1/R(T)]$ as a function of $1/T$, where $R(T)$ is the fraction of occupied electron traps at temperature T. $R(T)$ is related to the area $A(T)$ of the TSEE glow curve at temperature T by the relation

$$R(T) = 1 - A(T)/(dA/dR)$$

and

$$A(T) = \int_{T_i}^{T} N(T)\, dT$$

The slope of the resulting straight line is given by, according to the Balarin-Zetzsche formula[25,26]

$$\frac{E}{k_B} = \frac{d \ln [\ln 1/R(T)]}{d\,(1/T)} \tag{7.2}$$

where k_B is the Boltzmann constant $(= 0.862 \times 10^{-4}\ eV.K^{-1})$

7.6.   TSSE-TL correlation
7.6.1.   *Introduction*

Simultaneous measurement of both TSEE and TL signals is necessary to classify the TSEE glow peaks which are, and those which are not, related to TL. The TL glow curves can be affected drastically by small amounts of impurities and crystal defects[27] and the glow peak temperatures depend upon the heating rate and, in some cases, on the method of temperature measurement. Thus, it will be incorrect to discuss TSEE–TL correlation from the results of independent measurements.

7.6.2.   *Experimental technique for simultaneous TSEE and TL measurements*

We now describe the experimental set-up for carrying out studies at liquid nitrogen temperature (LNT) on TSEE, TL and TL emission spectra of pure LiF single crystals by Tomita *et al*[28]. The set-up is schematically shown in Figure 7.13 (a) and (b). Exoelectrons are detected by an electron multiplier (EM) tube; TL is detected by a photomultiplier (PM) tube (not shown in the figure) through the quartz window and spectral measurement on TL made by a Bausch and Lomb grating monochromator. The cryostat base can be rotated while remaining under vacuum so that the sample can face the Al window through which X rays enter the vacuum chamber. The TL received by the PM is not, however, normal to the sample surface but

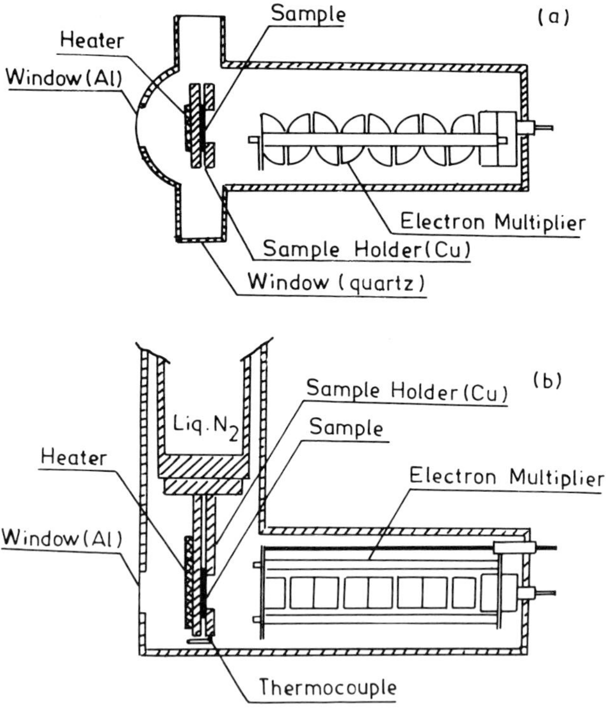

Figure 7.13. Schematic diagram of the device (cross sectional); (a) bottom view and (b) side view[28].

enters the photocathode after having been scattered. In this way the light intensity is reduced by a factor of several tens as compared with the intensity received normally to the sample surface. This fact, however, does not significantly alter the shape of the glow curves. The first dynode of the EM is held at a positive potential (50-80 V) with respect to the sample holder (Cu plate) which is grounded. The heater itself is covered with copper sheets to prevent any light from it reaching the PM. The temperature of the sample is measured by a copper-constantan thermocouple which is silver-soldered to the holder plate. The front surface of the holder, on which the LiF crystal rests, is covered with a thin gold layer by vacuum evaporation to improve conduction. The PM current can be measured by an electrometer and the EM pulses due to exoelectrons are amplified and counted. Alternatively, one can use a ratemeter to give an average signal output which can be fed to a y-t recorder in order to obtain a trace. The heating rates should be kept small ($0.5°C.s^{-1}$). To check on background signals, the sample holder, without the sample, is X irradiated at LNT for some time and then heated to record any TSEE and TL signals which can then be allowed for. An oil diffusion pump is used to create a vacuum in the chamber (10^{-6} torr). TSEE and TL glow curves of a pure LiF single crystal are obtained after the crystal is annealed at 400°C for 6 h in air and rapidly cooled to room temperature within a few minutes. This procedure restores the sample to its pre-irradiation state and the crystal can be used repeatedly. The results of simultaneous measurements are shown in Figure 7.14 and the emission spectra of each glow curve is shown in Figure 7.15. As we observe in Figure 7.14(a) and (b), TL and TSEE exhibit, below 0°C, closely related temperature dependent peaks. Above 0°C, while the TL glow curve consists of 5 peaks (numbered 0, 1, 2, 3 and 4), the TSEE glow curve is composed of a relatively intense peak at 100°C and an abrupt rise above 250°C. It is therefore evident that the traps responsible for TL below 0°C, are also responsible for exoelectron emission; the latter arising from a part of the released TL which excites the neighbouring electron traps (Auger process).

A similar study has been done in KCl, KCl:Tl and KCl:Cu samples by Kamada *et al*[29]. Several TSEE and TL glow peaks showed overlapping glow peak temperatures in the three samples of KCl:Tl, KCl:Cu and KCl (undoped) and this is shown in Table 7.1. Another significant feature of the TSEE and TL correlation is the decay profile

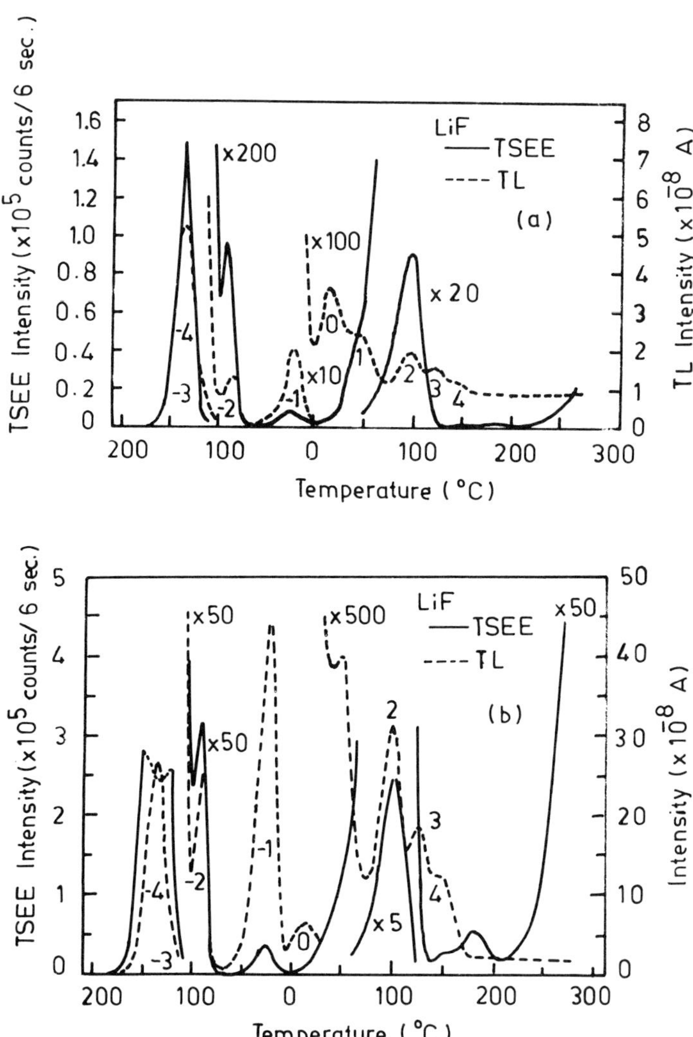

Figure 7.14. TSEE and TL glow curves for a pure LiF single crystal by simultaneous recording, where the crystal was X irradiated (W target, 30 kV and 10 mA) at LNT; (a) for 5 s and (b) for 30 s. The TL glow peaks are labeled from –4 to 4 in ascending temperature order. A factor is also shown by which the part of the curve is enlarged[28].

of the exoelectron emission and TL intensity. This is particularly illustrated for the case of the KCl:Tl sample. The decay for the 200 K and 280 K peaks is examined for both TSEE and TL. The decay profiles are shown in Figure 7.16. The TL intensity in the case of the 200 K peak decays with time but that of the exoelectron emission remains almost constant. This behaviour suggests that these exoelectrons are not the ones excited from shallow traps by the TL. Kamada *et al*[29] have proposed that the diffusion of V_k centres (holes) and the subsequent reaction of these centres with the F centre releases the excess electron from the latter: this is the emitted exoelectron. (A V_k hole interacts with an excess electron of the F centre to form an exciton. The energy released in the decay of the exciton is transferred to the other electron of this centre which is then released to become the exoelectron. Since any external radiative decay of the exciton is not observed, it is obvious that the

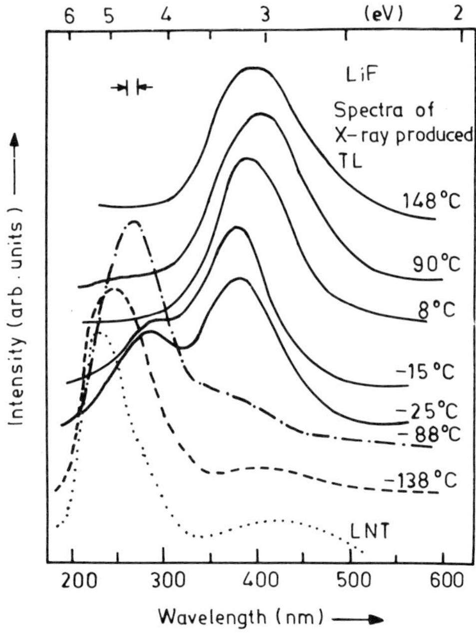

Figure 7.15. Emission spectra of X ray produced TL in LiF near each glow peak temperature. The full width at half maximum of mercury lines used for the calibration of wavelength is also shown for the reference of resolving power. The spectra are not corrected for change of spectral response of the photomultiplier[28].

Table 7.1. **Temperatures (K) at the maximum of the TSEE and TL glow peaks in KCl:Tl, KCl:Cu and undoped KCl, which were X rayed at LNT[29].**

KCl:Tl				KCl:Cu				KCl			
TSEE		TL		TSEE		TL		TSEE		TL	
A,	96		···	A,	124	a,	123	A,	123	a,	121
B,	121	b,	123	B,	165	b,	157	B,	207	b,	211
C,	156	c,	153	C,	211	c,	213	C,	241	c,	249
D,	209	d,	209	D,	247	d,	252	D,	293	d,	289
E,	270	e,	267	E,	305	e,	310	E,	340	e,	359
F,	300	f,	299	Γ,	336		···	F,	370	f,	373
G,	371	g,	373	G,	355	g,	353	G,	484	g,	469
	···		···	H,	377		···		···		···

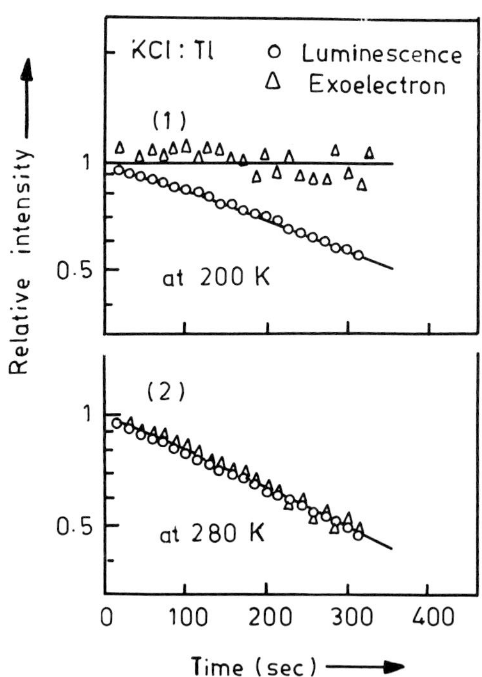

Figure 7.16. Decay of the exoelectron emission and luminescence intensities at about 200 K(1) and 280 K(2) in KCl:Tl[29].

exciton radiation is internally converted into the observed exoelectrons.) The corresponding TL peak at 200 K probably arises from the recombination of the diffusing V_k centres with the electron traps. The decay of the 280 K exoelectron and luminescence emissions are identical (Figure 7.16(2)). It is proposed that Tl^0 centres lose electrons to the conduction band at about 300 K. These electrons are then trapped by the Tl^{2+} forming the excited Tl^+ centres which then decay to give TL emission. Some of the thermally released electrons from Tl^0 centres may acquire sufficient kinetic energy to overcome the surface potential barrier to become exoelectrons.

7.6.3. *Improved TL-TSEE correspondence*

Much improved correspondence between TL and TSEE has been achieved by recording TL and TSEE glow curves from two separate samples of the same material[30]. The samples that have been studied are LiF (pure and with usual dopants, i.e. Mg, Ti) and Al_2O_3. One sample of the same material is admixed with graphite (with a view to improving conductivity) and is mounted on one heater. The other portion of the sample is mounted on another heater. Both the heaters are connected to the temperature programmer to keep the temperatures and heating rates identical

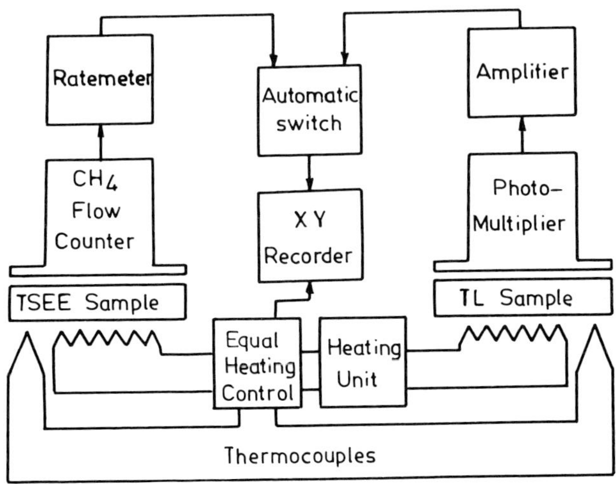

Figure 7.17. Circuit diagram for recording TSEE and TL[30].

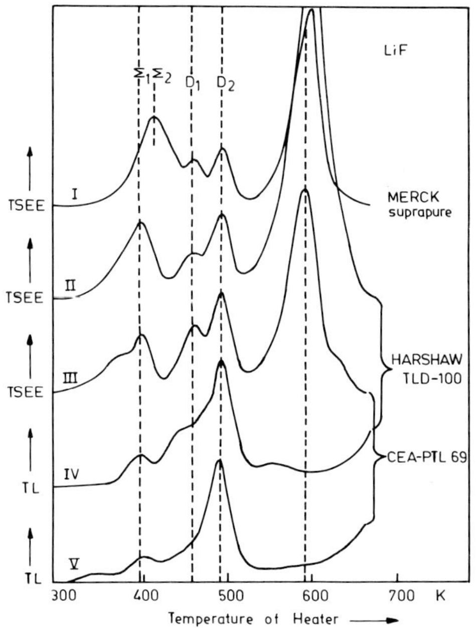

Figure 7.18. Simultaneously recorded spectra of different LiF materials[30].

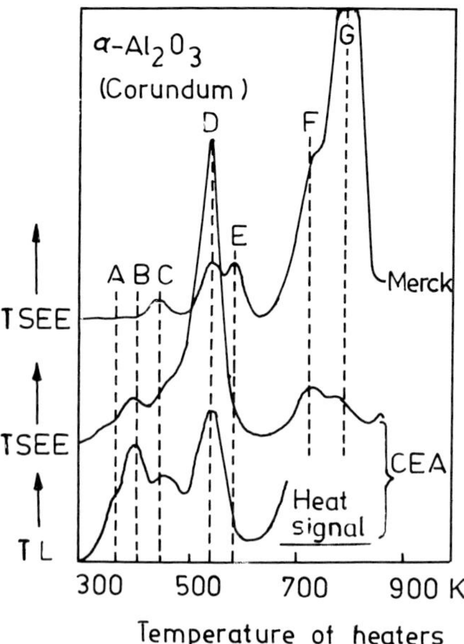

Figure 7.19. Simultaneously recorded spectra of different Al_2O_3 materials[30].

for both samples. The former material gives TSEE while the latter gives TL glow curves which can be recorded by the same *x-y* recorder through the alternative switch. The schematic diagram is shown in Figure 7.17. Figures 7.18 and 7.19 depict TL and TSEE correlations for LiF (pure and doped) and Al_2O_3 respectively.

Apart from the coincident TL and TSEE peaks that are visible in Figures 7.18 and 7.19, the striking feature of the study is that since TSEE glow curves show the electron peaks even in an ultrapure LiF powder sample without any dopant, the evidences that electron and hole traps are created by the dopants in LiF is not supported in these studies. On the contrary, it is suggested by Petel and Holzapfel[30] that the dopants act merely as activators.

Fukuda *et al*[31] have observed correspondence between several of the TSEE and TL peaks in the temperature range −200 to 400°C in the sintered and X irradiated CaB_4O_7 doped with Pb, Eu or Dy.

7.6.4. *Influence of excitation radiation on the TSEE spectrum*

Since TSEE has significant dosimetric applications[32], particularly in the low dose region (10^{-3}–10^{-4} Gy), it is pertinent, before we conclude TSEE-TL correlations, to point out the response of a typical TSEE material (SiO_2 layer) to a single and mixed radiation field. This is depicted in Figure 7.20 (a), (b) and (c). Note in (b) that the subsequent irradiation of the sample by γ rays washes out the electron-generated TSEE peak of (a); however, it

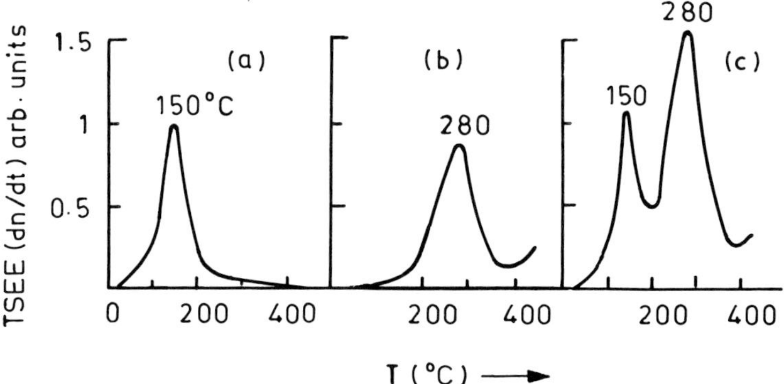

Figure 7.20. TSEE glow curves of SiO_2 layers after successive excitation by (a) electron, (b) γ-ray and (c) electron again; (a) 3.5×10^{-9} A flux for 10 min, $E=3$ keV; (b) 1.2×10^{5}R; (c) 3.5×10^{-9} A flux for 10 min, $E=3$ keV[50].

re-emerges if the sample is again irradiated with the electron beam to the same extent as in (a). It is surprising that the 65 keV electron bombardment of the sample does not induce any glow curve. Also, a 60 min UV excitation and a 2 min visible laser light excitation give identical TSEE glow curves. Similarly, an X ray exposure of the sample for up to 10^3 min is not sufficient to yield any glow curve.

7.7. TL-TSC correlation

7.7.1. *Introduction*

Thermally stimulated conductivity (or current) arises from the thermal de-trapping of carriers and thermally induced dipole orientation. Since TL also arises from the thermal de-trapping of carriers, a correlation between the two is expected, as some TSC peaks arising from trapped carriers are also accompanied by TL peaks. TSC measurements are thus utilised to determine the charge-

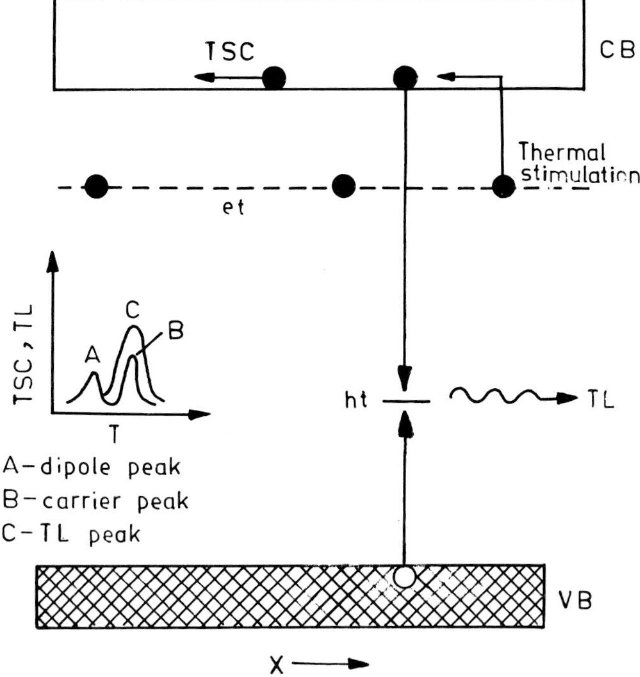

Figure 7.21. Mechanism of TSC and TL from trapped electrons. ●, electrons, ○ holes; ht, hole trap (recombination centre, RC); et, electron trap; X, electric field (DC).

carrier trapping details. As the two glow curves are usually found to be similar, the methods used for TL are also employed to analyse the TSC peaks. First, however, the TSC peak by dipole orientation should be distinguished from that arising from carrier de-trapping. As we know, a sample will not show a TL peak until excited by radiation and this also holds for the TSC peak originating from carrier de-trapping. Hence the emergence of a TSC peak in an unexcited sample corresponds to the dipole orientation and is not correlated with the TL peaks.

The mechanism of radiation induced TSC and TL are illustrated in Figure 7.21. The specimen pre-irradiated at some initial low temperature, is heated linearly under the action of a DC electric field X. The thermal release of electrons and holes from their traps gives rise to TSC. Many released electrons can also recombine radiatively with holes at the recombination centre, giving rise to TL.

7.7.2.   *Theoretical account of TSC*

If n_c is the concentration of electrons in the conduction band (or holes in the valence band), the current I when an electric field X is impressed on the sample, is given by

$$I = n_c\, e\, \upsilon = n_c\, e\, kX \tag{7.3}$$

where k is the mobility of the carrier. The conductivity σ is given by

$$\sigma = I/X = n_c ek \tag{7.4}$$

Referring to Figure 7.22, the rate at which the conduction electrons n_c and trapped electrons n_t change is expressed as

$$(-dn_c/dt) = (dn_t/dt) + (n_c/\tau) \tag{7.5}$$

and

$$-dn_t/dt = n_t\, s\, \exp(-E_t/k_B T) - n_c\, (N_t - n_t)\, \upsilon\, \Sigma_t \tag{7.6}$$

In Equation 7.5, the first term on the right hand side gives the increase in trapped electron population per unit time and the second term gives the number of conduction electrons recombining per unit time. In Equation 7.6 the first term on the right hand side gives the de-trapping rate from which the number of recaptured electrons in the trap (2nd term) is subtracted. Σ_t is the trap capture cross section, υ is the thermal velocity of the conduction electrons, s is the frequency factor, E_t is the trap activation energy and τ is the carrier recombination lifetime. We find next the expression for conductivity assuming slow re-trapping.

7.7.3.   *Slow re-trapping*

Under this assumption, the re-trapping term in Equation 7.6 can be ignored and we can write

$$dn_t/dt = -n_t\, s\, \exp(-E_t/k_B T) \tag{7.7}$$

using $T = T_i + q\,t$, $dT/dt = q$ and integrating, we get

$$\ln(n_t) = \left[-(s/q) \int_{T_i}^{T} \exp(-E_t/k_B T'')\, dT' \right] + \text{constant}$$

when $T = T_i$, $n_t = n_{to}$ (say), the initial number of trap electrons. Thus, with constant $= \ln(n_{to})$ we find

$$n_t = n_{to} \exp\left[-(s/q) \int_{T_i}^{T} \exp(-E_t/k_B T')\, dT' \right]$$

Substituting this in Equation 7.7

$$dn_t/dt = -(sn_{to}) \exp\left[-(s/q) \int_{T_i}^{T} \exp(-E_t/k_B T')\, dT' \right] \exp(-E_t/k_B T)$$

$$= -(sn_{to}) \exp\left[-(E_t/k_B T) - (s/q) \int_{T_i}^{T} \exp(-E_t/k_B T')\, dT' \right] \tag{7.8}$$

We now make a basic assumption that is valid in high resistivity

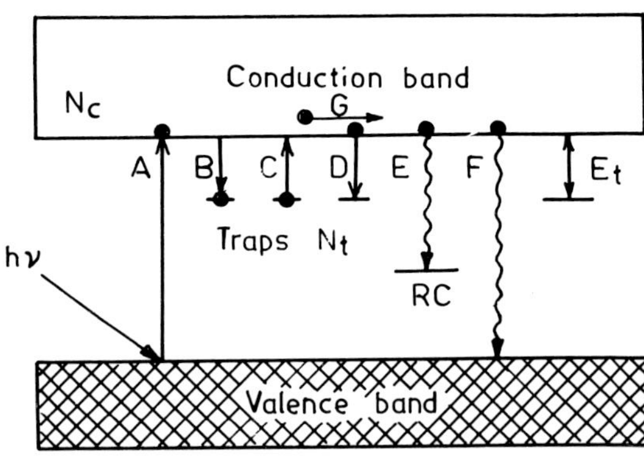

Figure 7.22. Simple band model for TSC analysis. Key: N_c, density of states in conduction band (CB). N_t, density of electron trap (ET) states. E_t, trap activation energy (eV). A, carrier generation process by radiation excitation. B, trapping process. C, de-trapping process. D, re-trapping process. E,F, recombination process. G, drift process.

crystals under normal experimental conditions and where the temperatures involved are not too high. This is $n_c \ll n_t$ and hence

$$(dn_c/dt) \ll (dn_t/dt) \tag{7.9}$$

Kelly *et al*[33] have criticised this assumption as being arbitrary and have emphasised that for $N_t > 10^{15}$ cm^{-3}, the assumption in Equation 7.9 and the solution outlined in Equation 7.8 are valid. As, for instance, at the start of heating cycle $dn_c/dt = -dn_t/dt$ and at high temperatures, above the TSC peak, $n_c > n_t$, they have defined the validity for the solution (Equation 7.8) as

$$N_c / (N_t + M) \le C \tag{7.10}$$

where M is the density of thermally disconnected traps (not contributing to the TSC peak being characterised by the activation energy E_t). We have taken M as zero here. C is a constant depending on the accuracy required of the final solution. For instance for an accuracy of the order of 1% of the peak value, $C = 10^4$.

Equation 7.5, with the assumptions given in Equation 7.9 becomes

$$n_c = -\tau \, (dn_t/dt) \tag{7.11}$$

Substituting Equation 7.8 in Equation 7.11 we find

$$n_c = s\,\tau\,n_{to} \exp\left[-(E_t/k_B T) - (s/q) \int_{T_i}^{T} \exp(-E_t/k_B T') \, dT'\right]$$

$$\sigma(T) = (s\,k\,e\,\tau\,n_{to}) \exp\left[-(E_t/k_B T)\right.$$
$$\left. - (s/q) \int_{T_i}^{T} \exp(-E_t/k_B T') \, dT'\right] \tag{7.12}$$

Since $I(T) = \sigma\,(T)\,X$

$$I(T) = I_o \exp\left[-(E_t/k_B T) - (s/q) \int_{T_i}^{T} \exp(-E_t/k_B T') \, dT'\right] \tag{7.13}$$

where $I_o = (s\,ke\,\tau\,n_{to}X)$ has been used.

Equation 7.13 is the standard equation expressing the thermally stimulated current as a function of temperature employing the constant rate of heating. The plot of the logarithmic current as a function of $(1/T)$, known as the Arrhenius plot, has been suggested by Maeta and Sakaguchi[34] as a method (initial rise method) to determine the trap depth for a given sample. Differentiating Equation 7.13 we can write

$$\frac{d \ln (I/I_o)}{d(1/T)} = (-E_t/k_B) + (sT^2/q) \exp (-E_t/k_B T) \qquad (7.14a)$$

For the maximum in the plot, i.e. when $T = T_m$,

$$\frac{d \ln (I/I_o)}{d(1/T)} = 0$$

which gives

$$(s/q) = (F_t/k_B T_m^2) \exp (E_t/k_B T_m) \qquad (7.14b)$$

Hence

$$\frac{d \ln (I/I_o)}{d(1/T)} = (-E_t/k_B) + (E_t/k_B) (T/T_m)^2 \exp [-(E_t/k_B T T_m) (T_m-T)]$$

$$(7.15)$$

Equation 7.15 gives the slope of the Arrhenius plot. In the region of temperature shortly after the heating starts, i.e. for $T \gtrsim T_i$, the second term on the right hand side of Equation 7.13 is negligibly small. We can therefore write for the slope of the plot in this region

$$\left(\frac{d \ln (I/I_o)}{d(1/T)}\right) = - E_t/k_B \qquad (7.16)$$

Thus, Equation 7.16 gives E_t. A typical Arrhenius plot for single crystal anthracene is shown in Figure 7.23. An easier expression that gives E_t values with errors within 3% has been given by Maeta and Sakaguchi[34]. We can write it as

$$E_t = \frac{(1.40) k_B T_m T_{m/2}}{T_m-T_{m/2}} (eV) \qquad (7.17)$$

where $T_{m/2}$ is the temperature on the rising side of the TSC-$T(K)$ curve at which the current is half of its peak value.

7.7.4. *Simultaneous measurement of TSC and TL*

A schematic block diagram of the experimental set-up is shown in Figure 7.24. If the specimen is in the form of a film or a compressed pellet then gold electrodes are evaporated onto both sides of a specimen at about 10^{-5} torr. When the sample is a powder, it can be pressed between two electrodes for a good ohmic contact. In that case the upper electrode may be of conducting glass so that

TL light may easily emerge from it to the PM tube. Both TSC and TL can be excited by X rays, keeping the specimen under short circuit conditions. In this manner, electron and hole traps are filled. A bias voltage is now applied to the specimen and, after the transient current is reduced to a negligible amount, the specimen is heated at a constant slow rate from the initial temperature T_i to some higher temperature to obtain TSC and TL glow peaks. These curves can be recorded by *x-y* plotters connected to electrometers.

From the occurence of TSC and TL peaks at the same temperature it can be concluded that electron traps are responsible for these two

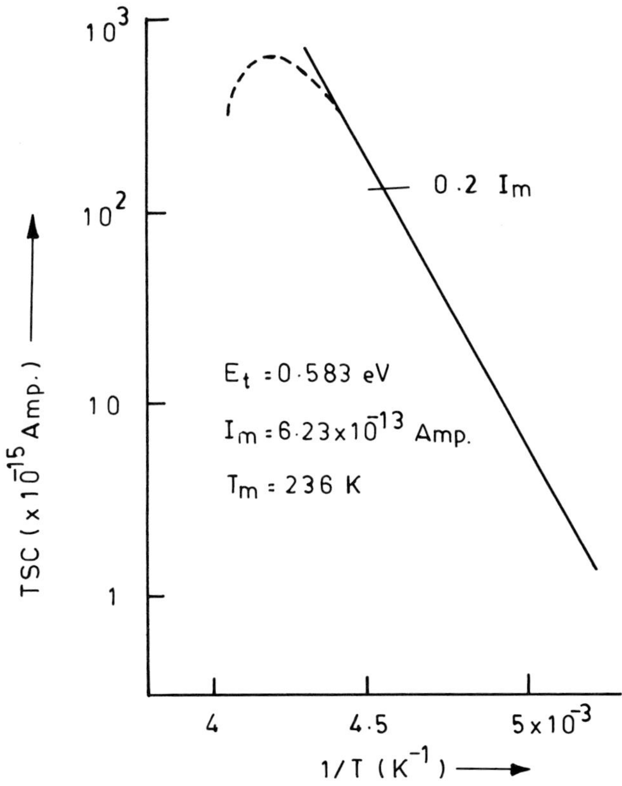

Figure 7.23. Arrhenius plot of TSC, calculated numerically. The range up to about 20% of I_m agrees with straight line[34].

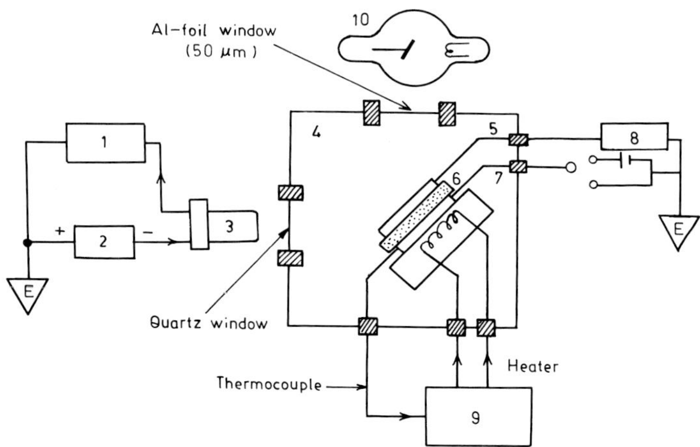

Figure 7.24. Schematic block diagram of apparatus for simultaneous measurement of TSC and TL. Key: 1,8, electrometers; 2, power supply; 3, photomultiplier tube; 4, vacuum chamber; 5,7, thin gold electrodes; 6, specimen; 9, temperature controller; 10, X ray tube.

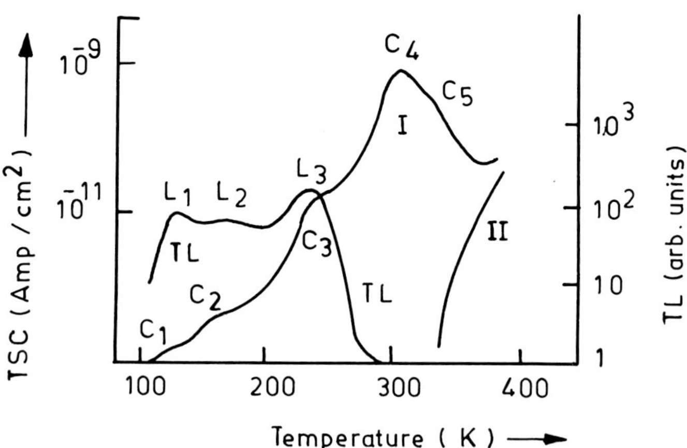

Figure 7.25. TL and TSC glow curves of low density polyethylene film[3]. I, conductivity spectrum of X ray irradiated specimen. II, without X ray irradiation, which shows only the steady state conduction current at about 320 K. L_1, L_2, L_3, TL peaks. Field applied = 90 kV.cm^{-1}.

processes. An example of an insulating polymer film specimen is shown in Figure 7.25. Another possibility can also exist in which the transition corresponds to the excitation of the carrier within the band gap by thermal stimulation and subsequent de-excitation, leading to TL emission but not current.

Complete coincidence on the temperature scale of the TSC and TL peaks should not, however, be expected as the TL peak usually precedes the TSC maximum[35,36]. This is illustrated in Figure 7.26 for the $CaSO_4$ specimen irradiated by X rays. The thermally stimulated current peak is shifted to the right relative to the TL peak.

Since the thermoluminescence intensity $I(T)$ is given by

$I(T) \propto - dn_t/dt$, using Equations 7.4 and 7.11, we find a correlation btween TSC and TL:

$$\sigma(T) = ekn_c = - e\,k\,\tau\,dn_t/dt \propto ek\,\tau\,I(T) \tag{7.18}$$

Thus, if k and τ were temperature independent, the $I(T)$ peak will coincide with the $\sigma(T)$ peak. Since this is not so, we can conclude that τ is temperature dependent. The mobility k in an insulator or a semiconductor increases with temperature as the conduction here is by the hopping process.

Another useful feature of TSC is shown in Figure 7.27. The electron and hole traps can be studied separately. If the illuminated surface is positively biased, the electron current which comes from electron de-trapping upon heating the specimen is predominant; when bias polarity is reversed, the hole de-trapping features are revealed.

7.8.   **Poole-Frenkel effect**

Another feature of TSC is the so-called Poole-Frenkel effect[37,38]. This involves the dependence of activation energy on the electric field strength applied to the specimen. If the trapping levels are charged when empty, the Poole-Frenkel effect will show. The trap depth is found to be field dependent[39,40].

$$E_t(X) = E_t(0) - \beta_{PF}X^{0.5} \tag{7.19}$$

where $\beta_{PF} = (e^3/\pi\varepsilon)^{0.5}$ is the Poole-Frenkel coefficient, e is the electron charge and ε is the dielectric constant of the medium of the specimen.

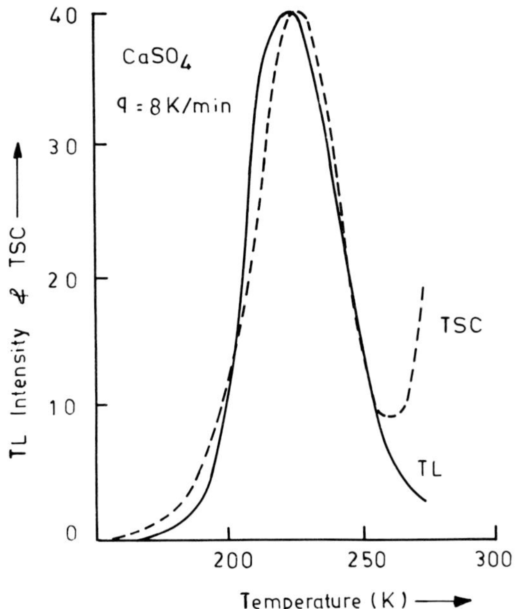

Figure 7.26. TL and TSC curves of X ray excited $CaSO_4$ crystal[51].

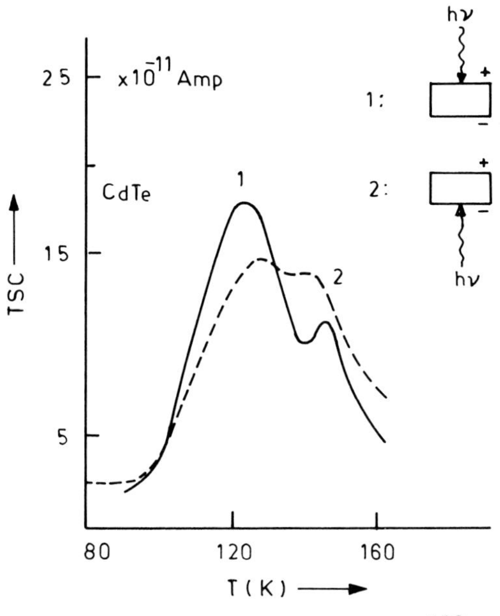

Figure 7.27. TSC current in CdTe[52]. 1, electron de-trapping. 2, hole de-trapping.

If we find E_t, using Equation 7.17, at two different values of the applied field strength, β_{PF} can be determined.
From Equation 7.14

$$T_m = E_t(X)/[k_B \ln (s \, k_B T_m^2/q \, E_t(X))]$$

Substituting from Equation 7.19

$$T_m = [E_t (0)/k_B d] - [\beta_{PF}/k_B d] \, X^{0.5} \tag{7.20}$$

where

$$d = \ln [s k_B T_m^2/q E_t(X)] = \ln [s k_B T_m^2/q(E_t(0) - \beta_{PF} \, X^{0.5})]$$

On account of the logarithmic function, d varies only weakly with X. Equation 7.20, therefore, gives an $X^{1/2}$ dependence of T_m which has been verified experimentally in poly(N-vinylcarbazole) (PVK)[39] and many other materials[41]. The experimental value of $\beta_{PF} = 3.8 \times 10^{-5}$ eV (m.V^{-1})$^{1/2}$ for PVK agrees well with the theoretical value (4.38×10^{-5}) using $\varepsilon = 3$ for PVK.

A similar, though not identical, field dependence of activation energy is also found to exist in electret depolarisation current measurements. In cellulose nitrate polymer foil, for example, a linear decrease in activation energy with increase in the polarising voltage has been observed[42]. Casserta and Serra[43] explain the lowering of activation energy in terms of the decrease of barrier height of traps due to the polarising field. One may assume that electron and hole charges trapped during polarising, form bound pairs whose binding becomes weaker as the polarising field strength increases.

7.9.   **TL—ITC/TSD correlation**
7.9.1.   *Introduction*
Ionic thermocurrent (ITC) or thermally stimulated depolarisation current (TSDC) (the first term was coined by Bucci *et al*[44] although the second is more illustrative and more frequently used) arises from (i) the relaxation of electric dipoles which are formed when an electric field is applied to the sample kept at elevated temperatures, and (ii) the traps which may be filled electrically in contrast to being filled by the ultraviolet or other ionising radiation as in thermoluminescence or TSC. The main features of TSD phenomena are of importance in the study of dipole relaxation processes and carrier trapping states. The specimen in

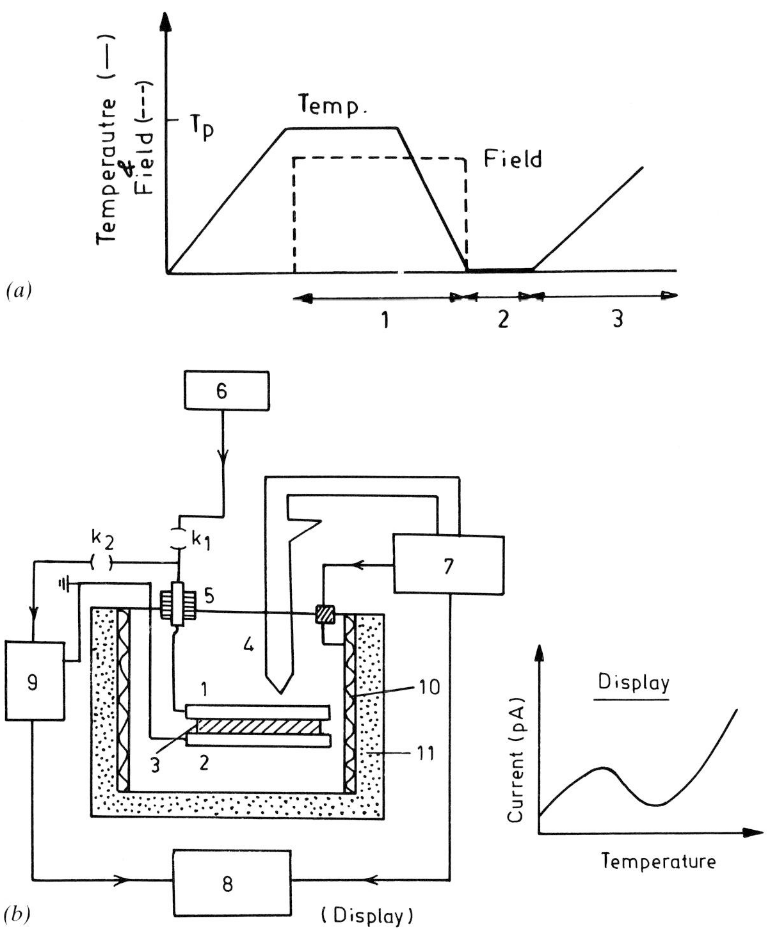

Figure 7.28. (a) Flow diagram of electret formation and discharge. 1, formation time. 2, storage time (optional). 3, TSD current. T_p, polarising temperature. (At this temperature the specimen is polarised. This is below the melting point but above the glass-rubber transition temperature.) (b) Experimental set-up (schematic). 1,2, electrodes (Cu-plates). 3, specimen in the form of thin polymer foil or pellet (in case of powder). 4, thermocouple. 5, amphenol cable connector. 6, DC power supply for polarising the specimen (K_1 is closed and K_2 open). 7, temperature controller. 8, *X–Y* recorder. 9, electrometer for recording TSD current (K_2 is closed and K_1 is open). 10, heater coil wound around a cylinderical can. 11, asbestos powder for insulation.

which dipoles have been orientated by the application of an electric field is known as an electret[45]. Currently, polymer electrets are of much practical interest as they possess good charge storage capability which makes it possible to use electrets as transducers (microphones), ear-phones etc. and as xerographic materials. More recently, electrets have been used as gas filters (charged porous electrets attract impurity particles from the flowing gas), in optical display systems and in radiation dosimetry[46]. The use of ultra-high voltage and DC power transmission demands the availability of highly insulating materials. Since space charge within the insulator limits the insulating capabilities, electret studies provide a means of investigating space charge effects which then enable us to develop materials having better electrical insulation properties.

7.9.2.   *Experimental technique*

The basic requirements for the experimental set-up are similar to those of TSC measurements. The electret forming and TSD measurement unit is schematically shown in Figure 7.28. As in TL and TSC, TSD glow curves are obtained at constant heating rates and similar methods are employed to determine the activation energy (for both the dipolar relaxation and the carrier de-trapping processes).

7.9.3.   *Theoretical description of TSD*

At elevated temperatures and under the influence of an external electric field (polarising field) the dipoles are produced and aligned. Upon cooling, with the field on, these dipoles in their state of alignment are fixed. The polarising field is then withdrawn and the dielectric acquires a permanent polarisation.

If there are N dipoles, each with a dipole moment p, per unit volume of the dielectric, then the polarisation P_o, when dipoles are oriented along the applied electric field, is given by

$$P_o = Np \tag{7.21}$$

During thermal stimulation, the dipole orientation begins to break, this is termed as depolarisation (or dipole relaxation). The rate of decay of polarisation, dP/dt, is given by

$$- dP/dt = P(t)/\tau(T) = \alpha(T)\, P(t) \tag{7.22}$$

where $\tau(T)$ is the dipole relaxation time which is a function of temperature. $\alpha(T)$ is, evidently, the relaxation frequency. The depolarisation current $I(T)$ is

$$I(T) = -dP/dt = \alpha(T)\, P(t) \tag{7.23}$$

Integration of Equation 7.22 gives

$$P(t) = P_o \exp\left(-\int_o^t \alpha(T)\, dt\right) \tag{7.24}$$

At low temperatures $\alpha(T)$ is small so that there is not much loss of dipolar orientation and the electret will stay charged for a considerable period of time. When the electret is heated, $\alpha(T)$ increases rapidly giving rise to a depolarisation current. From Equations 7.23 and 7.24

$$I(T) = P_o\, \alpha\,(T) \exp\left(-\int_o^t \alpha(T)\, dt\right)$$

or

$$I(T) = P_o\, \alpha\,(T) \exp\left(-(1/q)\int_{T_i}^T \alpha(T')\, dT'\right) \tag{7.25}$$

where $T = T_i + qt$ and T_i is the initial temperature and q is the linear rate of heating.

The increase in $\alpha(T)$ with temperature usually follows the Arrhenius equation:

$$\alpha(T) = \alpha_o \exp(-D/k_B T) \tag{7.26}$$

where D is the activation energy required to disorient a dipole and (α_o) is the natural relaxation frequency. D is a kind of potential barrier for the dipole which it must overcome in order to readjust its orientation. As in TL, the energy D can be estimated from the low temperature part (initial rise method) of the $I(T) - T$ data by plotting it as ln I (T) against $1/T$ and determining the slope of the resulting straight line.

It is possible to determine the dielectric polarisation $P_o(X)$ under the influence of uniform electric field X on the dielectric sample. If dN number of dipoles (each having a dipole moment p) have their orientations lying within θ and $\theta+d\theta$ with respect to the field direction,

$$P_o(X) = \int_{\theta=o}^{\theta=\pi} p \cos\theta\, dN$$

Where dN, following the Boltzmann distribution, is given by $dN = A \exp\left(-\varepsilon/k_B T\right) d\Omega = A \exp(Xp \cos\theta/k_B T)\, d\Omega$ where ε is the potential energy of the dipole oriented at an angle θ to the field and

$d\Omega$ is the element of solid angle between θ and $\theta+d\theta$ ($d\Omega = 2\pi \sin \theta\, d\theta$).

Thus,

$$P_o(X)/N = \int_{\theta=o}^{\theta=\pi} p \cos \theta\, dN / \int_{\theta=o}^{\theta=\pi} dN = p\, L(Xp/k_B T) \qquad (7.27)$$

where $L(Xp/k_B T) = \coth(Xp/k_B T) - (k_B T/Xp)$

is the well known Langevin function.

Typically, at low fields and at room temperature and above, $(Xp/k_B T)\ll 1$. Hence, by retaining only linear terms in the expansion of $\coth(Xp/k_B T)$, $L(Xp/k_B T)\approx(Xp/3k_B T)$

Using it in Equation (7.27) we get

$$P_o(X) = XNp^2/3k_B T \qquad (7.28)$$

Putting Equations 7.26 and 7.28 in Equation 7.25 we get

$$I(T) = \frac{Np^2 X}{3k_B T}\, \alpha_o \left\{ \exp\left[-(D/k_B T) - (\alpha_o/q) \int_{T_i}^{T} \exp(-D/k_B T')\, dT' \right] \right\}$$

$$(7.29)$$

7.9.4. *Correlation*

TL and TSD correlation is rather indirect. While TSD current can arise from both de-trapping of charge carriers and

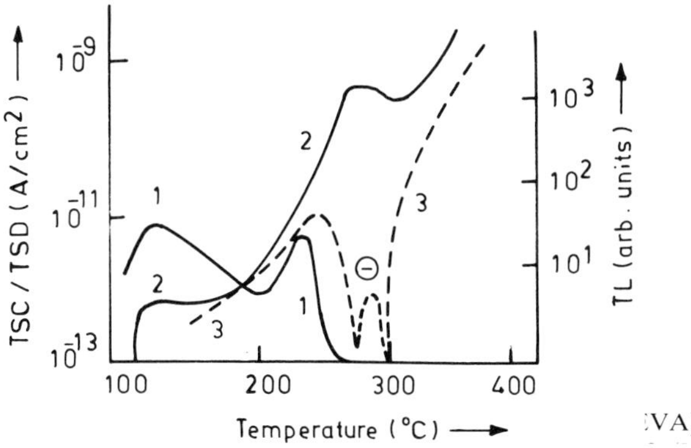

Figure 7.29. TSC, TSD and TL in ethylene vinyl acetate (EVA) copolymer with VA content = 22 wt%[3]. Curves: 1, TL; 2, TSC (X irradiated); 3, TSD (unirradiated). $\ominus$ current reversal. Bias field = 90 kV.cm^{-1}.

dipolar relaxation, TL corresponds only to carrier de-trapping. However, depolarisation current measurement experiments can separate out the two contributions. While the TL and TSD (carrier de-trapping) will arise only in an irradiated sample (a small amount of carrier trapping, caused by the application of an electric field may, however, occur during electret formation in an unirradiated sample), TSD (dipolar relaxation) shows in the unirradiated sample. Further, the current reversal at some temperature during the heating cycle is also exhibited experimentally and predicted theoretically in the thermally stimulated current thermogram. This is shown in Figure 7.29. An interesting application of the TSD phenomenon is a very simple electret dosemeter based on radiation induced thermally activated depolarisation (RITAD) reported by Podgorshak and Moran[47]. However, the one which is of practical use (at room temperature) has been described by Hobzová and Spurný[48]. This dosemeter uses a BeO ceramic disc with both faces metalised by silver. It is heated to a temperature of about 350°C and then cooled spontaneously to room temperature. During cooling a high DC field (6000 $V.cm^{-1}$) is applied to the sample. The basic features of the dosemeter are shown in Figure 7.30 while the dose response characteristics are shown in Figure 7.31. RITAD dosimetry has limited use as it can measure large doses only (5-50 Gy) so that only large accidental radiation exposures can be covered. There is, however, clear scope for further investigations and development in RITAD dosimetry with the aim of suppressing fading (TSD in an unirradiated BeO disc sample decreases to about a half in 250 h after polarising the sample) and improving upon

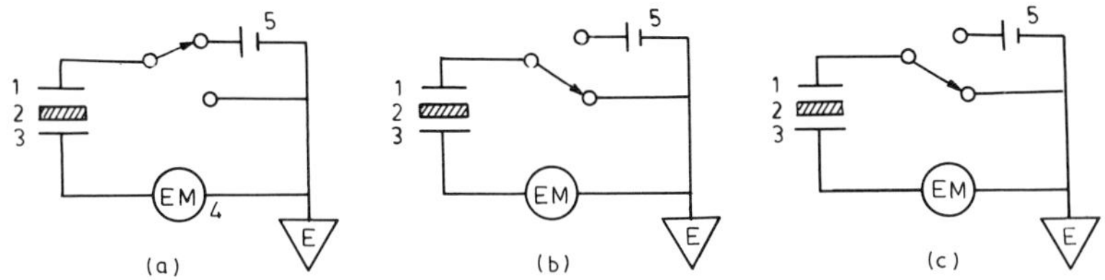

Figure 7.30 Electret dosemeter system[48]. (a) Sample polarising during spontaneous cooling from 350°C to room temperature. 1 and 3, electrodes; 2, BeO disc (sintered, 12.7 mm diam.×1.6 mm thick); 4, electrometer; 5, bias. (b) Radiation exposure at room temperature (large doses of few tens of $J.kg^{-1}$ are needed). (c) Readout as a function of temperature (see Figure 7.31).

material and techniques for lower dose response. The simplicity of the method and low cost of the set-up are, however, very attractive features of RITAD dosimetry. The other form of electret dosemeter[49] depends upon the formation of a surface charge on a polymer sheet sample by the usual polarising method (say by corona discharge) and then measuring the decrease in the surface field as a function of radiation dose. Improvements in the sensitivity of such a dosemeter from the kilorad to the millirad region have been achieved by utilising ionisation of the gas surrounding the charged surface so that the surface charge is neutralised by attracting ions of opposite polarity from the gas ionised by the incident dose of radiation.

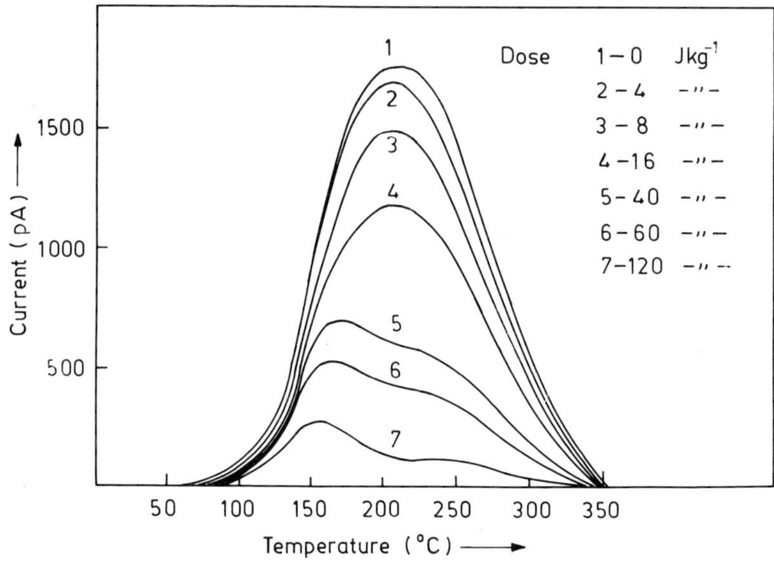

Figure 7.31. Dose dependence of RITAD curves[48].

References

1. Chen, R. and Kirsh, Y. *Thermally Stimulated Processes in Solids.* (London: Pergamon Press) (1981).
2. Takeuchi, N., Adachi, M. and Inabe, K. J. Lumin. **18/19** (II), 897 (1979).
3. Ieda, M., Mizutani, T. and Suzuoki, V. *TSC and TL Studies of Carrier Trapping in Insulating Polymers.* Memoirs of the Faculty of Engineering, Nagoya University, Nagoya, Japan, **32**, 173-219 (1980).

4. Lopez, F. J., Jaque, F., Fort, A. J. and Agullo'-Lopez, F. J. Phys. Chem. Solids **38**, 1101 and other references mentioned therein (1977).

5. Herreros, M. and Jaque, F. J. Lumin. **9**, 380 (1974).

6. Stoebe, T. G., Wolfenstine, J. B. and Las, W. C. J. Phys. (Paris) Suppl. **7**, C6-265 (1980).

7. Nambi, K. S. V., Bapat, V. N. and Ganguly, A. K. J. Phys. C: Solid State Phys. **7**, 4403 (1974).

8. Merz, J. L. and Pershan, P. S. Phys. Rev. **162**, 217 (1967).

9. Sastry, M. D., Dalvi, A. G. I., Page, A. G. and Joshi, B. D. J. Phys. C: Solid State Phys. **8**, 3232 (1975).

10. Gupta, N. M., Luthra, J. N. and Sastry, M. D. J. Lumin. **10**, 305 (1975).

11. Luthra, J. M., Gupta, N. M., Nambi, K. S. V. and Bapat, V. N. IN Proc. Natl Symp. on *Thermoluminescence and its Applications*, Kalpakkam, India, 12-15 Feb, pp. 147-165 (1975).

12. Prokic, M. J. Phys. Chem. Solids **40**, 405 (1979).

13. Becker, K. At Energ. Rev. **8**, 173 (1970).

14. Becker, K. CRC Crit. Rev. Solid State Sci. **3**, 39-81 (1972).

15. Becker, K. *Solid State Dosimetry*. (Cleveland, OH: CRC Press) (1973).

16. Samuelson, L. I. *Mechanism for Exoelectron Emission mainly from LiF*. Acta Radiol. Suppl. **359**. (1979)

17. Chen, R. and Kirsch, Y. *Thermally Stimulated Processes in Solids*. (Oxford: Pergamon) (1981).

18. Portal, G. and Scharmann, A. (eds) *Exoelectron Emission and its Applications*. Proc. 7th Int. Symp. Radiat. Prot. Dosim. **4** (3-4), (1983).

19. Proc. 8th Int. Symp. on *Exoelectron Emission and Applications,* Osaka, Japan, 1985. Jpn. J. Appl. Phys. **25** (Suppl.) (1985).

20. Holzapfel, G. IN 5th Polish Seminar on *Exoelectron Emission and Related Phenomena*, Karpacz, Poland, 1978. Ed. B. Sujak.

21. Fintlemann, D. IN Proc. 5th Int. Symp. on *Exoelectron Emission and Dosimetry*, Zvikov, p. 288 (1976).

22. Kamada, M., Furukawa, K. and Tutsumi, K. Jap. J. Appl. Phys. **20**, 71 (1981).

23. Bohun, A. Czech, J. Phys. **4**, 139 (1954).

24. Nassenstein, H. Z. Naturforsch. a **10**, 944 (1955).

25. Balarin, M. and Zetzsche, A. Phys. Status Solidi **2**, 1670 (1962) (in German).

26. Holzapfel, G. Z. Angew. Phys. **29**, 107 (1970).

27. Miller, L. D. and Bube, R. H. J. Appl. Phys. **41**, 3687 (1970).

28. Tomita, A., Hirai, N. and Tsutsumi, K. Jap. J. Appl. Phys. **15**, 1899 (1976).

29. Kamada, M., Furukawa, K. and Tsutsumi, K. Jap. J. Appl. Phys. **20**, 71 (1981).

30. Petel, M. and Holzapfel, G. IN Proc. 4th Int. Symp. on *Exoelectron Emission and Dosimetry*, Liblice, Czechoslovakia, 1973, p. 252.

31. Fukuda, Y., Tomita, A. and Takeuchi, N. Phys. Status Solidi a **99**, K135 (1987).

32. Spurný, Z. and Käämbre, H. IN *Techniques of Radiation Dosimetry*, eds. K. Mahesh and D. R. Vij (Chichester: John Wiley) (1985).

33. Kelly, P., Laubitz, M. J. and Braunlich, P. Phys. Rev. B **4**, 1960 (1971).

34. Maeta, S. and Sakaguchi, K. Jap. J. Appl. Phys. **19**, 597 (1980).

35. Fields, D. E. and Moran, P. R. Phys. Rev. B **9**, 1836 (1974).

36. Chen, R. J. Appl. Phys. **42**, 5899 (1971).

37. Frenkel, J. Phys. Rev. **54**, 647 (1938).

38. Ieda, M., Sawa, G. and Kato, S. J. Appl. Phys. **42**, 3737 (1971).

39. Zielinski, M. and Samoć, M. J. Phys. D: Appl. Phys. **10**, L105 (1977).

40. Samoć, A. and Samoć, M. J. Electrost. **8**, 121 (1979).

41. van Turnhout, J. and van Rheeuen, A. H . J. Electrost. **3**, 213 (1977).

42. Saini, K. K., Talwar, I. M., Lal, N., Mahesh, K. and Nagpaul, K. K. IN *Solid State Nuclear Track Detectors*, eds P. H. Fowler and V. M. Clapham (Oxford: Pergamon Press) (1982).

43. Casserta, G. and Serra, A. J. Appl. Phys. **42**, 3778 (1971).

44. Bucci, C., Ficschi, R. and Guidi, G. Phys. Rev. **148**, 816 (1966).

45. Sessler, G. M. (ed) *Electrets* (Berlin: Springer Verlag) (1980).

46. Ikeya, M. IN *Techniques of Radiation Dosimetry*, eds K. Mahesh and D. R. Vij (Chichester: John Wiley) (1985).

47. Podgorshak, E. B. and Moran, P. R. Science **179**, 380 (1973).

48. Hobzová, L. and Spurný, Z. Radiochem. Radioanal. Lett. **22**, 319 (1975).

49. Ikeya, M. Jap. J. Appl. Phys. **20**, 1615 (1981).

50. Wild, W. *et al.* IN Proc. 6th Symp. on *Exoelectron Emission and Applications*, Ahrenshoop, October 1979. p. 41.

51. Scharmann, A. IN *Einfuhrung in die Lumineszenz*, ed. N. Riehl (Munich: Thiemig), pp. 182-225 (1970).

52. Scharager, C., Muller, J. C., Stuck, R. and Siffert, P. Phys. Status Solidi a **31**, 247 (1975).

CHAPTER 8

Thermoluminescence Applications

K. Mahesh

8.1.   Introduction

During the past twenty years or so, very considerable progress has been made in the applications of the thermoluminescence technique for practical purposes. The most widely developed application, however, refers to its use in radiation dosimetry[1-5] which spans areas of health physics and biomedical sciences, radiation protection and control[6] (personnel monitoring of radiation). An equally impressive growth of the TL technique is reflected in its applications in dating of archaeological[7,8,9] and geological[10] materials. A recent application of TL concerns the dating of sand-dunes, which has implications in the study of their movement and some other aspects of the desert environment[11]. This study offers the interesting possibility of looking into the mechanism of spread of desert areas in the past and its relationship with future predictions for the pattern of desert spread. Interesting, but not yet sufficiently developed, applications have been mentioned in forensic science which is concerned with the characterisation of suspect materials and identification of forged objets d'art (ceramics, glasses and others which are amenable to TL investigations); in industry (study of burnt concrete of fire-damaged buildings and of industrial pollutants) and, via the TSC mechanism, in solid state ionography[12]. This chapter is intended to expand on these and other applications.

8.2.   Radiation dosimetry: personnel monitoring

8.2.1.   *Introduction*

Thermoluminescence has been effectively used as an experimental tool in this discipline. Here, we are mainly concerned with the radiation monitoring of personnel working with reactors and accelerators in nuclear establishments and those working in radiological laboratories handling and using a variety of radioactive sources. Over the years, TL dosemeters have almost replaced[20] the traditional use of film badge dosemeters since the latter have several inherent limitations such as fading with temperature and humidity, poor reproducibility, limited dose range and sensitivity, and need elaborate dark room facilities. The aspect of radiation protection and control concerning TL dosimetry (TLD) arises from the need to fulfil the statutory requirements for safety regarding radiation hazards of those personnel who are working with facilities using and involving radioactivity.

8.2.2.   *Principle of radiation dosimetry by TL*

Some of the incident ionising radiation energy absorbed by the insulating medium (i.e. a thermoluminescent material) is utilised in exciting the electrons from the valence band (VB) to the conduction band (CB) of the insulator. The mobile electrons in the CB have a reasonable chance of being trapped at a region of crystalline imperfection arising from the presence of, for example, an impurity atom, a lattice vacancy or a dislocation. The probability (p) of release per unit time of these trapped electrons back into the CB is dependent on temperature (T) and activation energy (E) and is given by

$$p = s \exp(-E/k_{\mathrm{B}}T) \tag{8.1}$$

where s, a constant for a given insulator, is the frequency factor. E characterises the energy depth of the trapping site relative to the bottom of CB.

By the suitable heating of this sample, the filled traps can be evacuated by the thermal stimulation of the trapped electrons to the CB. From here the electrons have a reasonable probability of recombining with a hole at some appropriate site where such an event will result in the emission of visible light. These sites are known as luminescent or recombination centres. Figure 8.1 illustrates this process. The number of TL glow curves and their peak temperatures reflect the trap types and their corresponding activation energies, while the wavelength spectra of the emitted light provided information about the luminescent centres.

If N is the concentration of empty traps in a TL dosemeter phosphor and, at any instant of time t, during the exposure of the dosemeter to a radiation of dose rate r, the remaining empty trap concentration is n, we may write for the rate of decrease of n

$$- \mathrm{d}n/\mathrm{d}t = Anr$$

where A is a constant of the phosphor which may be termed the radiation susceptibility. Simultaneous thermal ionisation of the traps being filled during irradiation has been ignored here, i.e. the filled traps are deep enough to resist thermal drainage. (The trap filling rate, simultaneously accompanied by the thermal ionisation rate of the filled traps, is determined by $d\eta/dt = A(N-\eta)r - s\eta \exp(-E/k_B T_r)$ where E is the trap depth and T_r is the temperature of irradiation of the sample). The above equation can be integrated to give

$$\int dn/n = \int -Ar dt + \text{constant}$$

Using the condition that at $t=0$, $n=N$, we get

$$n = N \exp(-Art) = N \exp(-A\rho) \tag{8.2}$$

where $\rho=rt$ is the radiation dose received by the dosemeter during

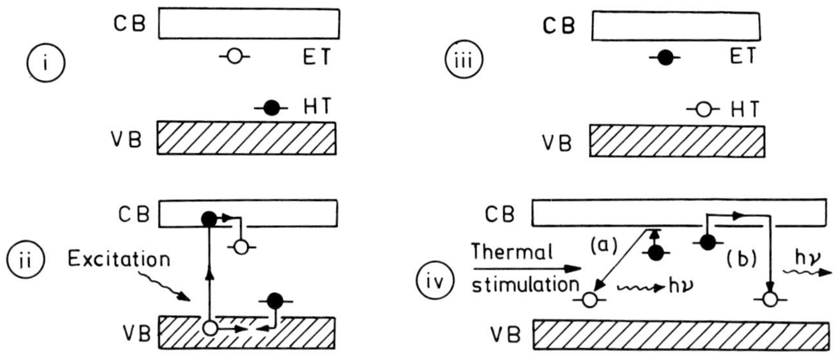

Figure 8.1. TL phenomena and basic mechanism of TL dosimetry.
    (i) Band model of insulator (TL phosphor) and empty electron traps (ET) and hole traps (HT).
    (ii) The effect of irradiation of phosphor by ionising radiation. The electron from VB is lifted to CB leaving behind a mobile hole in VB.
    (iii) The event following irradiation; an electron is trapped in ET and a hole in HT. Thus the consequence of ionising radiation is to populate electron and hole traps in proportion to the radiation dose.
    (iv) Thermal stimulation of electrons from traps and their subsequent recombination with holes at luminescent centres resulting in emission of light (hv). (a) Electron raised to the excited state of a luminescent centre and its eventual return to its ground state. (b) Electron lifted to the CB where it becomes mobile and its subsequent recombination with the hole at a luminescent centre.

the time t. From Equation 8.2 we can define A by postulating that if $\rho_{1/2}$ is the radiation dose needed to populate half of the empty traps, then from Equation 8.2

$$A = 0.693/\rho_{1/2} \tag{8.3}$$

The number of filled traps is evidently given by $\eta = N - n$, thus

$$\eta = N(1 - e^{-A\rho}) \tag{8.4}$$

If the irradiated sample is now heated to record thermoluminescence, we can write for the thermal ionisation of the filled traps

$$-d\eta/dt = p\eta = s\eta \exp(-E/k_B T)$$

and the thermoluminescence intensity is given by

$$I(T,\rho) = -cd\eta/dt = cp\eta = csN(1 - e^{-A\rho}) \exp(-E/k_B T)$$

For small values of ρ such that $A\rho < 1$, we may use the approximation

$$1 - e^{-A\rho} \simeq A\rho$$

hence

$$I(T,\rho) = csNA\rho \exp(-E/k_B T) \tag{8.5}$$

Equation 8.5 thus shows that the TL intensity at a given temperature (say at the glow peak temperature) is proportional to the radiation dose, provided the dose received by the sample is small. This equation is the basis of radiation dosimetry, and also of archaeological dating (section 8.9).

8.2.3. *Aims and objectives of radiation protection and monitoring*

For the purposes of providing necessary safeguards to personnel associated with nuclear industry and research, handling and using a complex variety of radioactive sources, clear objectives have been laid down through the joint efforts of international bodies such as ICRP[13], the World Health Organization (WHO) and the International Atomic Energy Agency (IAEA), Vienna, with a view to providing guidelines for the exercise of uniform standards in this matter. The aims, apart from providing for the routine monitoring of radiation areas, include the availability of adequate facilities for timely indication of any unforseen exposure to personnel in controlled access areas. In order to meet these requirements, especially in the case of high dose areas, the following are necessary[14]:

(i) daily determination of personnel doses and an estimate of the whole-body dose received;

(ii) frequent check measurements of the acquired dose for a cross section of people at the time of leaving the controlled access area;

(iii) access to a large computer facility for summing up the daily doses and for balancing the personnel dose data per week, per month and per year; and

(iv) presence and access checks for reasons of protection and in case of incidents.

In order to serve the defined objectives with regard to the protection of personnel against abnormal exposure to radiation, the monitoring of (a) the contamination of the working area, clothes and skin, and (b) the activity incorporation through inhalation or by other means of the radioactive gases/substances, is needed.

For these requirements, the different radiation monitoring devices such as GM counters, thin window proportional counters, scintillation and semiconductor junction detectors, as well as the thermoluminescent detectors, are available and should preferably be linked to computer supported data acquisition units. It will then be possible to store and display both data and worker's identification number and the monitoring information separately for the whole body, hands, feet, clothing and exhaled air; along with the built-in alarm check facility in case of excess exposure.

8.2.4.   ***TL dosimetry in personnel monitoring***

Thermoluminescence dosimetry is a reasonably sensitive technique to serve the stated objectives, especially for gamma ray exposure monitoring down to 10 mR or so. The typical dose ranges that can be covered by the TL dosemeters are 10^{-4} to 10^2 Gy[15]. The linear dose response is obtainable to about 1 Gy, above which supralinearity may be observed. The basic mechanism of TL dosimetry is depicted in Figure 8.1. The necessity for giving a heat treatment and anneal to the TLD prior to re-use is, however, a complication that has to be tolerated. LiF:Mg,Ti (TLD-100), for instance, requires a high temperature heating for 1 h at 400°C to erase any previous dose and to re-establish the sensitivity of the material. It is then followed by a low temperature anneal for 16-24 h at 80°C to remove the presence of the shallow traps that are responsible for substantial fading of the stored TL signal (alternatively a post-irradiation annealing for 10 to 15 min at 100°C can be used[16]). TL dosimetry is a suitable technique for local as well as postal monitoring services. It can be used for body as well as extremity (fingers, nails etc.) monitoring and lends itself to automation for maintaining dose records. For a good TL yield and reasonable accuracy of measurement, an appropriate thickness of the TL phosphor (see Figure 8.2)[17] and precision readout equipment is, of course, necessary.

Further, the choice of the phosphor should be such that it emits shorter rather than longer wavelength light in order to discriminate against the longer wavelength thermal radiation. Another common feature in many phosphors, that their glow peak temperature increases with the heating rate, should also be noted[18]. Table 8.1 lists the dosimetric features of some TLD materials[19].

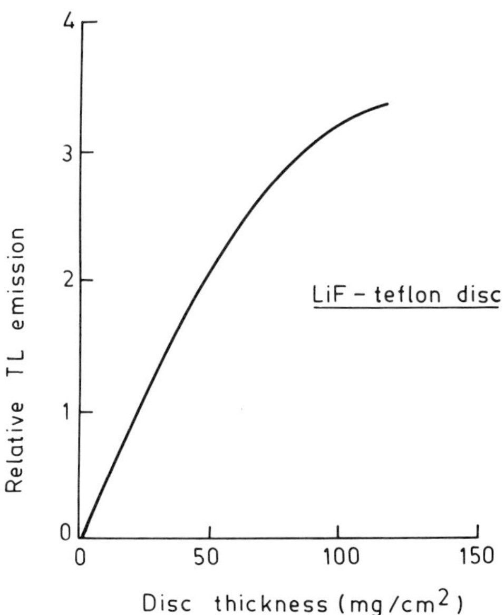

Figure 8.2. TL emission from LiF Teflon disc phosphors as a function of disc thickness.

Table 8.1. Dosimetric features of some TLD materials[19].

TLD phosphor	Gamma ray sensitivity relative to to TLD-100	TL emission spectrum peak wavelength (nm)	Dosimetric peak temp. ($^\circ$C)	Response per roentgen at 30 keV photon energy relative to ^{60}Co γ ray	Post irradiation fading
TLD-100	1.00	400	195	1.3	10% in 1 month
$Li_2B_4O_7$:Mn	0.40	600	200	0.98	10% in 1 month
$Li_2B_4O_7$:Cu,Ag	1.00	368	185	0.98	9% in 2 months
$CaSO_4$:Dy	38	480,570	210	11.27	3% in 1 month
CaF_2:Mn	5	500	260	15.51	10% in 1 month
Mg_2SiO_4:Tb	53	552	195	3.75	3% in 2 months
BeO	3.1	200–400	180–220	0.87	5% in 1 month

8.2.5. *Comparison of TLD with film badges*

Over the years, TLD has progressively replaced the use of the traditional film badge dosemeters in nuclear facilities[20] on account of several inherent limitations of the latter; such as the substantial fading with the temperature and humidity of the environment[21], poor reproducibility and sensitivity, limited dose range and the necessity for elaborate dark room facilities. The experimental set-up for a typical sealed-type TL dosemeter using CAF_2:Mn powder as the TL phosphor is schematically illustrated in Figure 8.3[22]. CaF_2:Mn powder ($\approx$90 mg) is cemented on to a nichrome coil or a carbon heater (16 mm$\times$16 mm$\times$0.38 mm) with the help of Dow Corning (No. 805) silicon cement. The sealed bulb dosemeter can be readily used as a personnel dosemeter because it is quite convenient to wear and easy to read and is also re-usable. The typical exposure measurement ranges can vary from 10^{-3} R to 5×10^3 R and self-background is about 2 μR.h^{-1}. The use of CaF_2:Dy (TLD-200) can give a still lower range. A comparison of the response (observed exposure/true exposure) of the CaF_2:Mn TL dosemeter with that of a film-badge in the mR range is given in Table 8.2[22].

It is evident from Table 8.2, that only in the higher dose range is the film badge reliable.

8.2.6. *Correlation of frontal exposure to critical body organs*

Dosemeters are generally worn on the chest and read the exposure (in air) received at that location. Since their response is

Table 8.2. Dose response comparison of CaF$_2$:Mn TLD with that of a typical film badge[22].

Gamma ray exposure (mR)	Average film badge reading/true exposure	Average TL reading/true exposure
4	~0	0.82
12	~0	1.02
35	0.83	0.98
94	0.81	0.99
235	0.84	1.00
422	0.90	1.02
938	0.88	1.01

generally dependent on the energy of the γ ray, a suitable dosemeter has to be found in which this dependence is minimal so that the given dosemeter can be used for larger dose ranges. This is a desirable feature of a dosemeter in so far as the exposure measurement is concerned because it always refers to the same absorbing medium (air). However, the dose absorbed by different body organs (some of which are located inside the body) can be energy dependent to a different extent (because of the compositional difference). Thus the rad/roentgen ratio as a function of γ ray energy will give the desired

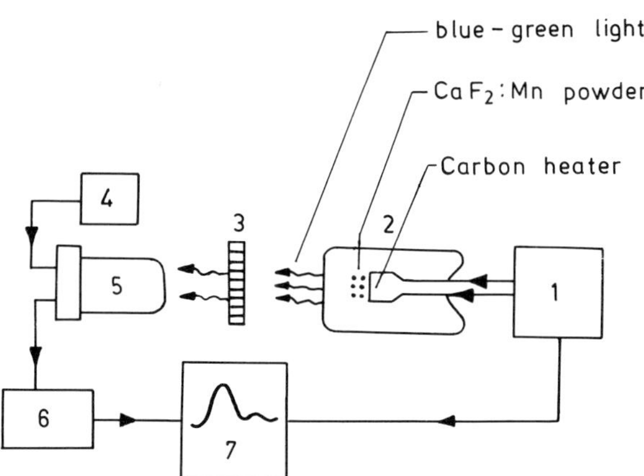

Figure 8.3. Experimental set-up of a typical sealed bulb dosemeter. 1, Heater and temperature controller. 2, Sealed glass bulb TL dosemeter. 3, Optical filter. 4, HV power supply. 5, Photomultiplier tube. 6, DC amplifier. 7, Chart recorder.

correlation between the actual dose received by an organ (measured by locating a dosemeter at the given organ site in a tissue-equivalent phantom) and the frontal exposure. These results are illustrated in Figure 8.4[22,23].

It will be seen from the figure that while the chest exposure adequately describes the absorbed dose for male gonads (testes) and gut mucosa, it over-estimates the doses absorbed by the female organ (ovaries) and the blood-forming organ (bone marrow), which are all buried within the body. Thus, as far as the chest exposure readings are concerned, these organs may be regarded as over-protected. At lower γ ray energies this mismatch factor is as high as 10. This brings us to the pertinent question of the desirability of having an energy insensitive TL dosemeter for medical purposes. As we see, because the absorbed dose in body organs is energy sensitive, the use of an energy insensitive dosemeter to read exposure will not have a corresponding response compatibility with the absorbed dose in the

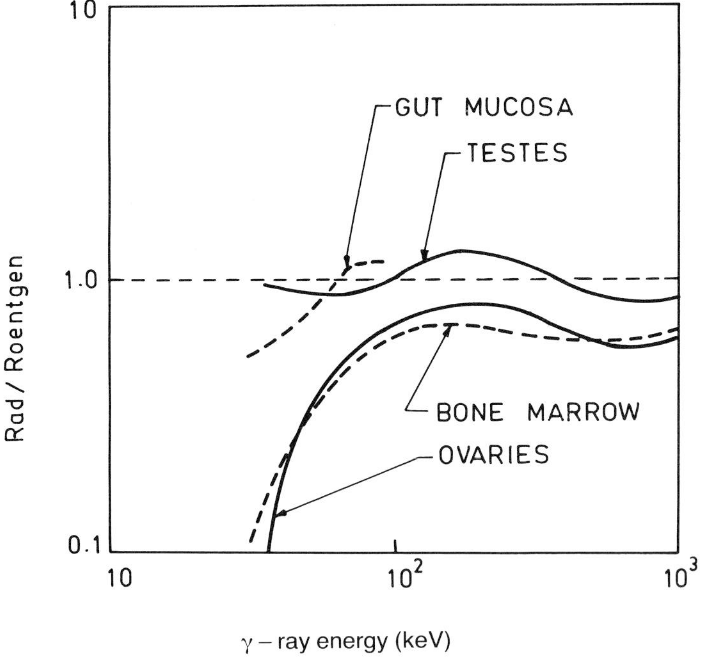

Figure 8.4. Absorbed dose in different body organs per roentgen of chest exposure as a function of γ radiation exposure energy.

tissue-equivalent medium. However, if male gonads are assumed to be the limiting critical organ, then, in case of personnel monitoring for the specific purposes of radiation protection, an energy insensitive dosemeter will be desirable. In diagnostic and therapeutic applications, on the other hand, where a knowledge of the energy relative to absorbed dose profile is needed, a dosemeter with some suitable shield (consisting of a combination of metals)[2] that can exhibit tissue-equivalence by showing the desired mismatch pattern for a given organ, will be preferable.

8.2.7. *Gamma ray dosimetry*

A general criterion for a good TLD for personnel monitoring services should be its stability (fading characteristics), sensitivity, dose response linearity, and energy independence.

Gamma ray dosimetry is, in fact, a dominant component of the personnel monitoring requirements and suitable dosemeters have been prepared for this purpose. From Table 8.1, it would appear that $CaSO_4$:Dy is a good choice since its sensitivity is 30 times as large as that of the more generally known TLD-100. At the same time, it suffers from the marked disadvantage that its response varies with energy, and it is not tissue-equivalent. However, because $CaSO_4$:Dy possesses an excellent lowest detectable dose limit (0.1 mGy), it is highly suitable in environmental dosimetry.

LiF, $Li_2B_4O_7$ and BeO have effective atomic numbers that match soft tissue (Z_{eff} ($Li_2B_4O_7$) = 7.4, Z_{eff}(soft tissue) = 7.5) and therefore absorb energy from photons and electrons in a tissue-equivalent manner. Of these, BeO is a highly toxic material and cannot, therefore, be used for personnel monitoring. While both LiF and $Li_2B_4O_7$ have overlapping and energy independent responses beyond 0.1 MeV photon energy, LiF has sharp and $Li_2B_4O_7$ only feeble energy dependence below this energy. It would thus appear that $Li_2B_4O_7$ is preferable to LiF for personnel dosimetric work. Coupled with this is the fact that LiF requires a very elaborate anneal procedure. $Li_2B_4O_7$ has a suitable lowest detectable limit (≈ 10 mR) and exhibits the onset of supralinearity at about 300R. The γ dosimetric features of another suitable TLD (CaF_2:Mn) (especially above 0.5 MeV) have been discussed by DeWerd[16].

8.2.8. *Beta ray dosimetry*

It was stated in section 8.2.6. that the tissue-equivalence

of a dosemeter is a desirable feature in therapeutic applications; $Li_2B_4O_7$:Mn is suitable from this point of view. Marinello *et al*[25] have used this dosemeter for estimation of total body dose of electrons ($E_\beta(max) \approx 4$ MeV) in the therapy of mycosis fungoides on skin. The dosemeter is used in the form of a Teflon disc of 9.5 mm diam. and 0.13 mm thickness with a composition of 95% Teflon, 4.9% lithium borate and 0.1% Mn (by weight). The dose estimation is, however, beset with some difficulties due to substantial absorption and scattering of β particles in the plastic foil covers as well as in the TLD material. BeO, which is also commercially available under the name of Thermalox-995, has also been used for β dosimetry, but it suffers from the disadvantage of substantial energy dependence of its dose response and a higher level of threshold dose[26]. The TL emission band maximum lies in the UV region, which is an additional disadvantage for efficient detection of the TL signal by a PM tube. Beta dose response of $CaSO_4$:Dy (with Al backing), LiF and

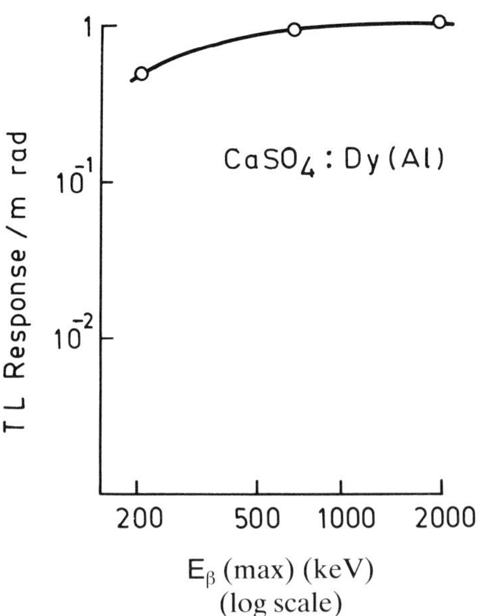

Figure 8.5. Energy dependence of the response of $CaSO_4$:Dy(Al) dosemeter for β rays[25].

CaF$_2$:Mn has been investigated in detail by Benkö *et al*[27], and of CaSO$_4$:Dy (Teflon backing) by Lakshmanan *et al*[28] using ^{147}Pm, ^{204}Tl, ^{32}P and ^{90}Sr sources. CaSO$_4$:Dy (Al backing) is found to be a better choice[26] on account of its excellent response as well as its independence of β ray energy (Figure 8.5). The phosphor is used in the form of a powder (30 μm grain size) and incorporated into a heat resistant resin. A layer about 10 mg.cm^{-2} thick is bonded onto an Al disc of 0.5 mm thickness. The dosemeter, before use, is given a 5 min anneal at 300°C. The threshold for the dosemeter is about 5 mrad.

8.2.9. *Beta dosimetry in mixed β–γ fields*

Piesch[29] has shown that it is not really necessary to estimate the β dose contribution in a mixed β–γ field for the simple reason that this contribution is significant only at the extremities (mainly fingers and hands). The rest of the body is essentially protected either by clothing or by virtue of its distance from the source (as in case of the face and eyes). This is particularly true for low values of mean energy $[\approx(1/3)E_\beta(\text{max})]$. This consideration then allows for the use of a somewhat thicker dosemeter medium for better γ response. In the usual situations of mixed β–γ fields, the β dose reading of the personnel dosemeter is found to be an order of magnitude less than the annual dose rate limit to the unprotected skin ($\approx$500 mSv). Thus, in mixed β–γ fields, the importance of β dosimetry lies essentially in the monitoring of extremities.

8.2.10. *Neutron dosimetry*

In neutron dosimetry, TLD has had only limited success owing to the ever present γ radiation along with the neutrons. As there is no way to suppress the γ sensitivity of TLD, neutron dosimetry always presents the problem of γ background. Fast neutron dosimetry is particularly difficult because most of the proton emitters (via n,p scattering), e.g. glucose or water which could be mixed in the dosemeter medium, are γ sensitive. Pearson and Moran[30] have used a neutron activation technique in which the neutron irradiation of some of the elements present in a given TLD material produces radioactivity. The irradiated TLD is then heated in order to bleach the TL induced by both neutrons and γ rays during the irradiation. Subsequently, the TL builds up again in the dosemeter due to its self-irradiation arising from the internal radioactivity of the dosemeter resulting from the neutron activation.

It is, therefore, a measure of the neutron dose. This technique, however, is not suitable for doses below 10 rad. Better sensitivity (down to few mrad) has been reported by Pradhan *et al*[31] by using $CaSO_4$:Dy powder and mixing it with sulphur powder. The neutron irradiation of such a dosemeter material gives rise to ^{32}P activity via the $^{32}S(n,p)$ ^{32}P reaction.

8.2.11. *Albedo neutron dosimetry*

Albedo neutron dosimetry has essentially been developed as a technique for routine monitoring of mixed n-γ fields. Those of the incident neutrons that are back-scattered into the dosemeter medium again are described as albedo neutrons. This additional registering of neutrons constitutes an error in the observed value of the incident neutron flux. The use of albedo measurement techniques in personnel monitoring has been described by Piesch and Burgkhardt[32] and this subject has been reviewed by Griffith *et al*[33] and, recently, by Lakshmanan[34].

Mixed fields of thermal and fast neutrons and γ rays can be separated by using a pair of ^{6}LiF (TLD-600) and ^{7}LiF (TLD-700) dosemeters placed under a shield of cadmium or boron foil to absorb thermal neutrons. The fast neutron flux is estimated via the albedo neutrons back-scattered through (n,p) scattering from the phantom surface in contact with the dosemeters. Thermal neutrons expend energy in a ^{6}LiF dosemeter via the ^{6}Li $(n,\alpha)^3$H reaction, whereupon the alpha particles cause the storing of TL signal in the dosemeter. TLD-700 is not sensitive to neutrons, while the two dosemeters are equally sensitive to γ rays. The albedo neutron flux is proportional to the incident fast neutron flux but the main disadvantage is the dependence of the dosemeter response on the neutron energy (Figure 8.6(a)). The dosemeter readings R_6 and R_7 with and without the shield (see Figures 8.6(b) and (c)) can be expressed as

(b) $R_6 = N_\gamma + N_a$

$R_7 = N_\gamma$

hence N_a (albedo neutron flux) $= (R_6 - R_7)$

(c) $R_6 = N_\gamma + N_i + N_a$

$R_7 = N_\gamma$

hence N_i (incident neutron flux) $= (R_6)_{no\ shield} - (R_6)_{Cd\text{-}shield}$. In the above, the effect of cadmium capture gamma rays has been ignored, although the response of TLD-700 is influenced by the capture γ rays from the cadmium shield[35].

8.2.12.   *TLD-SSNTD combination*

Tymons and Tuyn[36] have used a combination of LiF (as a TL dosemeter) and cellulose nitrate (CN) film (as a solid state nuclear track detector, SSNTD) to separate out the complex neutron fields that are found behind the shielding of high energy proton

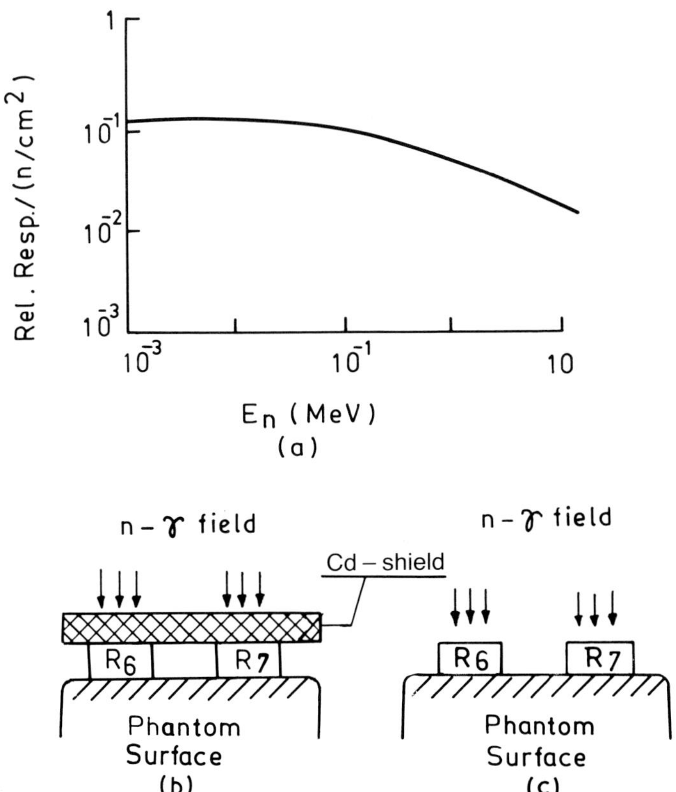

Figure 8.6. Albedo neutron dosimetry. (a) Albedo neutron response of TLD-600 dosemeter as a function of neutron energy. (b) TLD-600 and TLD-700 dosemeter pair under Cd shield. (c) The dosemeter pair without the shield.

accelerators. ^{6}LiF is used here as a radiator of α particles, which will form tracks in the CN foil placed under the TLD chip. Fast neutrons will induce recoil particle tracks directly in the foil[37]. Thus, while the ^{6}LiF chip will read the thermal neutron flux, the reading of foil tracks will give both the thermal and fast neutron flux together. An additional ^{7}LiF chip will read the γ field.

The tracks in the CN foil are revealed by etching it in a 10% NaOH solution at 60°C for 90 min and tracks appearing as holes are counted under a microscope. The CN film is calibrated for neutron dose by using a PuBe source. This calibration is compatible with 14 MeV neutrons but can be considerably different for neutrons of a different energy. The empirical formula[36] gives the neutron dose:

$$N \text{ (mrem)} = 5.2 \, (N_\text{u}/200) + N_\text{o} \tag{6}$$

where N_u = tracks cm^{-2} in the foil area under a ^{6}LiF disc, and N_o = tracks cm^{-2} in the outside area.

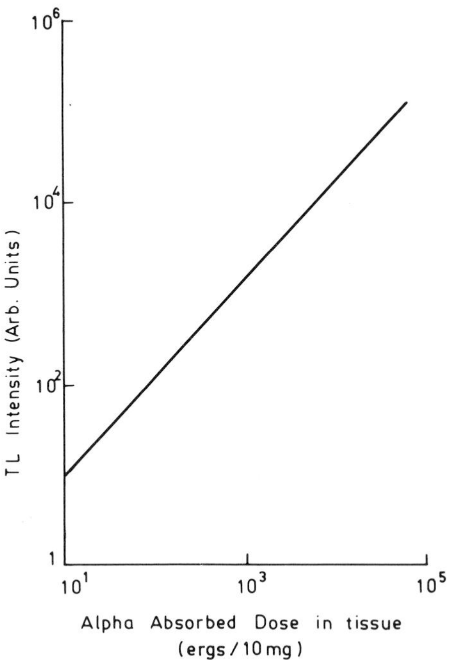

Figure 8.7. Integrated luminescence of TLD-100, plotted against absorbed dose in tissue for 910 MeV α particles[36].

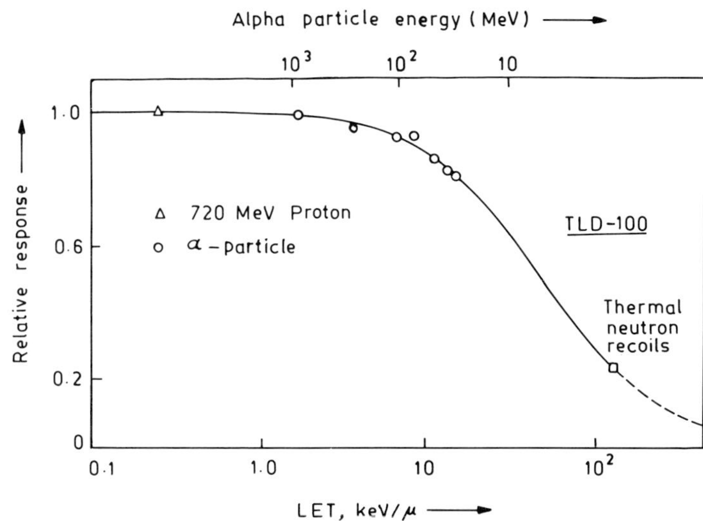

Figure 8.8. Relative response of TLD-100 plotted against alpha particle energy[36]. TL response for α particles is an order of magnitude lower than for β particles, due to very large attenuation of the former.

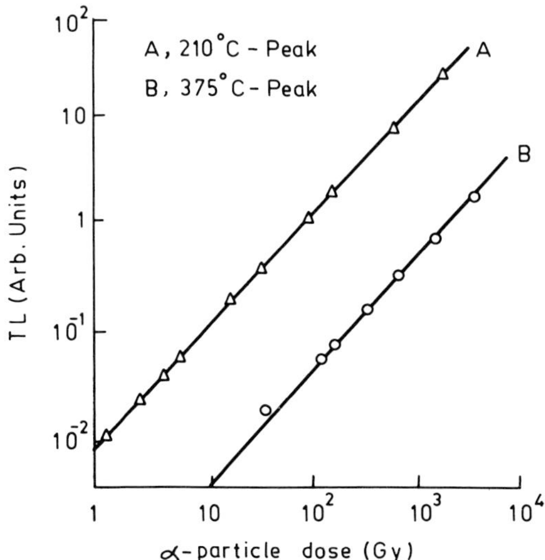

Figure 8.9. TL response of CaSO$_4$:Dy dosemeter to alpha particle dose[38].

8.2.13. *Heavy charged particle (HCP) dosimetry*

In work with accelerators, personnel monitoring in respect of radiation beams of alphas, protons and pions is called for. Wingate *et al*[38a] have studied the sensitivity of TLD-100 (containing only the natural isotopic ratio of ^{6}Li and ^{7}Li, i.e. 7.4 and 92.6%, respectively) to 910 MeV alpha particles from a cyclotron. Figure 8.7 shows the TL intensity of TLD-100 as a function of kerma in tissue for the direct 910 MeV α beam from the accelerator while the response of this dosemeter to α particles of different energies and also to the 720 MeV proton from the cyclotron is shown in Figure 8.8. More recent studies on HCP dosimetry have been reported by Kalef-Ezra and Horowitz[38b].

Aitken *et al*[39] have measured the light output of 230°C and 290°C glow peaks of TLD-100 as a function of absorbed dose of alpha particles from a ^{210}Po (50 mCi) source. It is found that the light output of the two peaks is nearly identical for different values of the alpha absorbed dose.

Lakshmanan *et al*[40] have studied the alpha dose response of a CaSO$_4$:Dy dosemeter and find it to be linear in the observed dose range of 1 to 10^4Gy. This response is shown in Figure 8.9. Negative pion beams, apart from their significance in personnel monitoring,

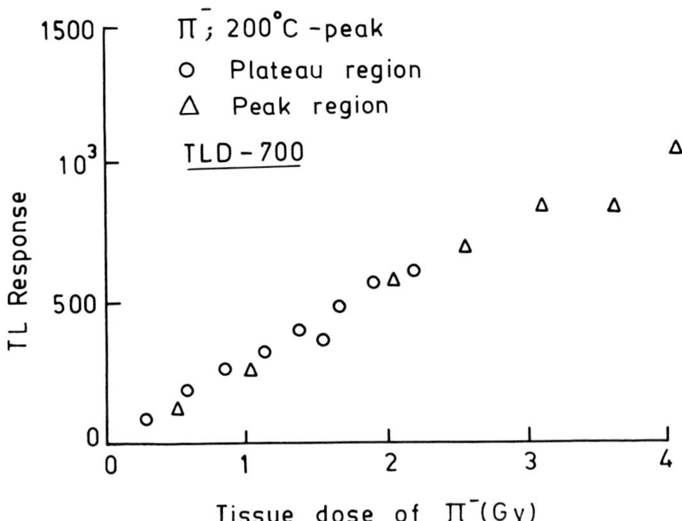

Figure 8.10. TL response characteristics of 200°C glow peak for the pion dose.

are of special interest in radiotherapy[41] and radiobiological investigations[42] owing to their easy capture in the malignant tissue nuclei followed by the consequent destruction of the latter. However, for negative pion beam dosimetry, TLD-100 and TLD-600 are not suitable because of the emission of neutrons in these materials following π^- capture. Cooke and Hogstrom[43] have found that TLD-700 is reasonably suited for pion dosimetry in the clinically useful range of 0.05 to 4 Gy. The appropriate dosimetric peaks are at 200°C and 260°C, although the TL response of the latter peak is relatively much lower. A powdered TL sample irradiated in a pion beam in the plateau and peak regions of the pion Bragg curve (5 cm and 15 cm depths respectively in a water-equivalent medium) accumulates the TL signal. The dose response is reasonably linear as shown in Figure 8.10.

TLD-700 is also found to exhibit a reasonably linear dose additivity behaviour[44] when irradiated with a mixed dose of low LET X rays and high LET negative pions. Prior to exposure, the powdered sample of the dosemeter is annealed for 1 h at 400°C, followed by a 24 h anneal at 80°C. Both the glow peaks, in the pion and X ray dose range from 10 to 10^3 rad, exhibit (within experimental error) that

$$\mathrm{TL}_\pi + \mathrm{TL}_X \approx \mathrm{TL}_{\pi+X} = \mathrm{TL}_{X+\pi}$$

$\mathrm{TL}_{\pi+X}$ denotes that the dosemeter is first irradiated with pions and subsequently with X rays while $\mathrm{TL}_{X+\pi}$ means the reverse order of exposure.

8.3. Environmental and space monitoring

8.3.1. *Environmental monitoring*

The surroundings in which one lives or works constitute the environment. There is, of course, the natural environmental radiation always and everywhere present but this is not of particular concern as far as the health and safety of the inhabitants is concerned. Increased radioactivity arising from atomic explosions, nuclear tests and nuclear reactor leakages can considerably increase the level of environmental radiation, which evidently requires to be monitored. For those workers occupied in nuclear plants, the local environment has the possibility of an increased radiation level caused by the radioactive material and by effluent. Radionculides in building materials contribute to a radiation level in domestic dwellings which may vary between about 60 to 300 mrem per year. Radiation exposure to humans in space arises from the higher cosmic ray intensity, solar UV radiation, mountain snows, and radioactive cargo (if any) in the aircraft.

Estimation of population doses is the main aim of environmental radiation surveys. Many techniques, such as the use of large scintillation counters located on airplanes flying at altitudes of 100-200 m for the aerial exposure rate, use of high pressure ionisation chambers for a quick measurement of the exposure rates at different locations, and the use of TLD (integrating dosemeters) for outdoor locations, have all been employed for this purpose. TL dosemeters have been found to be reasonably suitable, being accurate and inexpensive devices for the determination of environmental radiation doses. Outdoor measurements alone are not sufficient as a basis for determining population doses from environmental radiation fields. The indoor exposure of personnel can also be substantial and this must be considered. Indoor radiation levels can vary markedly from house to house because of the difference in building material used, being higher where cement, concrete, bricks and industrial wastes (such as fly ash and 'chemical' gypsum, a by-product of phosphate production) have been used in greater proportions.

It is important that measurements are continued over a longer period of time for reliable assessment of the population exposure. The results of short-term measurements will suffer from great uncertainty because of significant fluctuations in radiation level at any given location. TL dosemeters have the unique advantage of freedom from any need of care and attention once they are positioned at defined locations. Of course the chosen TL dosemeters should be of high quality, having low background (equivalent of less than 1 mrad) and low fading characteristics. For estimation of dose rates in air due to terrestrial sources, by using LiF-TLD, one can use the Lowder-de Planque formula[45,46]

$$\dot{D}_\gamma = 1.07 \left(0.81\, R' - \frac{\dot{D}_c}{1.26} \right) \tag{8.7}$$

where $\dot{D}_\gamma$ is the air absorbed dose rate in μrad.h^{-1}; 1.07 is the gamma ray conversion factor from rad in LiF to rad in air; 0.81 is the conversion factor from roentgen to rad in LiF; D_c is the cosmic ray dose rate in μrad.h^{-1} (for indoor measurements, 70% of the local cosmic ray dose is taken into account due to absorption of the soft component by the buildings); 1.26 is the conversion factor from rad in LiF to rad in air for cosmic ray muons and R' (μR.h^{-1}) is the corrected net response of the dosemeter. It is estimated from

$$R' = (R-B)\, C_i - \dot{D}_t T_t\, (C_f/T_e) \times 10^3$$

where R = dosemeter response (average reading of three detectors) in mR,

B = individual card background equivalent in mR, each card consisted of two separate TL dosemeters,

C_i = calibration factor of individual card,

$\dot{D}_t$ = equivalent of average transit exposure rate in mR.d^{-1} (transit in mail or otherwise from the place of readout facility to the location where card is positioned),

T_t = transit time in days,

T_e = net exposure time of the card in hours,

C_f = correction factor for TL fading,

10^3 = conversion factor from mR to μR.

As mentioned earlier in 8.2.7, $CaSO_4$:Dy is a highly suitable TLD because of its very low threshold and tolerable fading, thus enabling reliable measurements at the 1-2 mR level. Hsu and Chen[47] have studied the influence on TL response of $CaSO_4$:Dy, of various factors such as environmental temperature, relative humidity, visible light, mechanical shaking, etc. They have observed that the TL reading can fluctuate by as much as 30% due to a variation in the temperature of exposure from 0°C to 30°C. Thus, for accurate dose measurement, the temperature of the environment should also be recorded and corresponding corrections applied. Shaking of the phosphor occurs during transportation. The spurious TL in $CaSO_4$:Dy at temperatures 20-30°C is about 1.8 mR per month. Thus, during a normal exposure period of 3 months at a site, the spurious build-up (which can arise due to the presence of radioactive particles in the TLD) will be about 5-6 mR. The fading of the exposed dosemeter is about 30% during the first four days, after which it stabilises. Thus, an exposure dose of less than 10 mR is almost impossible to estimate directly. Low dose exposure can therefore be estimated by including an additional TLD pre-exposed to a known dose of 30 mR to 50 mR. After correcting for the influencing factors described above, small doses can be distinguished from the spurious TL. Weng[48] has made an intercomparison of the performance of various TLD materials for environmental dose measurement use. These are CaF_2 (chips), $CaSO_4$:Dy (powder), $CaSO_4$:Dy (Teflon disc), $CaSO_4$;Tm (powder) and LiF (Teflon disc). The environment monitored was close to the open-pool reactor building and the measured dose was about 18-20 mR. Wachsmann and Regulla[49] have also made an extensive survey of environmental radiation in West Germany. On the basis of measurements in and around more than 4000 dwellings, the average values for the outdoor and indoor exposure rates were found to be 5.7 and 7.6 μR.h^{-1} respectively, thereby showing that the indoor exposure rate was higher by a factor of 1.33 than that outdoors.

8.3.2. *Dose estimation from atomic bomb explosions*

TL dosimetry has been successfully applied[50] for the estimation of gamma ray fall-out from the atomic bomb explosions in Hiroshima and Nagasaki. The roof tiles of houses that absorbed gamma radiation during the explosion but were not exposed to fire, serve as the irradiated samples. From these samples the TL output is recorded as a function of temperature. These samples are then given a known γ dose to the extent that an overlapping glow curve is obtained. This value is the estimate of the γ ray fall-out. In these measurements, the contribution of neutrons and electrons are not known separately. In order to eliminate the problem of fading, only the part of glow curve above 200°C is considered for the purpose of dose matching.

In order to prepare the sample for TL study, a piece of exposed roof tile is ground in an agate mortar after removing a thin surface layer. The powder is then sieved to obtain a uniform grain size of 100-200 mesh. It is washed in distilled water and then in alcohol to remove dust. The powder is then separated into colourless and coloured minerals by means of a magnetic separator. The colourless portion is used for TL measurements. A sample of about 300 mg is spread out uniformly and then pressed onto the heater plate (2.5 cm diameter). The sample is heated at a rate of 75°C.min^{-1} to about 400°C and the glow curve is recorded. The γ doses received by roof tiles located at a distance of about 400 m from the hypocentre in the south west (Hiroshima) and west (Nagasaki) directions were about 6000 R.

8.3.3. *Space monitoring*

TL dosemeters have been used for radiation monitoring in manned space craft and satellites. These experiments are primarily aimed at estimating the cosmic ray intensities at those heights. Higher cosmic ray intensities, especially during the period of maximum solar activity, constitute radiation risk to space travellers. Particularly during space walks, when there is no protection from the walls of the space craft, the risk is quite substantial. Similar studies have been made to assess environmental radiation exposure to aircraft passengers and also to study the effect of radioactive air cargo at passenger seats located over the freight compartments.

Attix *et al*[51] have used a CaF_2:Mn dosemeter installed in recoverable satellites and space probes to measure radiation exposure. This dosemeter exhibits a linear γ dose range within 10^{-2} to 3×10^5 rad. The dosemeters in each flight are enclosed in an aluminium can filled with polystyrene foam for cushioning. A group of 5 dosemeters was contained in holes drilled in a Lucite sheet of 1.3mm thickness and 8mm $\times$ 16mm surface area. The dose measurements yielded values that varied from flight to flight (2.7, 4.5, 1.8 and 3.3 mrad per orbit). The radiation flux was identified to be a proton flux. The flight to flight variation in dose rate was attributed to the altitude and orientation of the orbits followed by the satellites.

Feher *et al*[51,52] have developed a small, portable, vibration and shock resistant TL reader (weight about 1 kg; volume 1000 cm^3 and electric power consumption of about 5 W), captioned as 'pille' (meaning 'moth' in Hungarian), for use in space flight experiments. It has been used aboard the Salyut-6 orbital station. The TL reader employs $CaSO_4$:Tm powder in a bulb dosemeter which has a dose linearity in the range 10 μGy to 100 mGy. Its TL response to X rays and directional dependence to ^{60}Co γ rays have been studied. The glow curve of the bulb dosemeter exposed to 2 mGy of ^{60}Co radiation obtained by Feher *et al*[52,53] is shown in Figure 8.11. The results obtained for the cosmic ray dose measurements aboard the orbital station are 15 to 30 mrad per day. The use of solid state nuclear track detectors also enabled them to measure the mean flux of particles with $Z \geqslant 6$ and LET$\geqslant$200 keV.μm^{-1} which was found to be 0.22 cm^{-2}.d^{-1}. The block diagram of the Pille and other details of the TL reader are given elsewhere[52,53].

Hsu and Weng[54] have made a survey of the radiation exposure on specific domestic and international air routes taken by Taiwanese air lines; namely the China Airline on international routes and the Far East Airlines for domestic air traffic. Most of the air routes are above the Pacific Ocean and the average flying altitudes vary from 9525 to 11,000 m. They employed $CaSO_4$:Dy TL dosemeters consisting of 40 mg of the sample as powder encapsulated in a black plastic vial. Each set of dosemeters consisted of 10 vials and two sets (one in the captain's cabin and another in the passenger cabin) were used in each measurement. The averages of measurements taken over a period of one year showed that air exosures aboard commercial aircraft (Boeing 747 and 707) flying above the Pacific Ocean are 10 times higher than on the ground. In the case of domestic flights operating at heights of 1830 to 3660 m, the exposure rate was found to be about 3 times higher than the ground value. Wachsmann and Regulla[55]

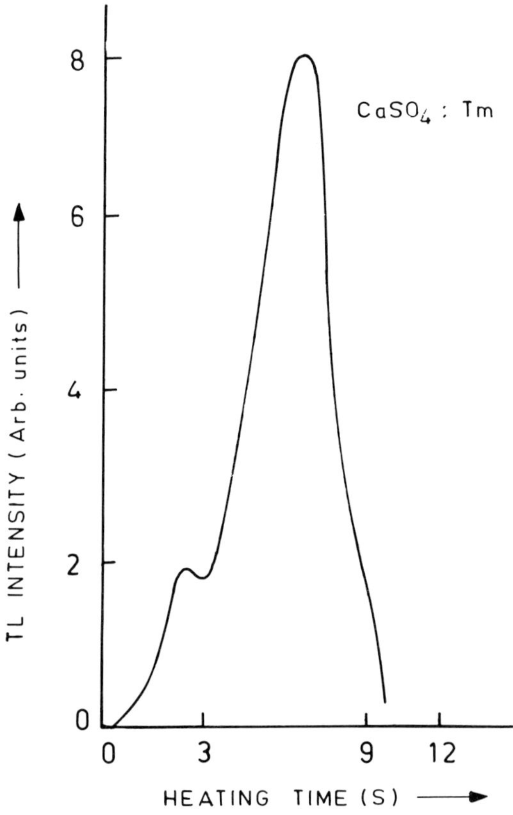

Figure 8.11. Glow curve of the bulb dosemeter used on board Salyut-6 orbital station. Different samples of dosemeter give almost identical glow curves.

using CaF$_2$:Dy crystals ($6 \times 6 \times 0.9$ mm^3) as TL dosemeters (surrounded by a 3 mm thick copper filter to obtain energy independence of their response for radiation of 80 keV energy and above) have made exposure measurements in passenger aircraft of the Lufthansa airlines and determined the effect of exposure from radioactive cargo on passenger seats located above the freight compartment. Their estimate of the mean exposure as equal to 5.15 mR.y^{-1} is negligibly small compared with that due to natural radiation background, and smaller still than the cosmic ray level of 25 to 140 μR.h^{-1} (depending on the flight altitude). For this reason, it may be concluded that there is apparently no risk to passengers on account of the dose received from the transportation of radioactive goods.

8.4. Health physics and biomedical applications
8.4.1. *Exposure measurements*

TL dosemeters have found uses in radiation diagnosis and therapy where dose measurements are required. The reasonable degree of accuracy that these dosemeters possess renders them quite suitable for practical use under conditions obtaining in hospitals. We see that a variety of radiations (electromagnetic, neutrons and charged particles) are used in these applications and, happily, there is available a selection of TL materials suitable for one or the other of these radiations.

For ^{60}Co gamma ray and orthovoltage X ray (from voltage operated X ray machines) dosimetry needs of the radiotherapy centres, the IAEA operates a postal dosimetry service[56] in which LiF powder (^{7}LiF:Mg,Ti, TLD-700) in plastic capsules is used as the dosemeter. When used under identical operating conditions of beam quality and other factors, the accuracy of the dosemeters used at different therapy centres lies within 10%. IAEA provides the readout facility for TL dosemeters received from different user centres.

Interesting use of the Li$_2$B$_4$O$_7$:Mn (TLD-800) TL dosemeter has been made by Regulla and co-workers[57] for measurement of very high doses of ^{60}Co radiation. Municipal waste (sewage sludge) is forced to pass through the annular region of two concentric cylinders of diameters 60 cm and 180 cm in order to hygienise it. The wall of the inner cylinder contains a distribution of ^{60}Co radioactive sources with a total activity of 150 kCi. For dose measurement of the radiation exposure of the sludge, TL dosemeters of 1 mm×1mm×6 mm size were also mixed with the sludge. Although the lithium borate dosemeters suffer from the effect of supralinearity above 1 kR, the careful annealing procedure allows a linear response up to about 400 kR, as shown in Figure 8.12. Sewage sludge, although predominantly water, also contains waste solid matter. The relative content of this matter in the sludge is found to influence the exposure level to the sludge. Up to about 4% of solid content, the influence is negligible, but at higher percentages, the presence of solid matter reduces the exposure level considerably. For instance, for the solid content at 6.3%, the exposure level is reduced by about 15%.

In the treatment of nasopharyngeal, uterine, cervical and breast tumours using a teletherapy ^{60}Co unit, it is important to measure dose distributions at locations around the tumour site with a view to checking these doses for scattered radiation that might be sufficient to cause leukaemia. This is done by placing a phantom in the gamma ray beam and placing TL dosemeters at various locations within the

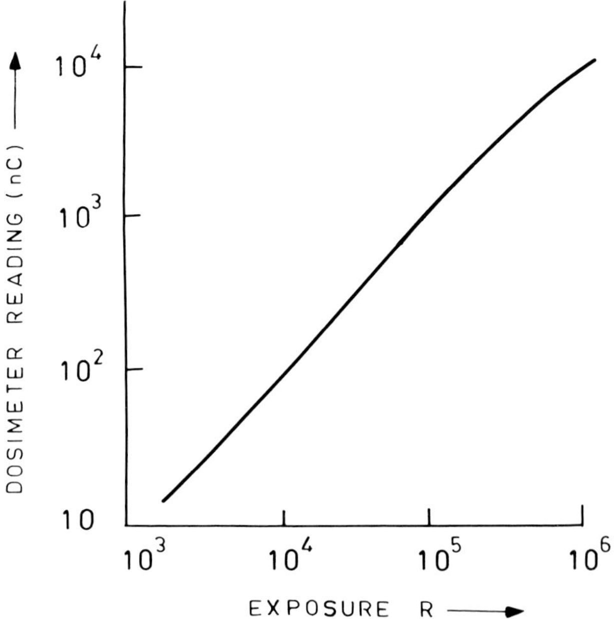

Figure 8.12. TL response of $Li_2B_4O_7$:Mn dosemeter to high exposure radiation of ^{60}Co[57].

phantom. Weng and co-workers[58] have carried out such a study using $CaSO_4$:Dy dosemeters in powder form, (75 mg samples) encapsulated in black polyethylene capsules with an internal diameter of 3 mm, wall thickness 1 mm and length 20 mm. The dosemeter is prepared in the mole ratio of 1:1000 of Dy_2O_3 to $CaSO_4.2H_2O$. The details of the treatment method for a nasopharyngeal tumour are illustrated in Figure 8.13, and the absorbed dose due to scattered radiation at different locations is shown in Figure 8.14.

In clinical diagnosis, the exposure dose of interest is usually in the range of a few mR to a few tens of roentgens and many TL materials are not only sensitive in this range but, equally important, are tissue-equivalent also. Accurate measurement in a wide dose range is ensured by the good linearity of dose response of some TL materials (e.g. $CaSO_4$:Dy, TLD-900)[59] and can be very useful for measuring low-level scattered radiation doses in organs lying outside the primary beam. Scarpa and others[60-62] have used LiF chips ($3.2 \times 3.2 \times 0.7$ mm^3) to measure radiation doses in areas outside the primary beam in a hind limb phantom of a mouse. The limb is represented by a polyethene phantom 7.8 mm in external diameter

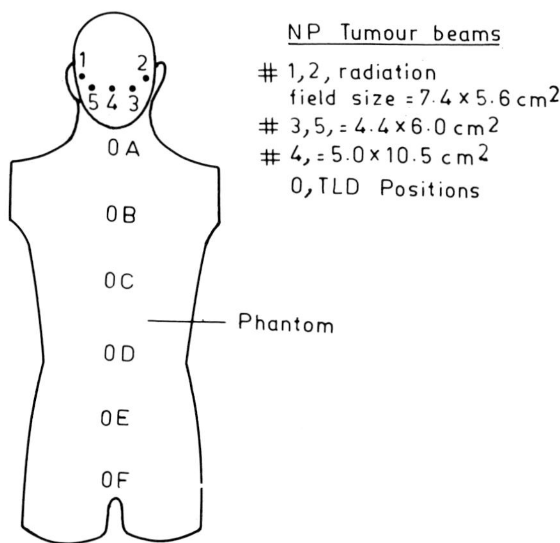

Figure 8.13. Radiation beam profiles for nasopharyngeal (NP) tumour treatment[58].

and 45 mm in length, and shielded with an open ended, 3 mm thick, hemicylindrical lead tunnel resting on a lead platform of the same thickness. Several dosemeters were located in a row inside this leg phantom. The bone is shown for reference only and was not actually present during dosimetric measurements. Leg phantom and the lead shield are shown in the insert of Figure 8.15. The area occupied by the animal was uniformly irradiated (within ±5%) by the uncollimated, vertical beam of X rays. The minimum dose is found in the central part, but increases rapidly near the two ends due to scattered radiation. The X ray quality used in the experiment is 250 kV, 0.5 mm Cu total filtration.

8.4.2. *Dentistry and clinical diagnosis*

In dental radiography (radiodontics) an evaluation of the radiation dose received not only at the required site but also at other neighbouring locations is needed to determine the benefit/risk factor. Regulla[63] has suggested that TL dosimetry will be more suitable and accurate for this purpose. The advantages that accrue to the TL dosemeters are:

(i) The possibility of using them in small sizes, thus practically enabling point to point or two dimensional (very thin dosemeter) measurement of dose.

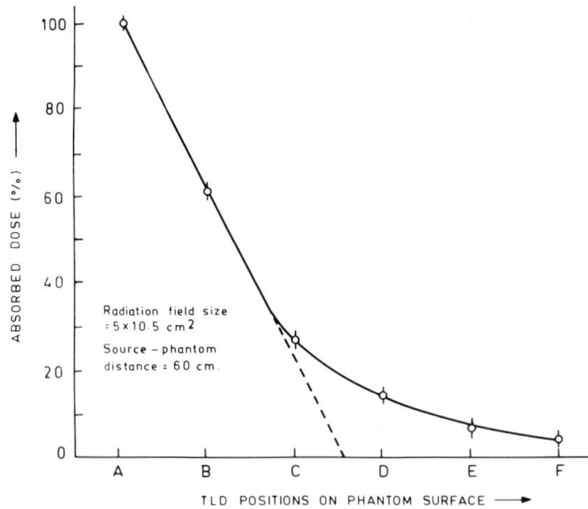

Figure 8.14. Absorbed dose measurement at different locations on phantom surface using $CaSO_4$:Dy TLD[58]. Field size 5×10.5 cm^2, source phantom distance 60 cm.

(ii) The near independence of dose response on the radiation energy. In this case the example of $Li_2B_4O_7$:Mn (Mn=0.34%) as a TL dosemeter is particularly notable as its response is quite independent over a large range of energy from 0.01 to 100 MeV.

(iii) The accurate and automatic readout equipment which is commercially available.*

(iv) The low detection threshold.

(v) The tissue-equivalence such as that of LiF, $Li_2B_4O_7$ and BeO.

---

*1. Harshaw Chemical Group, Cleveland, USA.

2. P.A. Pitman Ltd, Weybridge, Surrey KT13 8LE, UK.

3. Teledyne Isotopes, Westwood, NJ 07675, USA.

4. Eberline Instrument Corporation, Sante Fe, NM 87501, USA.

5. V/O 'Licensintorg', 31 Kakhovka Street, Moscow, B-420, USSR.

6. Institute of Nuclear Physics, ul. Radzikowskiego 152, 31-342 Kraków, Poland.

7. Indotherm Instruments Pvt. Ltd, P.B. No.5665, Dadar, Bombay 400 014, India.

8. Radiological Service TNO, Utrechtseweg 310, N-Arnhem, The Netherlands.

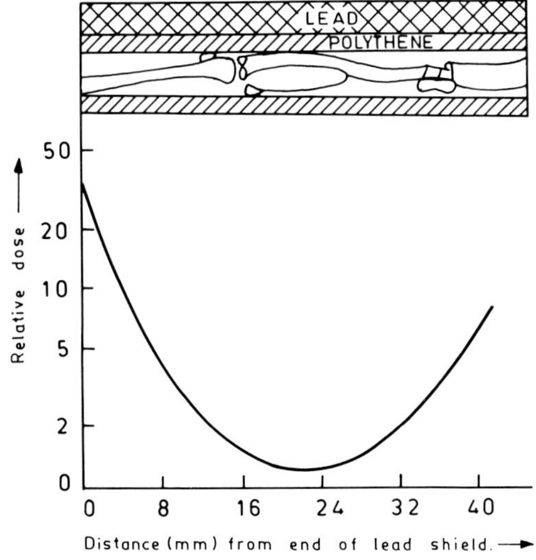

Figure 8.15. Distance (mm) from end of lead shield. Open beam X ray irradiation of shielded hind limb of a mouse (simulated by a polythene tube)[60].

Dental and other partial-body irradiation entails a knowledge of the sensitivity of individual organs to the radiation scattered from the collimated beam incident on the target in order to obtain data on the carcinogenic action of radiation. The desired zero dose to the shielded tissues is not achieved as radiation scattered from the unshielded targeted parts of the body is received in significant amounts in the shielded areas.

Table 8.3 shows gonadal dose measurements obtained using CaF_2:Dy dosemeters when dental irradiation is carried out. The effect of a lead apron (0.5 mm thick Pb) in drastically reducing the gonadal doses is clearly seen.

More recent and detailed studies on the radiation exposure of other organs of the human body as a result of dental radiography have been carried out by McKlveen[64,65]. Initially, the experiments were performed on volunteers. The TLD-100 dosemeters of size $0.125'' \times 0.125'' \times 0.035''$ were held in an elastic face-mask worn by the volunteers, and also fastened to the breast and abdomen. Because of difficulty in obtaining precise and statistically useful information, McKlveen subsequently experimented with a phantom made of a plastic resin (Polylite-32039, Reichhold Chemical Company, USA) which consists of propylene glycol, phthalic anhydride, maleic anhydride and styrene monomer. It has a specific gravity of 1.1 to 1.2 which compares favourably with the Reference Man values (for skin, epidermis, blood and, approximately, for skeletal muscles, brain and salivary glands). The phantom consisted of one inch thick slices which were drilled to accept TLD-100 dosemeters. Slices located in the vicinity of the eyes, upper and lower teeth and thyroid had two TLD's per location installed. A skeleton with a full set of natural teeth and encased in the phantom served as the full face exposure target. TLD packages were taped onto the surface of the phantom at other locations. The X rays for the full face exposure were obtained

Table 8.3. Gonadal dose measurement[63].

X-ray exposure of 1 s (Phillips Optident X ray unit)		Cross section upper jaw (μ rad)	Upper molar (μ rad)
Gonadal doses (μ rad)	1. Without shielding	700	22
	2. With lead shield plate (0.5 mm Pb)	5.7	2.9
	3. With lead apron (10.5 mm Pb)	0.6	0.5

from a machine working at 80 kV and 5 mA current. Table 8.4 shows the average exposure (mR) to various organs of volunteers receiving full face dental examination.

8.4.3.   *Dosimetry with human nails, teeth and hair*

In situations where radiation accidents occur and the victims are not likely to be wearing any kind of dosemeter, it becomes important to assess the accidental radiation exposure by some alternative means. A victim's nails, tooth and bone have been thought of as possible media of radiation dose accumulation. Wilhoit and Poland[66] have investigated thermoluminescence of irradiated tooth dentine and enamel. TL of teeth seems to hold considerable promise from the standpoint of radiation dose measurement on victims who have received large accidental doses. If the measuring technique, coupled with better preparation of samples, could be made more sensitive than it is at present, low doses such as those received during the lifetime of a person could also be estimated. These studies can also have significant implications in the forensic sciences. The dentine and enamel samples are prepared from the crown of the tooth by grinding to 200 mesh powder. The samples are pre-heated for 6 h at 800°C in a nitrogen atmosphere. A 7-9 mg sample of powder is then dispersed in CCl_4 and coated onto the sample holder and then irradiated to a known dose. It is found that the enamel sample exhibits a linear log-log response between the peak height (or glow curve area) and the dose up to about 10^5R. Dentine response, however, shows saturation effects at these exposure levels. At lower exposure levels (<600 R), the TL response

Table 8.4. Exposure of various organs due to full-face irradiation by X rays for dental examination[65].

Location	Exposure (mR)
Forehead centre	4
Upper lip centre	9
Lower lip centre	10
Cheek, right	177
Cheek, left	113
Mastoid process, right	349
Mastoid process, left	362
Back of head, centre	8
Thyroid	27
Chest	4
Abdomen	2

of dentine is close to that of TLD-100. The glow curves for enamel and dentine are compared with that of TLD-100 in Figure 8.16.

A more successful approach to this problem was developed by Watanabe and Nakajima[67] who have utilised the thermoluminescence of jewels from the watches of accident victims. The results give reliable estimates of the accident doses but this method makes it incumbent on the victim to be wearing a watch at the time of accident, a situation which comes close to that of wearing a dosemeter. Another disadvantage is the very small size of the sample which does not allow division into several samples for statistical reliability in the results. Another reliable technique, though not of thermoluminescence, has been tried by Watanabe and Nakajima[68]. This is based on the generation of free radicals by the radiation in human nail and hair and their detection by ESR. This method suffers from the very fast fading of the ESR signal. Garment buttons, however, exhibit only nominal fading and can be used for accident dose measurement within a few hours of an accident.

Grey and Bolt[69] have studied the TSC of irradiated nail samples exposed to 100 R of 65 kV X radiation. They found, however, that dark current increases rapidly with temperature and thus masks any

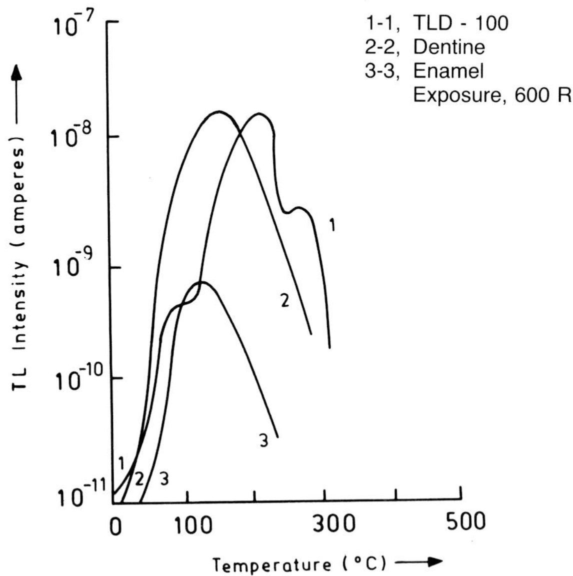

Figure 8.16. TL glow curves of TLD-100, dentine and enamel[66].

radiation induced TSC that might be produced during the heating cycle of the sample from 0 to 200°C (at which upper temperature the nail chars). It may therefore be concluded that, so far, there is no simple way of establishing the accidental dose received by a radiation worker if a dosemeter was not being worn at the time of the radiation accident.

8.4.4. *Pion therapy and radiobiology*

As mentioned in 8.2.13 there is a growing interest in the use of negative pion beams in radiotherapy[41] and in radiobiological investigations[42]. Negative pions are easily captured by the target nuclei of the malignant tissue which, thanks to the high deposition of energy (pion rest mass is large), are disintegrated (nuclear star production). TL dosemeters have been found useful in negative pion dosimetry. Cooke and Hogstrom[43] have found that TLD-700 is suitable for the dosimetry of π^- beams in the clinically useful range of 0.05 to 4 Gy. Powdered TL material is irradiated by the pion beam in the plateau (5 cm depth in water-equivalent) and peak (15 cm depth) portions of the depth-dose curve (Figure 8.17), by placing it at the equivalent depths in a polyethylene phantom. The TL response of the glow peak is determined by recording the peak height as a function of the pion dose. The characteristic features were shown in Figure 8.10 in respect of one of the two glow peaks (200°C) of a TLD-700 dosemeter. The features for the other peak (260°C) are similar except that the TL response is lower by a factor of about 10.

As already pointed out in 8.2.13, TLD-100, TLD-600 and TLD-800 dosemeters are not suitable for negative pion beam dosimetry

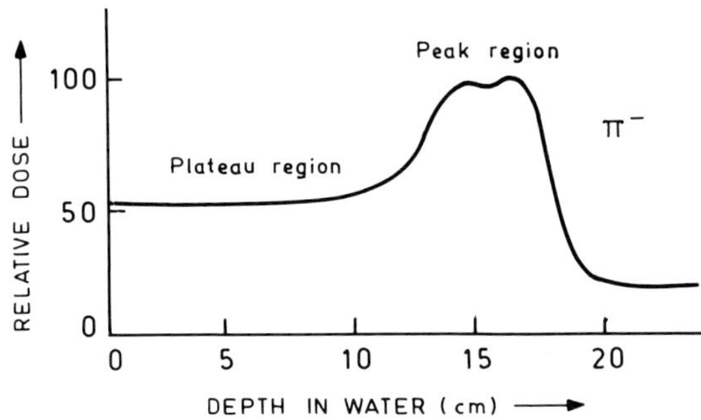

Figure 8.17. Negative pion depth-dose curve for irradiation of TL dosemeter[43].

because these materials emit neutrons following π^- capture. These neutrons will undergo a nuclear reaction, $^6Li(n,\alpha)^3H$, with the ^6Li component of the dosemeters and emit alpha particles. Cooke and Hogstrom[44] have also observed that the sensitivity (TL output/rad) for high LET (≥ 10 keV.μm^{-1}) compared with that for low LET radiation is greater for the 260°C peak than for the 200°C peak. It may be noted that the former peak (260°) also exhibits supralinearity at doses which are in excess of 200 rad for both high and low LET radiation while the latter peak does not exhibit it until about 500 rad for low LET γ radiation and about 5000 rad for high LET α radiation.

8.5. Agriculture

TL dosimetry in agriculture is mainly concerned with the irradiation of grains and seeds for enhancing their storage life and for elimination of agricultural pests that infest food grains. For this purpose sources such as ^{137}Cs and ^{60}Co having high γ intensity are used. In contrast to the chemical dosimetry which was employed in agriculture earlier, TL dosimetry is efficient and inexpensive. For irradiation purposes, a large irradiation facility is built and with the help of TLD chips mixed with the grains, the extent to which the grains are uniformly irradiated, as well as an average level of the dose received by the grains, is determined. Simon and Osvay[70] describe the KOLOS seed irradiator in which the average dose to the maize grains is determined. The irradiator employs a 3467 Ci ^{137}Cs gamma source. Al_2O_3:Mg dosemeters of 6 mm diam. and 1 mm thickness are

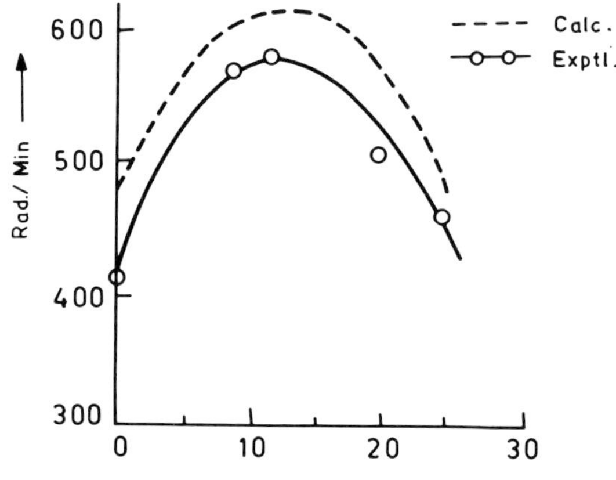

Figure 8.18. Dose rate distribution in a high level gamma irradiation facility[71,72].

packed in polyethylene capsules of a size similar to that of a maize grain. This ceramic dosemeter is especially useful on account of its good mechanical strength, so that when the capsules are mixed with the grain, they are not damaged during the handling of the grain by the irradiation facility. As many as 40 dosemeter pieces were mixed uniformly in the heap of grain. After the grain was fed into the facility, the first capsule appeared in 40 s and the last one in 100 s at the output aperture of the irradiator. Approximately 2000 grains per second (at the rate of 23 kg.min^{-1}) poured out of the aperture of the KOLOS irradiator. In the 40 dosemeters the dose distributions found are shown in Table 8.5.

It may be seen from Table 8.5 that 85% of the grains receive a dose between 2.5 to 4.5 Gy. The average dose is 3.5 Gy with a spread of 13%. Simon and Osvay[70] have further noted that the average value found is dependent upon the condition that the supply to the irradiating vessel is continuous, the latter is totally full of grains, and the passing speed of grain (at 23 kg.min^{-1} in this experiment) is maintained. They have also observed that a uniform feed of grain to the vessel and the latter's fullness are much more effective parameters influencing the average value of the dose than the passing speed of the grains. This value also depends effectively on the moisture content of the grain.

Another large and more sophisticated gamma irradiation facility[71,72] has been built in the Texas A&M University, USA, for carrying out biomedical and radiobiological investigations. It employs a ^{60}Co source of strength 2000 Ci. This activity is divided among 30 stainless steel pencils arranged in the shape of a hollow cylinder. Each pencil provides an activity of about 52.6 Ci. The central region of the source provides the most intense radiation field.

Table 8.5. Distribution of exposure-dose of maize grains.

Dose interval (Gy)	No. of dosemeters showing reading in this range	Per cent of dosemeters
2.00 to 2.49	2	5
2.50 to 2.99	4	10
3.00 to 3.49	17	42.5
3.50 to 3.99	9	22.5
4.00 to 4.49	5	12.5
4.5 to 5.00	3	7.5
	Total 40	100

The facility can provide a semicircular field with a radius of 150 ft for long-term irradiation of plants and animals stationed in the field. A sample basket (18 cm diam. and 25 cm height) that can be located in the central region allows samples to be placed in an intense radiation field. Larger samples, through the tunnel in the facility, can also be kept there.

LiF dosemeters are used to measure the dose in the central region. With the help of a plastic frame, dosemeters are positioned at different heights along the axis of the basket. The results of dose rate measurements are shown in Figure 8.18.

8.6.   Forensic sciences

Forensic science is concerned with the application of science in the detection of crime and the use in law courts of evidentiary material gathered in the process of investigations for the purposes of prosecution. The scientific investigation involves the identification, evaluation and interpretation of the physical materials associated with crime situations such as glass pieces, soils, nails, hair, lipstick marks, etc. These investigations are, therefore, potentially helpful in associating or dissociating a suspect with the crime. The well tried methods of spectrometry, neutron activation analysis, conventional chemical analysis, optical microscopy and gas chromatography are now routinely available for the study of suspect materials found at the scene of a crime, or items expected to be related to a crime and needing to be compared with items known to be associated with the suspect[73,74].

As an illustration, small quantities of lipstick may be smeared on the clothing of a man who has attacked a female. A lipstick generally consists of a moulded solid fatty base containing dissolved dyes and suspended pigments. Typical ingredients of a fatty base are bees' wax, hydrocarbons, castor oil, ethyl alcohol, lanoline, polyethylene glycols, etc. Thus, one can make a forensic examination[75,76] of the lipstick stains by UV spectrophotometry, paper chromatography and thin layer chromatography (TLC) to identify various dyestuffs present in the lipsticks. For the analysis of trace elements present in the lipstick, neutron activation analysis can be used. These results can then be compared with those of the lipstick recovered from the scene of the crime. Discriminatory evidence is, however, very important here as other brands of the lipsticks available in the market may give identical or similar results.

Another illustration of the use of physical methods in forensic investigations is in the detection of fingerprints by laser excited luminescence[77]. A small percentage of the incident laser light (argon laser) can be absorbed by the materials present in the fingerprint and be re-emitted at a longer wavelength. Thus the blue-green incident light is converted to yellow-orange. The Goggle filters transmit a larger fraction of yellow, orange and red light than blue or green and thus the finger print pattern is revealed as a yellow-orange image. This image may be recorded by conventional photography with optical filters over the camera lens.

The materials that are commonly encountered in crime situations are glass, soil, bits of paint, blood, hair, fibres, etc. The important aim of the scientific investigations is to convert suspicion into a reasonable certainty of either guilt or innocence. The characteristic nature of the TL glow curves is essentially related to various parameters in a given material, such as crystallisation history, impurities, structural imperfections and the radiation dose received by the material. Thus a particular material originating from different sources can give characteristically different glow curves. In forensic applications of TL, we wish to compare the glow curves of a sample lifted from the scene of crime with those obtained from samples recovered from a person suspected to be associated with the crime. In addition to the glow curve matching, a more sensitive and discriminating feature of the technique can be the matching of TL emission spectra, which can considerably decrease the chances of coincidental matching[78] which is a serious limitation of this technique in forensic investigations.

Ingham and Lawson[79] have made TL studies of glass and soil that have possible use in forensic investigations. For this purpose, the glow curves of glass of different brands of auto headlamps, and soils collected from different sites, have been compared. These can be evidentiary materials in a motorist's hit-and-run involvement. The glass samples are prepared by mild crushing or grinding and sieved to uniform particle size. The powdered samples are then exposed to a given radiation dose from either a ^{137}Cs or ^{60}Co radioactive source. About 7 mg of the sample heated at a constant rate of $8.8°C.s^{-1}$ gave a glow curve which was then compared with those of other samples obtained under identical conditions. If the curves of the given samples are exactly the same or very different, these can be compared visually. When, however, the glow curves of two given samples appear similar, then the statistical χ^2 method as used by Ingham and Lawson[76] gives the measure of the net difference between the glow curve intensities. Between the two curves, one is taken as 'expected' and the other one treated as observed. The curves are divided into 33 equal parts at 10°C interval. The value of χ^2 is then calculated as

$$\chi^2 = \sum_{i=1}^{33} \left(\frac{E_i - O_i}{E_i}\right) \tag{8.8}$$

E_i and O_i are the expected and observed values of the intensities at the ith temperature interval. As either of the two curves could be taken as expected, the events E_i and O_i are interchanged and a new value of χ^2 is found. Thus the χ^2 value for a sample is expressed as a pair. For statistically identical curves, χ^2 will be zero, while a larger χ^2 value represents a greater difference between the two curves. There should, however, be a certain threshold value of χ^2 for discriminating against coincidental matching or mismatching. While different samples of the same piece of glass may give a mismatch, the samples from two different glass pieces may give matching glow curves. In order to account for this, the two curves for which χ^2 will stand above the threshold value, will be regarded as different.

For determining the threshold, Ingham and Lawson have obtained glow curves for four samples that were different parts of the same lens

Table 8.6. χ^2 values (in pairs) for four samples of the same lens of an auto headlamp.

Pairs of glow curves	χ^2 values (in pairs)
1 and 2	0.101 0.096
1 and 3	0.429 0.597
1 and 4	0.472 0.421
2 and 3	0.494 0.366
2 and 4	0.151 0.165
3 and 4	0.871 0.587

Note: Glow curve numbers 1, 2, 3 and 4 refer to central, right, left and lower parts of the auto headlamp lens.

of an auto headlamp. Four glow curves make six different pairs for which six pair sets of χ^2 values are obtained. Thus, if glow curves are numbered as 1, 2, 3 and 4, the results of six pair-combinations of glow curves are shown in Table 8.6. From these data a mean $<\chi^2>$ and a standard deviation (σ) can be found. The threshold value χ^2_t $(=1.26$ for these data) is then given as

$$\chi^2_t = <\chi^2> + 3\sigma$$

because 99.7% of measurements lie in the range of $<\chi^2>\pm3\sigma$. Thus, for a χ^2 value of the test sample less than the threshold, a match is indicated. Ingham and Lawson investigated 10 auto headlamp glasses of different makes and found coincidental matching among only a few samples while the rest were clearly differentiated. From the above discussion, it is clear that at the present stage of development, TL results cannot be taken to establish the identity of two glasses, one recovered at the site of the crime and the other associated with the suspect, but can certainly be used to emphasise excluding evidence, i.e. the two samples for which the χ^2 values differ from each other by more than the threshold, are not related.

Values of χ^2 have also been found for window glass samples in which the incidences of coincidental match seem much less, probably because in the manufacture of these glasses there is less control and standardisation, unlike the auto headlamp manufacture.

TL differentiation of materials seems to be much more reliable in the case of soil samples collected from different sites, because no manual controls are involved in the composition of these samples. There can, however, be many other earth samples such as the surface samples from sand dunes or from desert areas which give matching glow curves.

8.7.   Industry

8.7.1.   *Introduction*

A fascinating application of TL was referred to earlier in which Regulla *et al*[56] describe the radiation treatment for rendering sewage sludge hygienic by using strong gamma sources and TL dosemeters for radiation dose measurement. Aramu *et al*[80], in another example, have studied the thermoluminescence of textile fibre. A host of interesting possibilities, therefore, seem to exist in which TL techniques can be exploited in industry. Some of these are described below.

8.7.2.   *TV viewing*

A useful application of TL dosemeters seems to be in the measurement of radiation exposure to viewers from television receivers. Weng[81] has studied X ray exposure from colour television sets of different makes by using highly sensitive CaSO$_4$:Dy TL dosemeters. According to the IAEA safety manual[82], the dose rate at a point 5 cm from the surface of a television set should not exceed 0.5 mR.h^{-1}. The effective energy of X rays emanating from TV sets is typically about 23-24 keV when a voltage of 25 kV is applied to the picture tube. For this energy the response of CaSO$_4$:Dy relative to BeO (its response is energy independent) is about six times as large. Although the measured exposures (0.001 to 0.255 mR.h^{-1}) are lower than the prescribed limits, the manufacturers will be helped by these measurements which should be incorporated in the user's pamphlet of information supplied by them.

8.7.3.   *Contaminated liquids*

A CaSO$_4$:Dy dosemeter has been used in the measurement of radioactive contamination of liquids by Hsu *et al*[83]. A 40 mg sample of CaSO$_4$:Dy powder encapsulated in a black plastic vial constitutes the dosemeter. Each satchet contains 6-10 vials and is immersed for a known time in a given volume of the contaminated liquid. In the case of ^{32}P contamination (β emitter), the TL output is

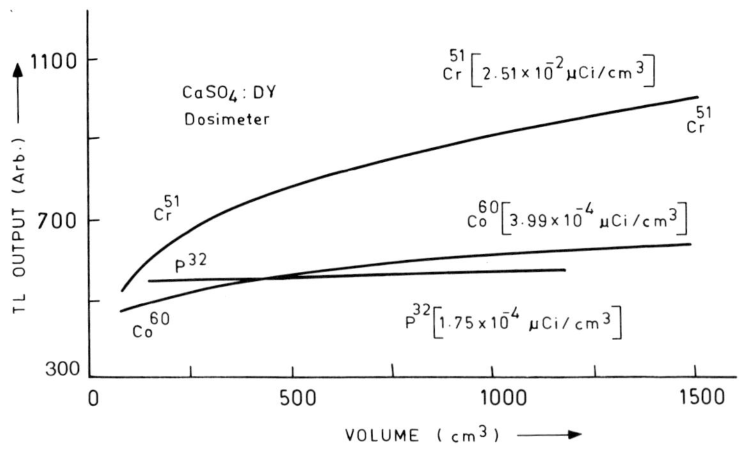

Figure 8.19. TL response as a function of volume of contaminated solution[83].

found to be low and independent of the volume of liquid while for γ emitters, the TL output is energy dependent and also increases with the volume. This is shown in Figure 8.19.

8.7.4.   *Factories using luminous materials*

In factories making luminous dials for watches, clocks and other display facilities, some sort of radioactive source in the printing solution is employed. As an illustration, ^{147}Pm (β emitter) is used in a luminous dial factory in Taiwan and TLD-700 phosphor has been used[71] to estimate the β dose from ^{147}Pm. Hands and fingers are especially exposed in handling this work and it is therefore required to monitor the dose for these parts of the body. Typical monitoring results of the fingers and hand in terms of dose equivalent in mrems are shown in Figure 8.20.

8.7.5.   *Fire damaged buildings*

Concrete is the main building material, having the advantage of great strength and reasonable resistance to fire. It has a lower thermal conductivity at high temperatures than at room

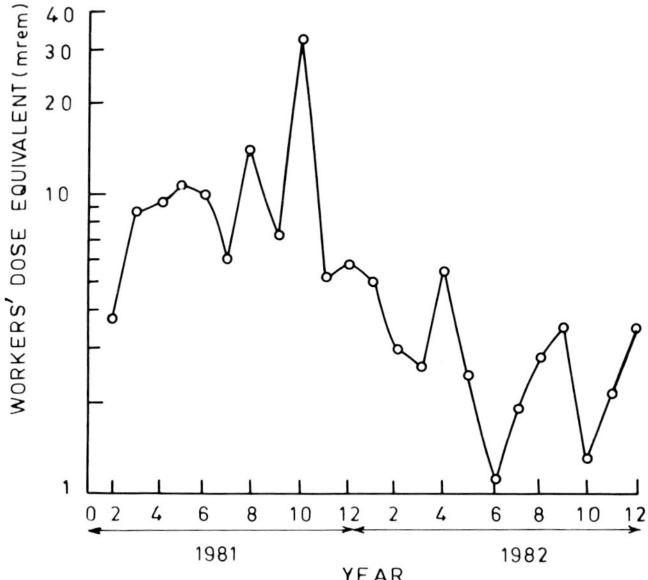

Figure 8.20. Time variation of dose (mrem) of finger and hand of luminous dial factory workers in Taiwan[71].

temperature roughly by a factor of 3. This property limits the depth of fire damage in concrete structures, which, if the fire has not continued for a prolonged period of time, can be retrieved. The extent of damage to the concrete of the building depends rather more on the duration of the fire than on the temperature. If a given concrete structure is to be retained after its exposure to fire, it is necessary to determine its residual post-fire strength.

Visual inspection of the surface and the deep concrete provides some acceptable criteria of the damage in terms of the colour. Some concretes retain pink colouration after exposure to temperatures in the range 300-600°C due to decomposition of iron compounds present in the concrete. Some limitations are inherent in this approach because many concretes do not exhibit pink colouration nor is it possible to determine precisely the depths the colouration has reached.

TL seems to be a very appropriate tool to make reliable estimates of the fire damage because the TL output is a sensitive function of the thermal and irradiation history of a material. Placido[84-86] has done pioneering work in this area. The basis underlying this technique is to measure the residual TL output left in a sample of fire damaged concrete. The sample consists of a small amount of sand (a few milligrams only) extracted from concrete elements at various depths by the following chemical process[84].

The concrete powder drilled from the site contains some TL bearing minerals other than quartz and these give spurious TL that can mask the TL output from the quartz in which we are interested. Quartz from the concrete sample is extracted in much the same way as from an archaeological pottery sample. The powder sample is treated for 10 min in a mixture of concentrated nitric, sulphuric and perchloric acids in the ratio of 1:1:2. The acid treated material is then washed in water and centrifuged several times to remove any remaining traces of acid. It is finally washed in methanol and acetone. A fine suspension in acetone is obtained which is thinly and evenly spread on a metal disc (1 cm diameter platinum disc). The acetone is then evaporated off in an oven at 50°C. Many disc samples are prepared in the same batch to give good reproducibility.

In order to simulate the effects of fire damage in concrete of a given composition (the temperature effect on the strength of concrete varies with its composition), Placido[85] cut 5 cm cubes from cast concrete beams using a water cooled carborundum wheel; the cubes were given thermal exposures of 1, 2 and 3 h at temperatures of 200 to 600°C and then cooled to room temperature. Using a hammer drill (with a 4.5 mm bit) at slow speed, powder samples were removed from the central location at each of the four faces which were vertical during heating in the furnace. The platinum discs coated with the extracted samples were heated ($q=20°C.s^{-1}$) in a nitrogen atmosphere and the TL output recorded on the X-Y recorder. Glow curves were thus recorded automatically after the temperature controller was switched on. The results obtained by Placido[85] are summarised in Figure 8.21. It is seen that the TL technique can

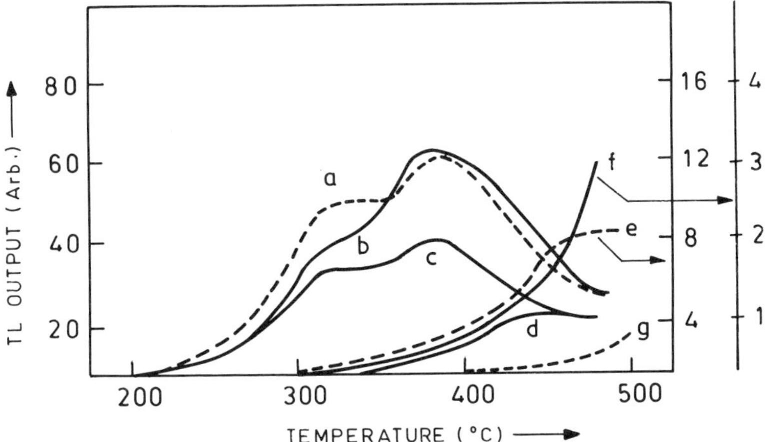

Figure 8.21. TL in unheated and fired concrete[85]. (a) Natural TL of unheated quartz sand (not treated with acid). (b) Natural TL of acid-treated concrete (unheated) showing clearly that the TL in concrete is mainly due to the quartz sand. (c) Unheated concrete sample (acid treated). The unexpected difference between curves (b) and (c) is due to the difference in the amount of active material (quartz sand) present in the studied samples which are, however, of the same mass. (Placido[85] has described the normalisation procedure to account for the difference in the quantity of active material. Low dose irradiation with a ^{90}Sr source gives a characteristic glow peak of quartz appearing at 110°C. The intensity of this peak can be used for normalisation of TL glow curves of samples from different concretes). (d) Sample from the concrete cube of origin common to (c) but cube heated for 3 h at 200°C. (e) Cube heated for 1 h at 300°C (note the vertical scale). (f) Cube heated for 1 h at 400°C (note the vertical scale). (g) Second readout cycle of the sample showing absence of TL and presence of blackbody radiation.

estimate precisely, through the loss of TL during a fire, the extent of fire damage in building concrete when compared with the unheated portion of the given concrete sample.

These studies reveal that fire damage is both time and temperature dependent. Heating at a lower temperature but for a longer duration will be more damaging to the concrete than heating at a higher temperature for a shorter period.

8.7.6. *Uranium mines*

Those who work in uranium mines run a greater risk of induced lung cancer due to the presence of a radon-charged atmosphere and the consequent inhalation of radon gas. Sufficient precautions are, no doubt, exercised against breathing in radon by using filter devices. The regular measurement of activity concentration (per unit volume) in the air in mines is, therefore, called for. Radon gas and its short-lived daughter products release alpha energy in air which is measured in units of WL (Working Level) $(1 \text{ WL} = 1.3 \times 10^5 \text{ MeV.l}^{-1})$. Regulla *et al*[87] have studied the use of a TL dosemeter which is in the form of a thin film (thickness=6 mg.cm^{-2}) of $CaSO_4$:Tm powder on a thin aluminium foil of 8 mg.cm^{-2} thickness and 8 mm diameter, and have determined its detection sensitivity for the radon daughters' activity concentration in the air filter. The glow curves show peaks at 200°C and the TL output bears a linear relationship with the α particle dose and exhibits discrimination against γ radiation. Thus the dosemeter can be operated in uranium mines even in the presence of a high γ field. The threshold detection limit of the dosemeter is 10^{-3} WL and it can also be used for long-term monitoring in dwellings.

8.7.7. *UV dosimetry in industry and medicine*

Ultraviolet radiation in sunlight is of immense biological benefit to mankind because of its sterilisation properties and for its role in photobiological processes in plant life. The sterilisation property which results from the ability of UV radiation to kill bacteria is exploited in medical applications and in industry, e.g. to sterilise surgical equipment and to prepare germ-free chambers and operation enclosures. Overexposure of sensitive human organs, such as the eyes, to UV radiation can be harmful for it can cause itching and burning sensations. The radiation also causes reddening of the skin (erythema); consequently UV radiation dosimetry is of interest to dermatologists. TL materials have many desirable features for use as UV dosemeters. The sensitivity of these materials as direct UV detectors is, however, low. For this reason, the TL dosemeter is first irradiated with a known dose of gamma rays which populates deeper trap states. Susbsequent irradiation with UV transfers carriers to shallow trap states in proportion to the UV dose. Thus the high temperature glow peak intensity of a given TL material is decreased in proportion to the UV dose. This is the well known PTTL (phototransferred TL) technique used in the TL dating of pottery and other objects.

Bassi and Busuoli[88] have arrived at a sensitisation technique through the heat treatment of CaF_2:Dy (TLD-200) which increases the direct sensitivity of the dosemeter to UV by a factor of 500 as compared with that of untreated TLD. The peak response of the dosemeter is found to be at the 250 nm wavelength of UV (Figure 8.22).

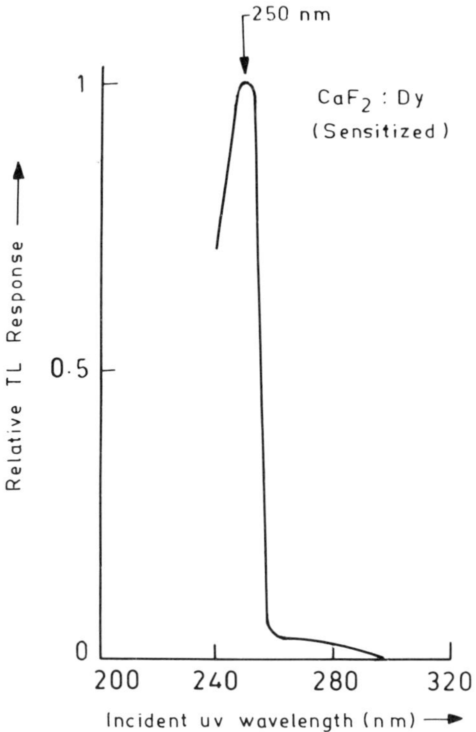

Figure 8.22. TL response of CaF_2:Dy relative to UV wavelength[88].

The significant feature of the dosemeter is its insensitivity to visible light which makes it particularly attractive for field work where there is a large component of visible radiation such as sunlight.

The procedure for heat treatment of TLD-200 is to heat the dosemeter to a temperature of 900°C for 15 min. It is allowed to cool to room temperature and then annealed (5 min at 600°C). The sensitivity of TLD is found to continue increasing for up to 14 such treatments, beyond which it falls suddenly. To improve the signal to noise ratio, the TLD should also be given a pre-readout anneal (15 min at 100°C) to bleach the low temperature glow peaks which contribute to fading. On account of the selective response of the CaF_2:Dy dosemeter to UV at 250 mm, it is not directly usable as a biological dosemeter. Wavelengths that are of interest in radiobiology have a typical spread in the range of 200 to 320 nm. However, if the spectral distribution of the given UV source (i.e. beam intensity relative to wavelength) is known, it is possible to normalise the radiation dose. CaS:Bi, on the other hand, appears to have a UV response devoid of such a sharp dependence on wavelength[89].

8.7.8.   *Polymers*

Except for some mention of TL in polymers with regard to its correlation with TSC and TSDC in section 7.9, we have seen that the process and technique of TL is mostly confined to inorganic solids. One may, however, recall that in many cases, Teflon

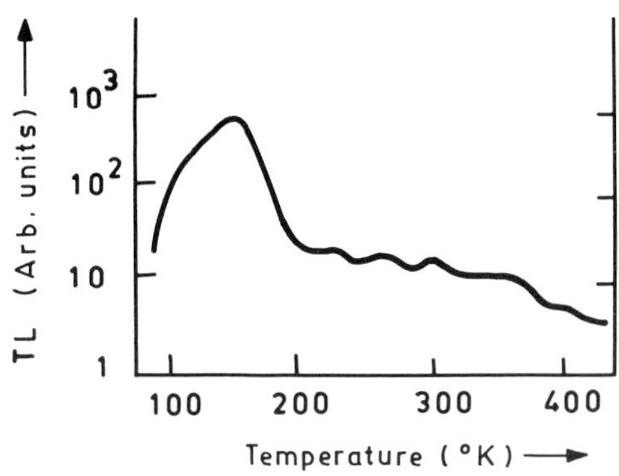

Figure 8.23. X ray induced TL in Teflon[90].

(polytetrafluoroethylene; PTFE) is used as a binder for some inorganic TL materials. This polymer gives X ray induced TL emission peaks even at temperatures higher than room temperature (see Figure 8.23). The implications of using Teflon as a binder for inorganic TL samples are, therefore, obvious.

In most polymers, TL emission is observed at below room temperature and therefore requires the use of liquid air temperatures. The modern electrical industry cannot be imagined without the use of polymers as insulating materials, particularly in high voltage applications. It is, therefore, pertinent to study factors that cause electrical breakdown of polymers. The space charge effect is one of the important factors. Both space charge and charge storage capability are found to relate closely to the carrier traps. Thermoluminescence, in combination with TSC and TSDC, is uniquely suited to the study and understanding of carrier traps in solids although, in polymers, the knowledge about the nature and origin of the carrier traps is scant. TL study of polymer insulators[90] will help in clarifying these aspects and aid in the development of still better polymer insulators for the electrical industry in particular and for many other important applications of polymers in general.

8.8.   Solid state ionography

Although the thermoluminescence process is not directly concerned in this technique, features of TL and TSC are involved in the ionographic imaging of objects by solid state ionic insulator dielectrics. The procedures involved are similar to the xerography process except for the reverse mechanism of charge interplay. In xerography, the amorphous selenium coating surface on an aluminium backing is first charged by means of a corona discharge. Subsequently, the light reflected from the printed paper is incident on the charged coating and, through the photoconducting mechanism, leaves a charged image on it (charge is allowed to leak away selectively). In ionographic imaging, the charge is built on the crystal surface in the form of a charge replica of the object (say, a copper wire mesh) by exposing it to X rays.

The solid state ionographic imaging technique, although only in the initial stage of development because of very poor resolution, was invented by DeWerd and Moran[91,92]. The principle underlying the imaging technique is that of TSC which shows peak values at definite temperatures characteristic of the host lattice and impurities in it.

While delibrate addition of selective impurities creates a population of optically active traps which help increase TL output, they, for the same reason, act to decrease the equilibrium value of TSC due to re-trapping of some of the carriers that became mobile earlier due to irradiation. High purity crystals (with only about 10 ppm impurity) are, therefore, required for a better yield of thermocurrent. DeWerd and Moran[91,92] have used pure crystals of Al_2O_3 of 1 mm thickness. A Mylar film (E I du Pont de Nemours, USA) in contact with the crystal surface receives the charge pattern of the object interposed between the X ray tube and the electrode-Mylar-crystal-electrode cassette. The charge print is taken by applying a suitable electric field while keeping the cassette at a temperature covering the low temperature TSC peak (100°C for Al_2O_3). For obtaining a visible image, the charged foil is kept in a xerographic box where the toner particles injected onto the foil are attracted to it and make a visible impression.

8.9.   **Thermoluminescence dating***

8.9.1.   ***Principles***

The thermoluminescence technique has been most widely applied in the dating of archaeological (mainly pottery) and geological materials (ocean sediments) and meteorites. The basis of age determination is to make a readout cycle of the sample and determine the TL output which we call the natural TL (NTL). If S is defined as the TL sensitivity of the sample then $S = (NTL)/D$ where D is the absorbed dose of radiation. If τ is the total time (years) since the formation of the sample then τ is the age of the sample. We may write $D = \dot{D}\tau$ where $\dot{D}$ is the annual dose rate received by the sample from its surrounding minerals and cosmic rays. We can write

$$\dot{D} = \dot{D}_\alpha + \dot{D}_\beta + \dot{D}_\gamma + \dot{D}_c \tag{8.9}$$

Hence

$$(NTL) = SD = S\dot{D}\tau = S\tau\,(\dot{D}_\alpha + \dot{D}_\beta + \dot{D}_\gamma + \dot{D}_c)$$

However, due to the very large attenuation of α particles in the outer layer of sample itself, the TL sensitivity for α radiation (S_α) may be

---

* Equipment for TL dating can be had from; (i) M/S Daybreak Nuclear and Medical Systems, Inc., 50 Denison Drive, Guildford, CT 06437, USA; and (ii) M/S Littlemore Scientific Engineering Co., Railway Lane, Littlemore, Oxford OX4 4PZ, UK.

taken as negligible in comparison with sensitivities S_β, S_γ or S_c (the latter three considered as nearly equal for simplicity). The α radiation affects only the outer surface of the object and can be scrubbed off to eliminate any contribution from α radiation. Thus, under the assumption that $S_\beta = S_\gamma = S_c$, we may write

$$\tau = (NTL)/(S_\alpha \dot{D}_\alpha + S_\beta \dot{D}_\beta + S_\gamma \dot{D}_\gamma + S_c \dot{D}_c)$$

or

$$\tau = \frac{(NTL)S_\beta}{(S_\alpha/S_\beta)\dot{D}_\alpha + \dot{D}_\beta + \dot{D}_\gamma + \dot{D}_c}$$

Putting $S_\alpha/S_\beta = k$ (say), which is quite small anyway ($k \approx 0.05$ to 0.3), and where $(NTL)/S_\beta$ is equal to the artificial β dose given to the sample under investigation to produce the same TL signal as the observed (NTL), we may write for τ in years

$$\tau = \frac{(NTL)/S_\beta}{k\,\dot{D}_\alpha + \dot{D}_\beta + \dot{D}_\gamma + \dot{D}_c} \tag{8.10}$$

Dose rates for different types of radiation are separately measured by placing TL dosemeters or ionisation chambers at the sample sites for a known time.

8.9.2. *Sample preparation for pottery dating*

Pottery contains many minerals, all or some of which may exhibit TL to a varying extent. For this reason, a suitable mineral from the point of view of its physical and chemical stability and reproducibility of TL glow curves, is chosen. Quartz fulfils these criteria well. Furthermore, quartz is a universal mineral component of all pottery collected from different culture sites. For routine dating, however, the simpler silk-screen technique of Ralph and Han for preparing samples can be adopted[93]. Powdered pottery is mixed with silicone oil which is then applied to an aluminium foil by means of a silk cloth. Thus a uniform coating of a mix of fine powdered sample and silicone oil is obtained on the foil which now serves as a sample for obtaining glow curves. Silicone oil inhibits the problems associated with oxygen, sunlight and moisture. Extraction of quartz mineral from the given pottery sherd (fragment) for the standard estimation of natural TL signal is, however, more elaborate as described below.

The sherd is crushed after removing about a 2 mm thick surface layer. Care is observed not to crush large mineral inclusions. Clay and grains of diameter less than 75 μm and 75-105 μm respectively are selected by sieving. Clay particles may be used for the estimation of β dose rate. The grains are then separated into colourless minerals (quartz, feldspar of various types, and calcite) and coloured minerals (mica, amphibole and haematite) using a magnetic separator. The colourless mineral is treated for 50 min with diluted (5%) HF after IICl and NaOH treatment in order to remove minerals other than quartz. This treatment also etches away the surface layer of quartz grains wherein α particles are mostly stopped. These quartz grains finally form the sample. Figure 8.24 shows the relative TL output of the bulk sample, the magnetic extract, and the quartz extract.

8.9.3.   *Pottery dating*

In dating ancient pottery finds, we date the event when the pot was fired by its original potter at temperatures between 750° and 1000°C. Whatever thermoluminescence had been acquired by the constituents of the pot earlier, during geological times, is rightly assumed to be removed at these temperatures and from then on the thermoluminescent clock begins to tick and the material acquires its

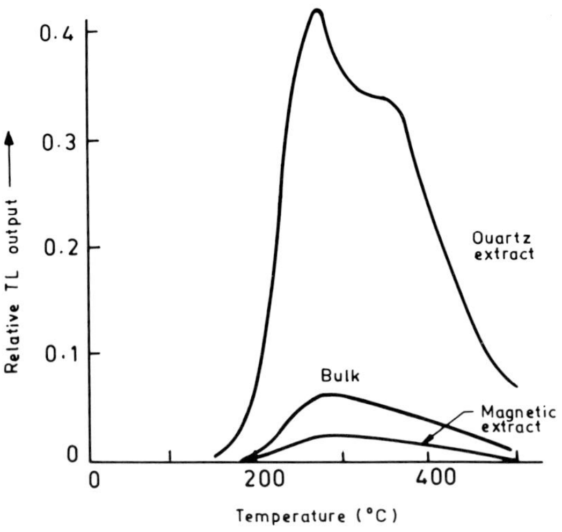

Figure 8.24. Natural TL from pottery sherd and its extracts[126].

TL dose via the natural irradiation of the pot by nuclear radiation (α, β and γ) emanating from within, from the surroundings at the burial site and, to a lesser extent, from cosmic rays. Typical annual dose estimates for a pottery fragment are depicted in Figure 8.25[94]. The internal dose arises from the radioactive impurities occuring in the sherd and the external radiation dose comes from the soils covering the sample during its history. There are essentially two methods for the determination of the age of pottery: (a) the quartz inclusion method, and (b) fine-grain method, which we now describe.

For the quartz inclusion method we recall that the main constituents of a pottery sherd are found[95] to be quartz (SiO_2) (55-70%), Al_2O_3 ($\sim$20%), Fe_2O_3 ($\sim$4%), and K_2O ($\sim$2%) in addition to other insignificant components such as MgO, CaO, Na_2O, TiO_2. The sherd is gently broken by pressing it in the jaws of a vice but avoiding crushing it, so that the grains of quartz of about 100 to 200 μm size included in the clay matrix are not broken. The TL observations are made on these quartz grains. The advantage of this technique, developed by Fleming[96a], is that quartz is free of radioactive

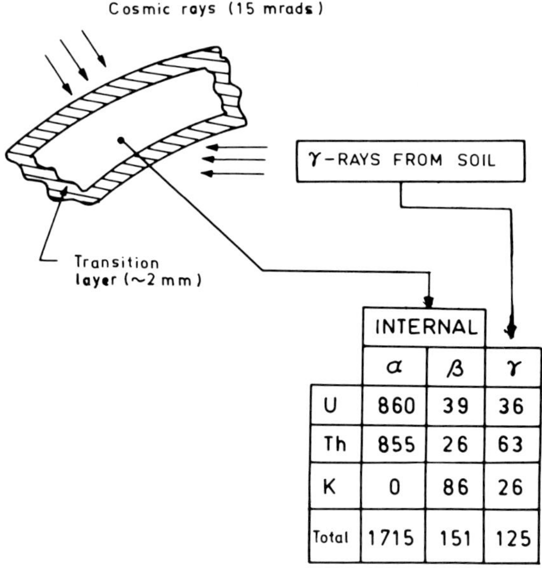

	α	β	γ
U	860	39	36
Th	855	26	63
K	0	86	26
Total	1715	151	125

Figure 8.25. Annual dose (mrad) for a pottery fragment with typical radioactive content (U, 3 ppm; Th, 12 ppm; K, 1%) buried in soil of the same radioactivity[94].

impurities, possesses a much higher TL response, and the alpha exposure from the burial soil and the clay matrix of the sherd affects only marginally the thin outer skin of the quartz grain. This skin can be easily removed by etching the grain with HF acid. The glow curve for this sample (heated in a nitrogen atmosphere) is recorded and may be ascribed to the accumulation of natural doses of radiation (β, γ and cosmic rays, the response of quartz minerals to β rays is nearly identical with its γ ray response). Call the area under the glow peak G_n. If the sample is now exposed to gamma rays from an artificial ^{60}Co source to a dose D_a equal to 1000 rad (say) the corresponding glow curve area is G_a. Then the naturally accumulated dose D_n over time τ (in years) is given by

$$D_n = D_a \, (G_n/G_a) \tag{8.11}$$

and

$$\tau = D_n/R_n \tag{8.12}$$

where R_n is the natural radiation dose rate (per year) received by the quartz inclusion from the surroundings as well as from the clay matrix of the sherd.

Before breaking the sherd to obtain inclusions, its surface layer, at least 1 mm thick, is removed. This is the transition layer (Figure 8.25) which may have had its TL affected by exposure to sunlight during some period of its history. After this layer is removed, all the subsequent handling of the sherd samples is done in dim red light.

The quartz inclusion method has recently been used by Takamiya and Nishimura[96b] to determine TL dates of volcaniclastic materials. Accurate TL measurements of quartz separated from these materials have been possible by using the photon counting technique and microcomputer controlled experimental set-up. Volcaniclastic materials provide an unique time marker in sediments. The ages of the samples dated by them are in the range $15\text{-}120 \times 10^3$ years.

For the estimation of annual dose, we need to determine the internal dose (which is β dose only; the α dose is excluded) originating within the sherd on account of the presence of its content of uranium, thorium and potassium, and the external dose (the γ dose) coming from the burial site of the pottery. Ichikawa and Nagatomo[97] have determined these two doses by using a $CaSO_4$:Tm TL dosemeter (Matsushita Electric, Japan) which has high sensitivity for both β and γ radiation. A dosemeter grain of the same size as the inclusion grain, is enclosed in a 40 μm thick polythene bag to shield it from the alpha particles. The bag containing the dosemeter is embedded in a clay matrix of the pottery under investigation. This fabrication is placed for 50 days inside a lead box with a 5 cm thick wall so as to shield its contents from any external radiation. This dosemeter is already calibrated for β and γ ray dose input and TL output, and therefore this dosemeter set-up measures the β dose which comes from the radioactive contents of the pottery. For measurement of cosmic ray and γ ray dose from the surrounding earth, a 2 mm thick copper capsule containing a $CaSO_4$:Tm dosemeter grain (copper shields the dosemeter from beta and alpha radiation) is buried for 50 days at the centre of a 30 cm radius sphere of soil excavated from the site from where the pottery was recovered. From the observed TL output we can find the annual γ and cosmic ray dose. Some results are tabulated in Table 8.7.

There are limitations to the inclusion method. As enough grains are needed to obtain a proper average of TL ages, larger pottery samples are required. The sample preparation also is not entirely simple. Further, as the alpha dose contribution is excluded, the dependence on soil gamma dosage becomes more important and involves a dependence on the conditions of the burial sites. There are, however, other environmental factors to be considered, for example the potsherd should have been buried to a depth of at least 30 cm during most of its antiquity.

In the fine grain method, a piece of pottery sherd is broken either by the method outlined above or by scrapping with a metal scrapper[98] by taking care that larger size grains of quartz are not broken to avoid mixing them with the fine grains. Grains in the size

Table 8.7. TL dates of some Japanese pottery sherds[97].

Samples, identification numbers		Accumulated dose (rad)	Annual dose $(rad.y^{-1})$	TL age (y)	Archaeological age (y)
Jomon	1	492	0.192	2560	BC 590
Jomon	2	487	0.213	2290	BC 310
Haji	1	342	0.214	1600	AD 380
Haji	2	358	0.212	1690	AD 280
Haji	3	351	0.214	1640	AD 330
Haji	4	375	0.230	1630	AD 340
Haji	5	387	0.231	1680	AD 300
Haji	6	350	0.225	1560	AD 420

range 1-5 μm are separated from the crushed mass by a sedimentation technique. These fine grains have not caused attenuation of α particles and therefore the α dose can also be evaluated in addition to β and γ doses by this method. Grains of the desired size are obtained by their settling rate in acetone and are allowed to deposit on 1 cm diam. and 0.010 inch thick aluminium discs. Grains smaller than 1 μm are discarded because of the larger amount of spurious TL in them. About six samples (with an average thickness of 4 μm and 1 mg mass of the sample layer) are prepared by evaporating acetone. The rest of the clay material (about 2 g) is ground with pestle and mortar and used for radioactivity measurements. Because the TL sensitivity of the sample may change after the heating cycle for reading its natural TL, the artificial dose accumulation must be carried out on the other remaining samples. Thus, natural thermoluminescence is measured on one, the natural plus beta induced TL on a second, and natural plus alpha induced TL on the third sample. The beta and alpha sensitivities are obtained by subtraction. ^{210}Po is used as an artificial source of α radiation while ^{90}Sr (0.5 Ci) is the β source. Age is determined from the formula

$$\text{Age (years)} = \frac{G_n}{(G_\beta \, D_{\beta\gamma}) + G_\alpha \, D_\alpha} \tag{8.13}$$

where G_β and G_α are the artificially induced TL glow curve areas per rad of beta and alpha particles respectively and $D_{\beta\gamma}$ and D_α are the annual dose rates of β, γ and α particles. It is assumed that TL sensitivity for β and γ rays is equal.

Annual doses for alphas and betas can be determined by the estimation of uranium, thorium and potassium content in sherds by neutron activation analysis. Alternatively, the α dose may be found by scintillation counting of alpha activity from a layer of powdered sample using ZnS scintillator. The β dose may be determined by a TL dosemeter as in the inclusion technique.

8.9.4. *The dating peak and plateau test*

During antiquity, the thermal decay of carriers even from deeper traps can take place. For this reason, it is observed that in natural samples there is almost no glow at temperature below 300°C whereas the artificially irradiated sample of the sherd may show considerable glow in this region. The glow in the 400°C temperature region is related to the traps filled during the period of antiquity and

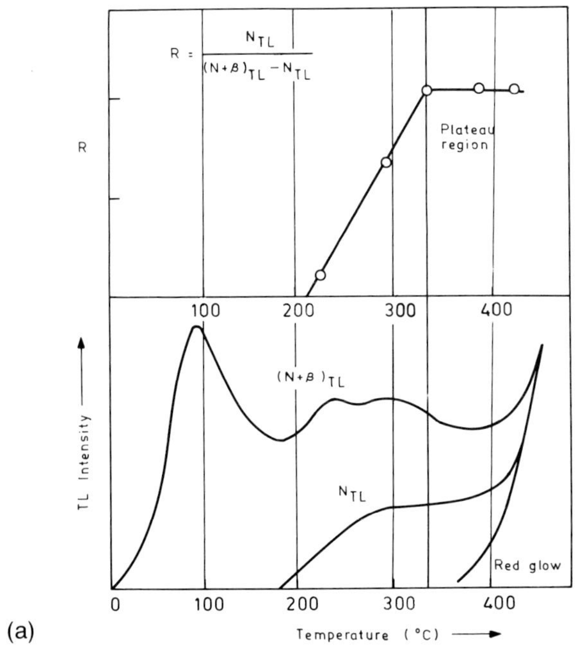

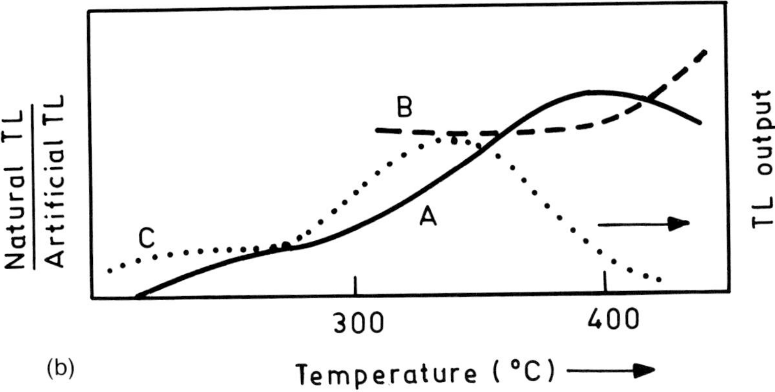

Figure 8.26. (a) Plateau test. N_{TL} = natural TL. $(N+\beta)_{TL}$ = (Natural + β dose) TL[99]. (b) Plateau test for a quartz extract from a geological sample. Sample A, without annealing. Sample B, annealed at 280°C for 150 s. Dotted curve C is the natural TL glow curve[101].

the ones that have remained stable. As first pointed out by Aitken, if we therefore plot the ratio of the natural glow of the antique sample to that of the artificially irradiated one (of the same sherd) as a function of temperature, then the ratio starting from a zero value will attain a saturation (plateau) in the temperature region 300-400°C, as shown in Figure 8.26(a)[99]. The occurence of a plateau, therefore, establishes that we are actually dating the antique sample. This is known as the plateau test.

Quartz, in this temperature region, exhibits two glow curves, one peaked at 325°C and the other one at 375°C. Frequently the presence of the 325°C peak interferes in such a way that the plateau is not observed, the fact that the sample does not exhibit any fading or spurious TL notwithstanding. Fleming[100] has shown that the suitable peak to use for dating is the 375°C peak as its growth curve as a function of dose is not altered by the first heating of the sample. Valladas[101] has shown that the interfering glow peak at 325°C can be erased by annealing the sample for 150 s at 280°C. This aspect is schematically depicted in Figure 8.26(b). The 'hidden' plateau, after this thermal annealing, is unfailingly revealed.

8.9.5. *Pre-dose technique for archaeological dose estimation*

Fleming[102] has analysed the sensitivity enhancement of the quartz grains as a result of giving them a fixed dose of β radiation and their subsequent heating to 500°C (or vice versa). The quartz grains after some irradiation (by say 1 rad of β radiation) exhibit a strong glow peak at 110°C. Pre-dosing and heating enhances substantially the sensitivity of this peak and by making measurements on the enhancement of this sensitivity, known as the pre-dose technique, Fleming has determined the archaeological dose accumulated by the quartz grains of the sherds during burial. Figure 8.27 shows the occurence of a prominent glow peak at 110°C in a nominally irradiated quartz sample. The advantage of this technique is that while only the old samples (500 years or more) can be dated by utilising the 375°C glow peak for archaeological dose determination as the light output in younger samples will be unreasonably small, we can date samples of more recent origin because the 'pre-dose' increase in sensitivity of the 110°C glow peak is quite high (up to about $28\%.\mathrm{rad}^{-1}$ in quartz samples extracted from terracotta). Archaeological dose estimation can be made by either of the following procedures:

(i) The sample is pre-dosed by a laboratory source of β radiation and then heated to 500°C. Two portions of quartz extract, ideally of equal weight, are required.

(ii) The sample is heated and then pre-dosed by a laboratory source of β radiation. Only one quartz sample is needed. As the condition of ideally equal weights of two samples in (i) is not fully met, the latter procedure for making a dose estimation is preferable.

In procedure (i), a test dose (beta radiation) of 1 rad is given to one of the two quartz grains, and it is given a heat cycle to record the 110°C glow peak. The peak height is noted which is, say, S_o. This is the sensitivity per rad. A pre-dose of β ($= 335$ rad) is then added to this sample followed by 500°C heating. When the sample is cooled, a test dose of 1 rad is given and the TL is recorded. Let the new height of the 110°C glow peak be $S_{N+\beta}$ (we find that $S_{N+\beta} >> S_o$). The second grain is heated to 500°C and when cooled, is given a 1 rad test dose.

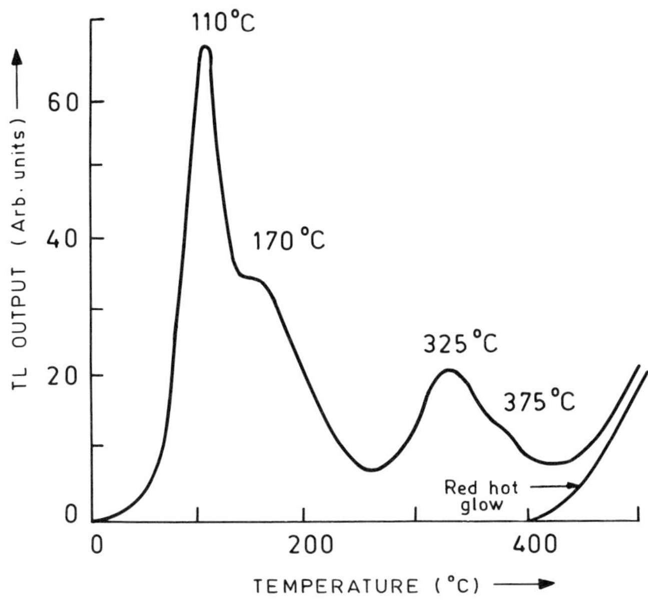

Figure 8.27. Glow curve of an irradiated quartz[102].

The corresponding 110°C glow peak height is designated as S_N ($S_N < S_{N+\beta}$). The sensitivity enhancement rate p (per rad) is defined as

$$p = \frac{S_{N+\beta} - S_N}{S_o} \frac{1}{\beta} \tag{8.14}$$

where β is in rads. The accumulated archaeological dose D_A in rad is now given by

$$D_A = \frac{S_N - S_o}{pS_o} \tag{8.15}$$

In the second procedure, S_o is determined as before. The sample is then heated to 500°C, allowed to cool and then given a test dose of 1 rad. The height of the 110°C glow peak gives, as before, S_N. The sample is reheated to 500°C, allowed to cool and then receives the test dose. This determines S'_N (say). Now a pre-dose of $\beta (=335$ rad) is applied to this sample followed by a 500°C heating. After cooling, the test dose is given and the 110°C glow curve recorded. The glow peak height now determines $S'_{N+\beta}$. The equality of S_N and S'_N shows that heating alone does not give rise to sensitivity enhancement q (per rad) which, in this procedure, is given by

$$q = \frac{S'_{N+\beta} - S'_N}{S_o} \frac{1}{\beta} \tag{8.16}$$

and

$$D'_A = (S'_N - S_o)/(q\, S_o) \tag{8.17}$$

It is observed that usually, as expected, D_A and D'_A are about equal. However D'_A is regarded as the more reliable, in view of the unfeasibility of stringent requirement of equal grain weights in procedure (i).

The β dose mentioned is not unique but is one that corresponds to the onset of saturation in the value of $S'_{N+\beta}$ in the sample. The premise of the pre-dose technique rests on the heating to 500°C as a simulation of the pottery firing in ancient times and the laboratory pre-dose as a simultation of the accumulated dose by the sample during the period of antiquity, causing enhancement in sensitivity of the 110°C peak.

8.9.6.   **Subtraction technique**
Fleming and Stoneham[103] have described the so-called

subtraction technique of combining measurements for accumulated doses made on both the quartz inclusions and the fine grains extracted from the pottery sherd. If A, B and G are doses accumulated by the quartz grains due to alpha, beta (from the uranium, thorium and ^{40}K content of the clay matrix of the fired pottery) and gamma (from the surrounding burial soil) and F, I are the total doses determined in the fine grains and inclusions and α, β are the annual dose rates for alpha and beta activity, then

$$\left.\begin{array}{c} A + B + G = F \\ fB + G = I \end{array}\right\} \qquad (8.18)$$

where the factor f accounts for the attenuation of β radiation due to grain size (f lies typically between 0.90 to 0.95) and the α contribution is negligible in the case of quartz grains,

$$F - I = A + B\,(1-f)$$

It is also reasonable to assume that

$$(A/B) = (\alpha/\beta)$$

hence

$$F - I = A\,[1+(1-f)\,(\beta/\alpha)]$$

and

$$T(\text{age}) = (A/\alpha) = \frac{(F-I)}{[\alpha+(1-f)\beta]} \qquad (8.19)$$

The annual dose rates α and β can be determined from the radioactive analysis of the pottery clay[104,105]. The advantage of the subtraction technique is that since G measurement is not involved, the age determination is not dependent on the conditions of burial soil and samples even from archaeological museums, for which no burial soils are stored, can also be dated.

8.9.7.   ***Dating of meteorites***

In some ways the basic features of TL dating in meteorites are not necessarily similar to those in the archaeological material[106]. The exclusive reason is the fact that meteorite ages are much greater (10^4 years and above). As a result the anomalous TL fading becomes too significant to be ignored. (The drainage of traps, which is not accounted for by the kinetics equations of Randall and Wilkins and is only weakly dependent on temperature , takes place, leading to the loss of TL which is known as anomalous fading. It is understood that the tunnelling of carriers from the traps and their migration to the recombination centres gives rise to this fading.) The active material in meteorites is feldspar. The problem associated with chronology in meteorites actually relates to their period of orbiting in space, their nearness to the sun (i.e. the orbiting temperature) and exposure of the meteorites to space radiation causing trap filling.

The other important information likely to emerge from TL studies in meteorites is, therefore, the temperature of the meteoroid (parent body of which the meteorite reaching the earth was once a part) and the consequent bearing on its orbiting history in space; as well as the shape and mass of the meteorites before their entry into the atmosphere[107-109]. As the trap filling rate is proportional to the dose rate and the TL intensity is proportional to $-d\eta/dt$ (see section 8.2.2), this forms the basis of archaeological dating and radiation dosimetry by the thermoluminescence technique. It is further understood that the levels of exposure to radiation are far below the saturation limit of TL($\eta=N$), because any date after the traps have all been filled will not be registered. For meteorites, the saturation levels appear to be reached with doses $\approx 10^6$ rad which, considering an average exposure of 1 rad per year, corresponds to exposure ages in excess of 10^6 years. Christodoulides *et al*[110] and Durrani and Christodoulides[111] have used archaeological methods to determine the ages of some stony meteorites and have fixed 10^7 years as an upper limit of the exposure age.

There is also the anomalous thermal fading in meteorites. When longer periods are involved, the thermal fading of TL corresponding to even the deeper traps (responsible for higher temperature regions of the glow curves) may become significant. The thermal de-trapping of these traps will, therefore, also take place along with their filling due to irradiation. In the case of a meteorite of sufficient age, it may become possible that the rate of trap filling ($-dn/dt$) equals the rate of thermal de-trapping so that an equilibrium condition will be reached in which the TL no longer increases with the radiation dose although the saturation level has not yet been reached. This state of equilibrium will, naturally, depend upon the E and s values characterising the trap as well as r and T_r. Thus at equilibrium

$$A(N-\eta)r = s\eta \, \exp(-E/k_{\mathrm{B}}T_r)$$

which gives (when $\eta = \eta_{\mathrm{eq}}$)

$$T_r = (E/k_{\mathrm{B}})/\ln \left(\frac{s}{Ar[(N/\eta_{\mathrm{eq}})-1]} \right)$$

where T_r denotes the temperature at which the meteorite is being irradiated. As the value of T_r is related to the orbit described by a meteoroid during the past few thousand years, the importance of knowing T_r is evident. Before applying the TL method, however, it is important to decide whether the TL in a meteorite sample at T_r is in equilibrium or still building up with dose. For this, it is to be noted that, under the prevailing conditions of temperature and radiation exposure in the original orbits of the parent meteoroid (representing a typical distance of the parent body from the sun as 2.5 astronomical units, ($1\mathrm{a.u} = 1.5 \times 10^{11}\mathrm{m}$), the TL levels are saturated. But, if the event which perturbed the meteoroid from its original orbit to the earth intersection orbit is also assumed to have fragmented it, then accompanying shocks and temperatures may have possibly drained off the TL, thus setting the TL clock to zero. In the new orbit the exposure levels and temperatures are likely to cause equilibrium TL in about 10^5–10^6 years, which are the typical exposure ages found for many meteorites.

The terrestrial age of meteorites is the number of years elapsed since the meteorite fell on earth. Many meteorites are observed when falling to the earth and are entered into records. These are termed 'falls'. There are also those which are 'finds' and, as such, their terrestrial ages are not known. Although these ages have been determined by radiocarbon dating methods[112], thermoluminescence dating is also quite interesting and reliable[113,114]. The TL method consists of determining the residual TL signal in the high temperature region ($\sim 400°C$) of the glow curve of the meteorite samples ('falls') and correlating this with the known age. A reasonably well defined correlation is observed (Figure 8.28). When a meteorite falls to the earth it receives negligible cosmic ray dose on account of its shielding by the earth's atmosphere. The accumulated TL will now, therefore, decay slowly during the terrestrial residence of the meteorite. The results obtained by McKeever and Sears[113] are shown in Figure 8.28. The high temperature TL intensity seems to correlate strongly with the terrestrial ages of the meteorites – 'falls' as well as 'finds'. The radiocarbon dating results in the case of Estacado meteorite (the names indicate their terrestrial locations; Estacado is in Hale county, Texas, USA, where this meteorite was found) seem to be 'anomalous' in view of its high TL intensity. The recorded evidence of its fall, found subsequently, suggests a more recent terrestrial arrival, nearly the same as would be predicted by the correlation of Figure 8.28. This rather points towards the better reliability of TL methods in determining terrestrial age.

8.9.8. *Dating of geological materials*

There are essentially three types of rock formations: igneous rocks, metamorphic rocks and the sedimentary rocks. Of

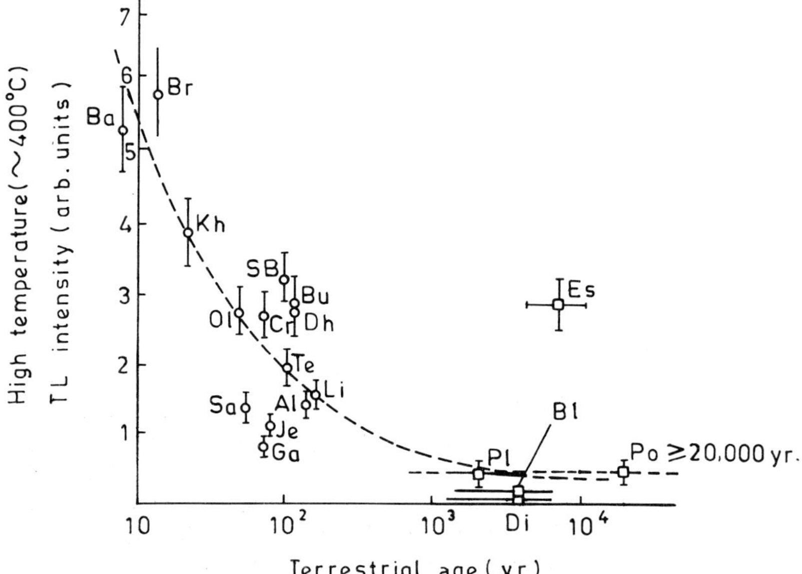

Figure 8.28. TL intensity and terrestrial age correlation. ○, observed 'falls' (precise ages are known from records). □, 'finds' (ages determined from radiocarbon dating). The dashed curve is an estimation of the decay curve. Key: Ba, Barwell; Br, Bruderheim; Kh, Khanpur; SB, Soko Banja; Bu, Butsura; Ol, Olivenza; Cr, Crumlin; Dh, Dhurmsala; Es, Estacado; Te, Tennasilm; Li, Limerick; Al, Aldsworth; Sa, Saratov; Je, Jelica; Ga, Gambat; Pl, Plainview; Bl, Bluff; Di, Dimmit; Po, Potter. The radiocarbon age of Estacado (Es) is shown, which, for its corresponding TL intensity, is anomalous.

these only the sedimentary rocks are of recent origin and as such are amenable to TL dating[115]. The other two varieties are very old and saturation effects will not permit correct age estimation. Sedimentary rocks and ocean sediments contain thermoluminescent rock minerals which are mostly calcite, quartz, dolomite, aragonite and magnesite[116]. Mechanical breaking of grains, exposure to ultraviolet from sunlight and surface temperatures during weathering of rocks may be considered as the zeroing mechanisms of the TL clock in the rock-forming minerals. When these zeroed surfaces are buried under subsequent sediment layers, the presence of uranium, thorium and ^{40}K radioactivity in the rock matrix slowly builds up the TL. From the measurement of natural TL and the rate of annual dose received by the samples (the silt attached to the sediment almost entirely gives the observed TL)[117-119]. Figure 8.29 illustrates the possible main role of sunlight exposure of a sedimentary sample in zeroing the TL[120]. The event being dated in these rocks is, therefore, the last exposure to sunlight. The natural dose (n) in the sediment samples, has been determined by Wintle and Huntley[117] from the measurement of reduction in TL due to standard sun-lamp exposure (for 15 min) as a function of additive gamma dose. Different samples receive varying γ doses in addition to the natural dose. The reduction, R, in TL after these are given a sun-lamp exposure for 15 min is determined. A

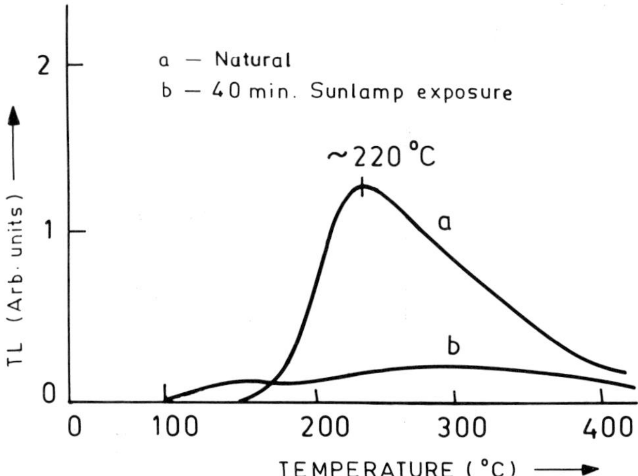

Figure 8.29. TL glow curves of an Antarctic ocean sample taken from a depth of 580 cm below the deep sea core (water depth 4330 m).

schematic plot between R and γ dose is shown in Figure 8.30, from which the natural dose (the gamma equivalent dose) is found, as shown in the figure. The age in years is, as in case of pottery, given by

$$\tau = \frac{n}{\text{dose rate per year}}$$

The annual dose contribution giving rise to the TL acquired by the samples come from the decay of ^{40}K and the decay chains of ^{238}U and ^{232}Th found in the sediment. Additional contributions to TL activation come from the decay of excess $^{230}_{90}$Th and $^{231}_{91}$Pa which are continuously precipitated from the ocean[121]. The annual dose determination is rather involved owing to the time dependence of dose rate, but has been outlined by Wintle and Huntley[117,118].

8.9.9. *Dating of sand dunes*

The same principles are applicable for finding the ages of sand-dunes in deserts[9,99]. In fresh sand-dunes the sun exposure erases any accumulated TL, which starts building up again as soon as it is buried under new sand as time passes. The accumulated TL will, in turn, be a function of the depth below the surface of the desert at a given location. The depth profile of the TL age will then provide a history of movement of the sand-dunes as well as throwing light on the dynamics of desert advancement.

The natural dose has been measured in a different way in these

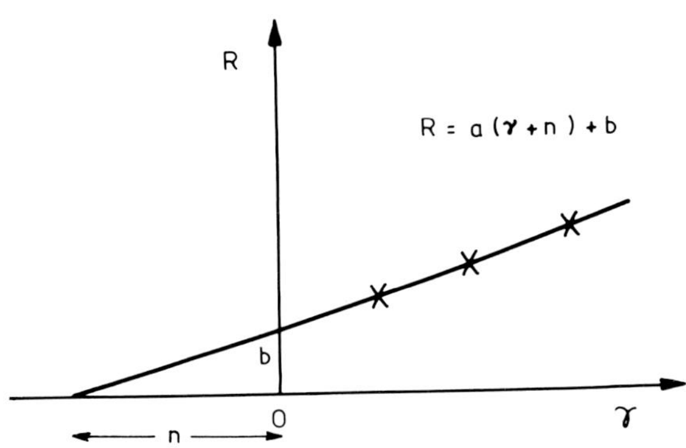

Figure 8.30. *R–γ plot to determine equivalent natural dose in ocean sedimental rocks.*

samples. This approach, outlined below, seems to circumvent the problem of the possible changes in the sensitivity of samples after the sun-lamp bleach. The natural dose (n) determined here is the equivalent of the β dose (β_n) and not the gamma dose as in case of ocean sediments. It is given as

$$n = [1-(I_b/I_n)]\beta_n \qquad (8.20)$$

where β_n is the β dose that gives the same TL intensity as the natural TL (N_{Tl}) and is determined from the additive β dose procedure (Figure 8.31); I_b is the TL intensity after the sample having N_{Tl} is given a 1000 min sun-lamp bleach (from a 300 W lamp at 30 cm from the samples); and I_n is the natural glow curve intensity (N_{Tl}). The measurement of I_b and I_n is illustrated in Figure 8.32.

8.9.10   *Dating of biological materials*

Jasinska and Niewiadomski[122] first observed TL in fossil bones and suggested the possibility of dating them. Earlier, low temperature peaks in archaeological dental samples were reported by Wilhoit and Poland[66], and Driver[123] observed TL in irradiated thin slices of bone and marine shell. Benko and Kozorus[124,125] have dated fossil teeth, which is particularly successful on account of the

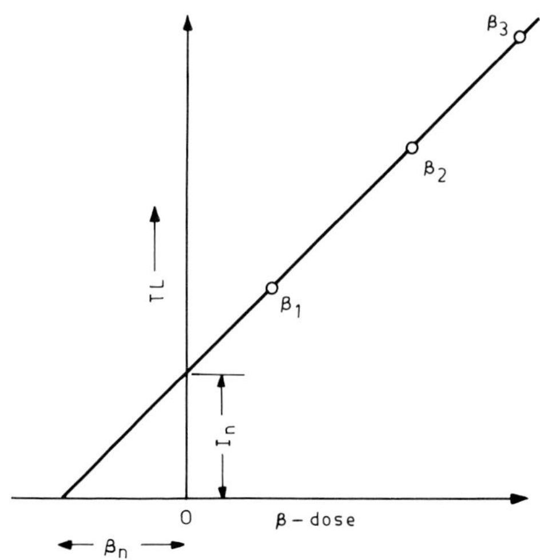

Figure 8.31. Additive dose procedure for determining β_n in the natural sample.

low content of organic matter and, as a result, a low level of spurious TL. The main mineral constituent of bone and teeth is hydroxyapatite which is TL sensitive. Because the glow curve peak in these samples appears at around 70°C, the only possibility is to use the pre-dose technique of pottery dating in order to determine the archaeological dose in the dental samples. The TL mineral is extracted from the tooth using the following procedure:

1. After removal of the roots, the teeth are rinsed in ethanol.
2. The material, after etching in 30% phosphoric acid to remove any light-induced TL, is crushed in low red light conditions and grains of 30 to 50 μm size are taken.
3. The samples are dried in vacuum at ambient temperature.

The test dose employed in determining the archaeological dose D_A in dental enamel grain samples (10 mg each spread on a Pt-Ir planchet) extracted from human teeth a few centuries old was 0.01 Gy (3.7×10^8 Bq, ^{90}Sr source of β radiation) while the pre-dose was 2.4 Gy. The second sample is pre-heated to 200°C for determining S_N. Glow curves, shown in Figure 8.33, were recorded in a nitrogen atmosphere. The annual dose rate has to be found in the usual manner, in the surrounding soil of the burial site.

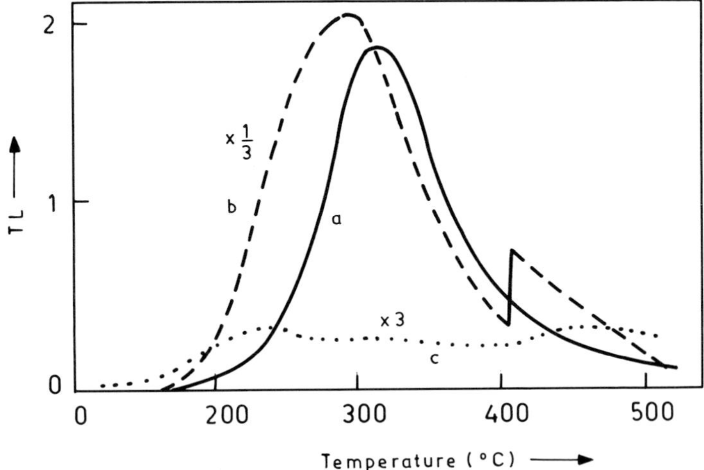

Figure 8.32. Typical glow curves for 90-100 μm quartz fraction of a sand-dune sample. (a) natural TL (N_{TL}). (b) N_{TL}+5 krad β exposure. (c) natural TL bleached with a 10^3 min sun-lamp exposure.

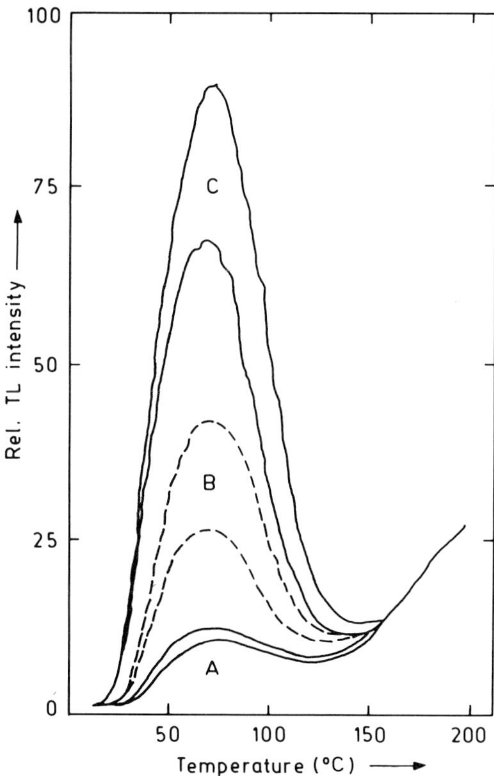

Figure 8.33. Sensitivity enhancement; S_N (lower curves in A, B and C) and $S_{N+\beta}$ (upper curves) of the 70°C peak in dental enamel samples of archaeological origin: A, 19th century AD; B, 14th; C, 10th. Archaeological doses are 4.45, 4.03 and 6.79 Gy, respectively[125].

References

1. Cameron, J. R., Suntharalingam, N. and Kenney, G. N. *Thermoluminescent Dosimetry*. (University of Wisconsin Press) (1968).
2. McKinlay, A. F. *Thermoluminescent Dosimetry*. (Bristol: Adam Hilger) (1981).
3. Oberhofer, M. and Scharmann, A. (eds) *Applied Thermoluminescence Dosimetry*. (Bristol: Adam Hilger) (1981).
4. McLaughlin, W. L. (ed) *Trends in Radiation Dosimetry*. (Oxford: Pergamon Press) Int. J. Appl. Radiat. Isot. **33**, (11) (special issue) (1982).

5. Horowitz, Y. (ed) *Thermoluminescence and Thermoluminescent Dosimetry.* Vol. II (Florida: CRC Press) (1984).

6. Mahesh, K. *Thermoluminescence Dosimetry in Radiation Protection and Control.* Joint Symp. on Radiation Safety and Dosimetry, University of Rome, Italy, 6-8 June 1984. Nucl. Sci. J. (Taiwan) **22**, 211 (1985).

7. Aitken, M. J. *Physics and Archaeology.* (Oxford: Clarendon Press) (1974).

8. Fleming, S. J. *Thermoluminescence Technique in Archaeology.* (Oxford University Press) (1980).

9. Aitken, M. J. *Thermoluminescence Dating.* (New York: Academic Press) (1985).

10. PACT. J. Eur. Study Group on Physical, Chemical and Mathematical Techniques Applied to Archaeology **9** (I and II) (1983).

11. Singhvi, A. K., Sharma, Y. P. and Agrawal, D. P. Nature **295**, 313 (1982).

12. Proc. SPIE **96**, 158 (1976).

13. ICRP. *Recommendations of the International Commission on Radiological Protection.* Publication 26 (Oxford: Pergamon Press) (1977).

14. Piesch, E. Kerntechnik, **20** (8/9), 391 (1978).

15. Greenslade, E. and Marshall, T. O. J. X-Ray Technol. **1**, 2 (1981).

16. DeWerd, L. A. IN Thermally Stimulated Relaxation in Solids. ed. P. Braunlich (Heidelberg: Springer-Verlag) p. 275 (1979).

17. Bjarngard, B. E., McCall, R. C. and Berstein, I. A. IN Luminescence Dosimetry, ed. F. H. Attix. AEC Symp. Ser. No. 8 (Springfield, VA: NTIS) (1967).

18. Attix, F. H., West, E. J., Nash, A. E., Gorbics, S. G. and Johnson, T. L. NRL progress report (Office of Naval Research, Washington, DC) (1968).

19. Pradhan, A. S. Bull. Radiat. Prot. **1**, 10 (1978).

20. Attix, F. H. Health Phys. **22**, 287 (1972).

21. Becker, K. Health Phys. **23**, 734 (1972).

22. Schulman, J. H., Attix, F. H., West, E. J. and Ginther, R. J. Rev. Sci. Instrum. **31**, 1263 (1960).

23. Jones, A. R. Health Phys. **12**, 663 (1966).

24. Bjarngard, B. and Jones, D. *Nato Summer School on Dosimetry, Brussels* (1967).

25. Marinello, G., Boret, C. and LeBourgeois, J. P. Nucl. Instrum. Meth. **175**, 198 (1980).

26. Busuoli, G. and Julius, H. W. IN Proc. 5th Int. Conf. on Luminescence Dosimetry, Sao Paulo, Brazil (1977).

27. Benkö, L., Uchrin, G. and Biro, T. IN Proc. 4th Int. Cong. International Radiation Protection Association, Paris, 24-30 April 1977, Vol. 4.

28. Lakshamanan, A. R., Chandra, B., Pradhan, A. S., Kher, R. K. and Bhatt, R. C. Int. J. Appl. Rad. Isot. **31**, 107 (1980).

29. Piesch, E. IN 3rd Information Seminar on the European Radiation Protection Dosimeter Intercomparison Programme, Grenoble, France 6-8 October 1980.

30. Pearson, D. W. and Moran, P. R. United States Energy Research and Development Authority (USERDA) Technical Report, COO-1105-227 (1976).

31. Pradhan, A. S., Bhatt, R. C., Lakshmanan, A. R., Chandra, B. and Shinde, S. S. Phys. Med. Biol. **23**, 723 (1978).

32. Piesch, E. and Burgkhardt, B. Nucl. Instrum. Meth. **175**, 180 (1980).

33. Griffith, R. V., Hankins, D. E., Gammage, R. B., Tommasino, L. and Wheeler, R. V. Health Phys. **36**, 335 (1979).

34. Lakshmanan, A. R. Nucl. Tracks Radiat. Meas. **6**, 59 (1982).

35. Lakshmanan, A. R. and Bhatt, R. C. Int. J. Appl. Rad. Isot. **28**, 665 (1977).

36. Tymons, B. J. and Tuyn, J. W. N. Health Phys. **32**, 447 (1977).

37. Becker, K. and Abd-el-Razek, M. Nucl. Instrum. Meth. **124**, 557 (1975).

38. a Wingate, C. L., Tochilin, E. and Goldstein, N. J. X-ray Technol. **1**, 421 (1981).

38. b Kalef-Ezra, J. and Horowitz, Y. S. Int. J. Appl. Radiat. Isot. **33**, 1085 (1982).

39. Aitken, M. J., Tite, M. S. and Fleming, S. J. J. X-ray Technol. **1**, 490 (1981).

40. Lakshmanan, A. R., Bhatt, R. C. and Supe, S. J. J. Phys. D: Appl. Phys. **14**, 1683 (1981).

41. Smith, A. R. *et al*, Med. Phys. **4**, 408 (1977).

42. Baarli, J., Bianchi, M., Hill, C. K., Sullivan, A. H. and Tuyn, J. W. N. Radiat. Environ. Biophys. **16**, 283 (1979).

43. Cooke, D. W. and Hogstrom, K. R. Phys. Med. Biol. **25**, 657 (1980).

44. Cooke, D. W. and Hogstrom, K. R. Health Phys. **39**, 568 (1980).

45. Lowder, W. M. and de Planque Burke, G. *The Response of LiF-TLD to Natural Environmental Radiation*. USERDA Report HASL-313 (1977).

46. Niewiadomski, T., Koperski, J. and Ryba, E. Health Phys. **38**, 25 (1980).

47. Hsu, P. C. and Chen, M. N. Nucl. Sci. J. (Taiwan) **12**, 29 (1975).

48. Weng, P. S. Nucl. Sci. J. (Taiwan) **12**, 61 (1975).

49. Wachsmann, F. and Regulla, D. F. IN Proc. Symp. on the Natural Radiation Environment, Houston, 23-28 April 1978, eds. T. F. Gesell and W. M. Lowder, Vol. 2, p. 1022 (US DOE, Washington) (1980).

50. Ichikawa, Y., Higashimura, T. and Sidei, T. Health Phys. **12**, 395 (1966).

51. Attix, F. H ., West, E. J. and Price, W. E. NRL Report 5938 (US Naval Research Laboratory, Washington, DC) (1963).

52. Feher, I., Deme, S., Szabo, B., Vagvolgyi, J., Szabo, P. P., Csoke, A., Ranky, M. and Akatov, Yu. A. Adv. Space Res. **1**, 61 (1981).

53. Akatov, Yu. A. Adv. Space Res. **1**, 67 (1981).

54. Hsu, P. C. and Weng, P. S. Health Phys. **31**, 522 (1976).

55. Wachsmann, F. and Regulla, D. F. Kerntechnik **7**, 318 (1978).

56. IAEA. *Biomedical Dosimetry: Physical Aspects, Instrumentation, Calibration.* STI/PUB/567 (Vienna: IAEA) (1981).

57. Regulla, D. F., Schurmann, G. and Suess, A. IN Radiation for a Clean Environment. STI/PUB/402 (Vienna: IAEA) p. 465 (1975).

58. Weng, P. S., Hsu, P. C., Tsai, F. L. and Huang, C. C. IN Proc. 15th Int. Congr. of Radiology, Brussels, Belgium (1981).

59. Furetta, C. and Gennai, P. Health Phys. **41**, 674 (1981).

60. Scarpa, G., Moscati, M. and Furetta, C. Br. J. Radiol. **52**, 75 (1979).

61. Bojtor, I. *et al* Acta Radiol. **13**, 32 (1974).

62. Bojtor, I. *et al* Acta Radiol. **16**, 341 (1977).

63. Regulla, D. F. IN Proc. 3rd Int. Congress of Maxillofacial Radiology, Kyoto, Japan (Japan Science Press) p. 191 (1974).

64. McKlveen, J. W. J. Am. Ind. Hyg. Assoc. **41**, 864 (1980).

65. McKlveen, J. W. Health Phys. **39**, 211 (1980).

66. Wilhoit, D. G. and Poland, A. D. Health Phys. **15**, 91 (1967).

67. Watanabe, S. and Nakajima, T. J. Nucl. Sci. Tech. (Japan) **10**, 202 (1973).

68. Watanabe, S. and Nakajima, T. J. Nucl. Sci. Tech. (Japan) **11**, 575 (1974).

69. Grey, L. J. and Bolt, C. Phys. Med. Biol. **23**, 759 (1978).

70. Simon, J. and Osvay, M. *Dosimetry of the KOLOS seed Irradiator.* Report of Budapest Institute of Isotopes, Budapest, Hungary (1979).

71. Weng, P. S. and Furetta, C. IN Techniques of Radiation Dosimetry, eds K. Mahesh and D. R. Vij (Chichester: John Wiley) (1985).

72. Weng, P. S. Hoken Butsuri (Japan) **2**, 160 (1967).

73. Shaw, D. Contemp. Phys. **17**, 307 (1976).

74. Mahesh, K. J. Forensic Med. Toxicol. **1**, 56 (1984).

75. Andresko, J. Forensic Sci. Int. **17**, 235 (1981).

76. Mesh, S., Palma, G., Moauro, A. and Capannesı, G. Arch. Kriminol. **161**, 137 (1978).

77. Salares, V. R., Eves, C. R. and Carey, P. R. Forensic Sci. Int. **14**, 229 (1979).

78. Sekharan, P. C. IN Proc. Natl Symp. on Thermoluminescence and Applications, Kalpakkam, Madras, India, 12-15 Feb. 1975, p. 638.

79. Ingham, J. D. and Lawson, D. D. J. Forensic Sci. **18**, 217 (1973).

80. Aramu, F., Maxia, V. and Rucci, A. J. Lumin. **3**, 438 (1971).

81. Wang, Y. S. Health Phys. **28**, 78 (1975).

82. IAEA. *Basic Safety Standards for Radiation Protection.* Safety Series no 9 (Vienna: IAEA) (1967).

83. Hsu, P. C., Weng, P. S. and Tseng, C. L. Int. J. Appl. Rad. Isot. **27**, 725 (1976).

84. Placido, F. Mag. Concrete Res. **32**, 112 (1980).

85. Placido, F. FIRE, p. 464 (February 1981).

86. Smith, L. and Placido, F. IN Proc. American Concrete Institute Convention, San Juan, Puerto Rico, Sept. 1980.

87. Regulla, D. F., Haider, B. and Jacobi, W. Nucl. Instrum. Meth. **175**, 233 (1980).

88. Bassi, P. and Busuoli, G. Nucl. Instrum. Meth. **143**, 195 (1977).

89. Mahesh, K., Vij, D. R., Singh, N., Lal, N. and Nagpaul, K. K. Nucl. Tracks Radiat. Meas. **9**, 139 (1984).

90. Ieda, M., Mizutani, T. and Suzuoki, Y. *TSC and TL Studies of Carrier Trapping in Insulating Polymers.* Memoirs of Engineering Faculty of Nagoya University, Japan. **32**(2), 173-219 (1980).

91. DeWerd, L. A. and Moran, P. R. Med. Phys. **5**, 23 (1978).

92. SPIE, Optical Instrumentation in Medicine V, **96**, 158 (1976).

93. Ralph, E. K. and Han, M. C. Nature **210**, 245 (1966).

94. Aitken, M. J., Zimmerman, D. W. and Fleming, S. J. Nature **219**, 442 (1968).

95. Ichikawa, Y. Bull. Inst. Chem. Res. Kyoto Univ. **45**, 63 (1967).

96. a Fleming, S. J. Archeometry **12**, 135 (1970).

96. b Takamiya, H. and Nishimura, S. Nucl. Tracks Radiat. Meas. **11**, 251 (1986).

97. Ichikawa, Y. and Nagatomo, T. Bull. Inst. Chem. Res. Kyoto Univ. **53**, 11 (1975).

98. Zimmerman, D. W. Archaeometry **10**, 26 (1967).

99. Sharma, Y. P. PhD Thesis (unpublished), Kurukshetra University, Kurukshetra, India (1984).

100. Fleming, S. J. Archaeometry **12**, 133 (1970).

101. Valladas, G. IN Proc. 18th Int. Symp. on Archaeometry and Archaeological Prospection, Bonn, Germany, 14-17 March, 1978. p. 499.

102. Fleming, S. J. Archaeometry **15**, 13 (1973).

103. Fleming, S. J. and Stoneham, D. Archaeometry **15**, 229 (1973).

104. Aitken, M. J. and Fleming, S. J. IN Topics in Radiation Dosimetry, ed. F. H. Attix (New York: Academic Press) (1973).

105. Huxtable, J. and Aitken, M. J. Monograph III, J. Hong Kong Archaeol. Soc. p. 116 (1978).

106. Sears, D. W. PACT **2**, 231 (1978).

107. McKeever, S. W. S. and Seers, D. W. Meteoritics **14**, 29 (1979).

108. Sears, D. W. Earth Planet. Sci. Lett. **26**, 97 (1975).

109. Sears, D. W. ICARUS **44**, 190 (1980).

110. Christodoulides, C., Durrani, S. A. and Ettinger, K. V. Mod. Geol. **1**, 247 (1970).

111. Durrani, S. A. and Christodoulides, C. Nature **223**, 1219 (1969).

112. Boeckl, R. Nature **236**, 25 (1972).

113. McKeever, S. W. S. and Sears, D. W. Nature **275**, 629 (1978).

114. Sears, D. W. and Mills, A. A. Meteoritics **9**, 47 (1974).

115. Singhvi, A. K. and Mejdahl, V. Nucl. Tracks Radiat. Meas. **10**, 137 (1985).

116. Medlin, W. L. Phys. Rev. A **135**, 1770 (1964).

117. Wintle, A. G. and Huntley, D. J. Quarternary   Sci. Rev. **1**, 31 (1982).

118. Wintle, A. G. and Huntley, D. J. Can. J. Earth Sci. **17**, 348 (1980).

119. Berger, G. W. and Huntley, D. J. PACT **6**, 495 (1982).

120. Wintle, A. G. and Huntley, D. J. Nature **279**, 710 (1979).

121. Ku, T. L. Ann. Rev. Earth Planet. Sci. **4**, 347 (1976).

122. Jasinska, M. and Niewiadomski, T. Nature **227**, 1159 (1970).

123. Driver, H. S. T. PACT **3**, 290 (1979).

124. Benko, L. and Kozorus, L. Revue d'Archemetrie **4**, 149 (1980).

125. Benko, L. and Kozorus, L. Nucl. Instrum. Meth. **175**, 227 (1980).

126. Aitken, M. J., Fleming, S. J. and Zimmerman, D. W. IN Radioactive Dating and Methods of Low-level Counting. STI/PUB/152 (Vienna: IAEA) p. 523 (1967).

CHAPTER 9

Recent Developments in Thermoluminescence Dosimetry

P. S. Weng

9.1   Introduction

Application of TL methodology in radiation dosimetry continues to attract world-wide attention of researchers and practitioners. Proceedings of the recent international conferences[1,2] provide a detailed survey of the current development in this area. Attention is being paid to the improvement in TL materials with regard to their intrinsic response, calibration sources, equivalent dose estimation, regulatory procedures, and performance quality programmes. Laser heating of TL phosphors (see Appendix IV) has added a new dimension to the methodology of thermoluminescence and its possible ramifications in radiation dosimetry. This chapter gives an account of some typical recent developments in TL dosimetry of radiation.

9.2.   Thermoluminescence dosimetry of lasers and ultraviolet radiation

The increasing application of non-ionising radiations such as laser light and ultraviolet radiation in medicine and industry requires added attention with regard to the risk associated with their use. Techniques similar to those used in the dosimetric methods for ionising radiation are being developed and they also allow exposure limits to be measured. In some instances TLDs and other methods are used[3,5].

It is possible to use the TL methodology in laser and UV dosimetry. However, not all the TL phosphors used in estimating gamma, beta, neutron and X radiation doses are found suitable since they have poor sensitivity to lasers and UV radiation. Some TL phosphors such as $LiF:Mg,Ti$, $CaSO_4:Tm$ and natural CaF_2 have been made sensitive to lasers and UV radiation with, in some cases, pre-irradiation to high gamma doses and/or partial annealing. The preparation of these TLD materials and the mechanism of TL emission are rather complex and time consuming. These reasons tend to limit any wider use of such phosphors.

Recently, a few TL phosphors with high intrinsic TL sensitivity and direct response to UV stimulation have been reported[6] allowing UV dose evaluation to be simple and practicable.

9.2.1. *NaCl:Ca(T)*

The powder specimens of Ca-doped NaCl (10^{-3} molar fraction) are obtained by recrystallisation from aqueous solution. The specimens are annealed at different temperatures, 550°C and 750°C, in air for 2 h each in a silica boat and subsequently quenched to room temperature. The microcrystalline TLD-grade powder of mesh size <80> 120 is then collected. The TLD-grade material quenched from 750°C is designated as NaCl:Ca(T).

A quantity of 2×10^{-5} kg can be spread uniformly on a kanthal plate. The specimen can then be excited by using the standard UV wavelength at 253.7 nm. The TL glow curves of NaCl:Ca(T) phosphors irradiated by UV can be recorded by the TL reader using a heating rate of 180°C.min^{-1}.

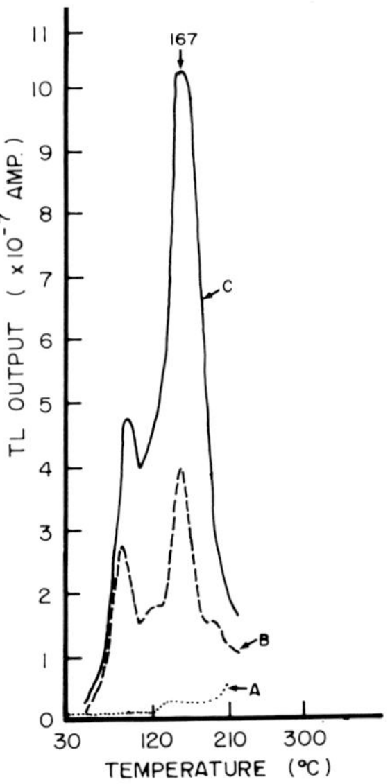

Figure 9.1. Typical TL glow curves of untreated NaCl:Ca. Curve A, UV dose: 2.4×10^{4} J.m^{-2}; Curve B, typical glow curve of 550°C quenched NaCl:Ca; and curve C, typical glow curve of 750°C quenched NaCl:Ca(T). Reproduced from the journal Health Physics, Ref. 7, by permission of the Health Physics Society.

Figure 9.1 shows the TL glow curves of untreated and thermally treated NaCl:Ca specimens after a test dose of 2.4×10^4 J.m^{-2} at room temperature. It is observed, in Figure 9.1 that NaCl:Ca(T) phosphor displays a well defined peak around 167°C (Peak 2) along with a weaker peak at a lower temperature. This indicates the high sensitivity of the phosphor to UV radiation[7].

Peak 2 grows with the increase in magnitude of the UV dose, and its position as well as shape remains unaltered. The intrinsic TL sensitivity is a function of UV dose. The response is linear for the dose range of 10^2 to 10^4 J.m^{-2}. The glow curves observed for untreated, 550°C quenched NaCl:Ca and NaCl:Ca(T) as shown in Figure 9.1 illustrate the fact that the TL glow curves result from direct interaction of UV radiation with the phosphor.

In divalent impurity-doped alkali halides a large fraction of the impurity ions in solid solution is used up in the formation of impurity–vacancy dipoles. Such dipoles are presumed to exist in the crystal along with other crystalline defects such as positive and negative ion vacancies, dislocations, etc. The centre responsible for peak 2 is a complex formed by the association of a dipole with a negative ion vacancy in the dislocation region. Because of thermal quenching, these centres in NaCl:Ca(T) are expected to be more numerous. High concentration of these centres leads to enhanced TL output.

Ultraviolet radiation is incapable of producing electron–hole pairs directly since the energy required ($\sim$12 eV) is much higher than the energy of the incident UV radiation (5-6 eV). The TL exhibited by NaCl:Ca(T) possibly results from the removal of the bound electron from its parent activator ion due to multiple excitation by UV radiation. The liberated electron can get trapped at the nearby negative ion vacancy in the complex centre. Heating the specimen to the temperature of peak 2 releases the electron from its trap. The released electron returns to its parent activator ion in which process it first drops to the emission level of the activator and finally returns to the ground state with TL emission.

9.2.2. α-Al_2O_3

When unirradiated α-Al_2O_3 is illuminated with 253.7 nm UV radiation at 77 K (-196°C) , almost no TL is subsequently produced: only one very weak peak is observed at 295 K as shown in Figure 9.2, curve a. When it is irradiated with X radiation at room temperature, its TL glow curve, registered from 77 to 675 K, exhibits three main peaks situated at 365, 408 and 531 K and a weak shoulder around 450 K. If the irradiation at room temperature is immediately followed by cooling to 77 K, two very weak peaks appear at 113 and 238 K as shown by curve b in Figure 9.2. When unirradiated α-Al$_2$O$_3$ is irradiated at room temperature with X radiation, then cooled to 77 K and illuminated with 253.7 nm UV radiation at this temperature and finally heated from 77 to 675 K, its TL glow curve presents five main peaks situated at 113, 238, 365, 408 and 531 K designated respectively as A, B, C, D and E on curve c as shown in Figure 9.2, and a weak shoulder around 460 K (D')[8].

The intensity of peaks A and B depends strongly on the UV exposure time. This behaviour is analogous to that of other materials such as LiF or CaF$_2$. During the UV illumination, the filling of low temperature traps is accompanied by their simultaneous bleaching. At the beginning of illumination, trap filling predominates and the efficiency of the photo-induced phenomenon is enhanced, then the optical bleaching becomes predominant and the efficiency decreases. The photo-induced thermoluminescence of α-Al$_2$O$_3$ is induced only with excitation wavelengths lower than about 450 nm.

Considerable work has been done in the study of various TL phosphors as UV detectors and dosemeters. Certain phosphors are sensitive to UV radiation, including Mg$_2$SO$_4$:Tb, MgB$_4$O$_7$:Dy,

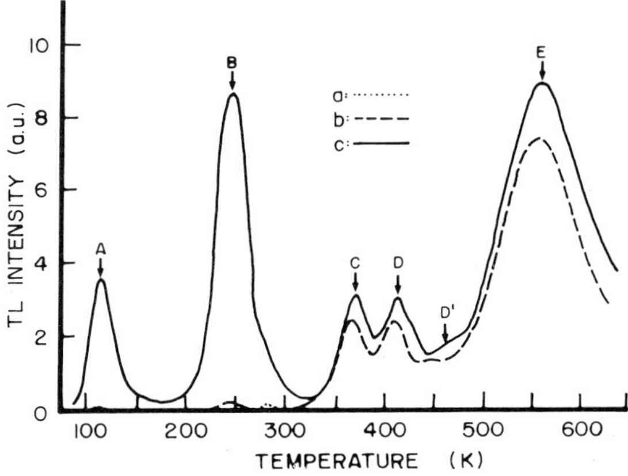

Figure 9.2. TL glow curves obtained with α-Al$_2$O$_3$ after (a) illumination with 253.7 nm UV radiation at 77 K; (b) X irradiation at room temperature; (c) X irradiation at room temperature followed by ilumination at 77 K with 253.7 nm UV radiation. (Courtesy Nuclear Technology Publishing, Ref. 8.)

Al_2O_3:Si:Ti, MgO, $CaSO_4$:Tm, $CaSO_4$:Dy, CaF_2:Dy(intrinsic TL), LiF:Mg:Ti, $CaSO_4$:Tm, natural CaF_2 and BeO. When they are pre-irradiated with a high dose of gamma radiation (of the order of 10 Gy) and partially annealed (transferred TL) as mentioned previously, they can be used as a UV dosemeter to a certain extent. The complexity of phototransferred TL poses serious disadvantages for practical application of this technique to UV dosimetry. A dosemeter which responds directly (intrinsic TL) to UV radiation would be more practical and hence more effort is now being put into dosimetric systems utilising phosphors exhibiting intrinsic TL such as NaCl:Ca(T) mentioned above. The phosphor Mg_2SO_4:Tb is another TL material having high sensitivity for intrinsic TL.

9.3. Thermoluminescence dosimetry of X and gamma radiation

The copper-doped LiF phosphor is very sensitive to X and gamma radiation, and is becoming more popular nowadays. However, glow curve deconvolution provides another approach to reach the same goal. Glow curve deconvolution is a software-implemented analytical process which enables the well known glow curves exhibited by many TLD materials to be separated into their individual glow peaks. By analytically manipulating the resulting peaks and determining relationships between the areas under the peaks, it is possible to significantly expand the dosimetric information currently available from the dosemeters, particularly:

1. Reduction in minimum measurable dose by as much as a factor of ten[9,10].

2. Determination of time elapsed between a major accident dose and readout[11].

3. Elimination of the pre-read anneal portion of the readout cycle.

4. Reduction of dose estimation inaccuracies due to fading.

A typical glow curve for LiF is shown in Figure 9.3 along with some of its major component peaks. Deconvolution enables these peaks to be individually displayed, analysed, and their areas quantified, so that absorbed dose patterns can be identified. A measure of the deconvolution can be obtained by adding the areas under the several peaks to reconstruct the composite glow curve, then comparing the sum with the original curve. This measure of closeness is called the figure of merit (FOM) and is typically less than 2%.

Each peak of the glow curve shown in Figure 9.3 has its own half-life. For example, the half-lives for peaks 2, 3, 4 and 5 are 10 h, 0.5 y, 7 y and 80 y, respectively. Peak 1 fades with a half-life of a few minutes, and peak 6 and above are outside the temperature regime of most automatic readers. In Figure 9.3 peak 5 has by far the largest area, as well as the longest half-life in the above group; therefore, the use of peak 5 alone provides a stable measure of absorbed dose.

The minimum measurable dose that a TLD material can detect is determined by its intrinsic sensitivity and the signal-to-noise (S/N) ratio. The major limiting factor of the S/N ratio is the background which, in turn, is made up of three key components: (1) low temperature peak 2 and its fading characteristics; (2) glow due to infrared heating of the dosemeter, an effect that is most pronounced at higher temperatures; and (3) electronic system noise of reader. Glow curve deconvolution utilises, in real time, an analytical automatic peak searching technique adapted from the field of nuclear spectroscopy to extrapolate the leading edge of peaks 3, 4, and 5 into the peak 2 region, and thereby eliminate peak 2 completely. In addition, deconvolution operated by establishing and subtracting the infrared component and reader noise from the total absorbed dose signal can lower the minimum measurable doses.

It was pointed out earlier that the glow peaks that contribute to the glow curve have different half-lives. These half-lives lead to decay periods which can be used to determine the time elapsed between the occurrence of a significant overdose and readout.

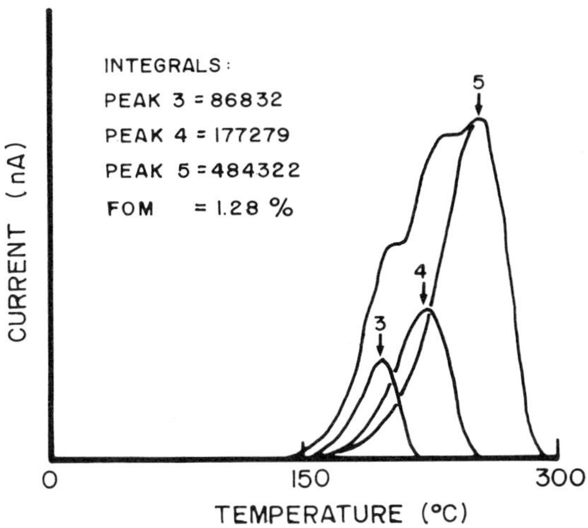

Figure 9.3. LiF glow curve and its three major component peaks. (Courtesy Harshaw-Filtrol, U.S.A.)

9.4.   Thermoluminescence beta radiation dosimetry

A recent study has shown that surface sensitive graphite-mixed MgB_4O_7:Dy and LiF measure beta doses nearly independently of energy, the detection limit being 0.1 mGy[12]. These types of dosemeter lend themselves to automation. LiF 0.2 mm thick on an aluminium holder is a sensitive beta element with flat energy response. Multi-element dosemeters suffer from practical difficulties although such types of dosemeter may give information on the energy spectra of the beta radiation if surface sensitive detectors are employed[13].

A new method of beta dosimetry based on multi-element dosemeters has been developed[14,15]. This method employs a three-element TLD system. However, a fourth element could be added to correct for the gamma dose distribution to each of the three TLD elements. The method focuses specifically on determining the effective incident beta energy distribution from the responses of the three TLDs, from which the dose distribution in LiF or tissue can be derived. The approach uses electron transport theory to determine the dose distributions for well defined incident beta energy distributions. This method has been tested experimentally using a variety of beta sources, and the results are in good agreement with those measured with an extrapolation chamber.

The specific experimental techniques consist of setting up, calibrating, and obtaining backscattering and resolution corrections for a beta spectrometer based on a plastic scintillator. The computation techniques involve (1) adapting a Monte Carlo electron transport computer code to use measured beta energy distributions as input data, and (2) using the code to calculate the energy deposition of these distributions of electrons in a slab of material. Computer codes that have been developed to calculate the energy deposited by photons in LiF are used to derive a theoretical value for the TLD response calibration factor.

Optimisation techniques are used to analyse the responses of a three-element TLD system based on electron transport theory as mentioned above. Specifically, the method determines the effective beta energy distribution incident on the dosemeter system, and thus the system performs as a beta spectrometer. Electron transport theory provides the mathematical model for performing the optimisation calculation. In this calculation, parameters are determined that produce calculated doses for each of the chip-absorber components in the three-element TLD system. The resulting optimised parameters describe an effective incident beta distribution. This method can be used to determine the beta dose specifically at 0.07 $kg.m^{-2}$ or at any other depth of interest. The doses at 0.07 kg.m^{-2} in tissue determined by this method can be compared with those experimentally determined using an extrapolation chamber. The results are also compared with those produced by a commonly used empirical algorithm.

9.5. Thermoluminescence beta dosimetry in mixed beta–gamma fields

In mixed beta–gamma fields, particularly for personnel dosimetry, it is generally agreed that the dosimetry techniques should differentiate the beta dose from the gamma dose. Specifically, the beta dose should be resolved into the skin dose component at 0.07 kg.m^{-2} and the deep dose component defined as the integrated dose deposited below the skin.

For the purposes of personnel dosimetry, the most common approach is to use a dosemeter system consisting of several TLDs each of which is covered by an absorber of a different thickness. Here, the skin dose and the deep dose are derived from the TLD responses by using an empirical algorithm. Although many investigators have used this particular approach, it remains, at best, purely empirical.

A special two-element dosemeter has been investigated as a possible candidate for the measurement of beta–gamma doses in a mixed field[16]. The idea is to have a two-element dosemeter which can be read out within one heating cycle and the responses of the two elements would identify doses due to the beta and hard beta plus gamma radiation, respectively. A sandwich type dosemeter has been developed in which two TL elements are separated by a transparent, heat-resistant layer having poor thermal conductivity, such as a silicon rubber layer. The dosimetric peaks of the two TL elements can be evaluated from the common glow curve.

Two TLD-100 (LiF) hot pressed chips are attached to a 0.6 mm thick (density $= 1.2 \times 10^3$ kg.m^{-3}) transparent silicon base resin irradiated with ^{90}Sr–^{90}Y source. The glow curves are shown in Figure 9.4 where separation of glow curves may be attained.

Another approach is to use TLD-200 (CaF$_2$:Dy) for the discrimination of alpha, beta and gamma radiation. It is based on the peak height ratios of the TLD-200 glow curve, particularly peak 2 (P$_2$) and peak 3 (P$_3$) ratio, and peak 2 and peak 4 (P$_4$) ratio. In summary, the following experimental results are obtained[17].

1. P$_2$/P$_3$ and P$_2$/P$_4$ peak height ratios can be used for photon energy identification in the region where the TLD-200 sensitivity to photon energy is not very linear, i.e. below roughly a hundred keV and above several MeV.

2. The range of average beta energy which can be identified by TLD-200 peak height ratios is below ~0.1 MeV.
3. The alpha radiation can easily be identified due to the high P_2/P_3 peak height ratio.
4. The P_2/P_3 and P_2/P_4 ratios are generally higher for beta (including electron) radiation than for photon radiation. Thus, the two types of radiation are generally distinguishable.

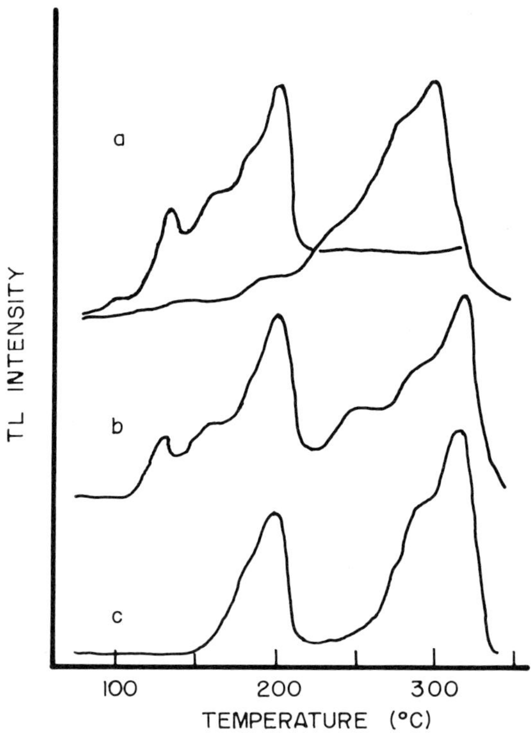

Figure 9.4. Glow curves of a sandwich type dosemeter. (a) glow curve of a one element detector heated from the detector side (left) and the same detector heated from the rubber layer side (right); (b) glow curve of a two-element dosemeter read-out with special heating profile; (c) same as (b) but a 115°C pre-heat is applied in the reader. The maximum temperature is 350°C and the heating rate is 3.5°C.s^{-1} for all cases. (Courtesy Nuclear Technology Publishing, Ref. 16.)

5. If the radiation field is not complex, the glow curve distribution can firstly be used for a rough estimation of the radiation type, energy, and the dose range. Then, a more detailed investigation of peak height ratios can be used for a more accurate evaluation of characteristics of the radiation field.

6. Like the other types of detector based on integral response, deconvolution (unfolding) techniques may become necessary if the detection involves a more complex radiation field.

9.6.   **Thermoluminescence neutron dosimetry**

The fast neutron response of most of the TLDs is generally low due to (1) luminescent yield decreases with linear energy transfer (LET) of particles in their sensitive volume; and (2) energy transferred from neutrons in these detectors is, due to the lack of hydrogen in them, generally much lower than energy transferred in tissue. While the first factor is a consequence of internal properties of a detector, the second can be influenced by adding an external proton radiator. In this case enhancement of response depends on the neutron energy spectrum, dimensions of a detector (particularly the thickness normal to the direction of fast neutron impact), dependence of luminescent yield on LET of particles and also on light transmission in a TLD.

Three types of external proton radiator have been investigated: Teflon (PTFE), polyethylene (PE), and Plexiglass (PMMA). Their thicknesses are 2, 3, and 4 mm, respectively. Three types of TLD have also been investigated: (1) Al_2O_3 in a sintered disc, diameter 6 mm, thickness 0.8 mm; (2) $CaSO_4$:Dy in a Teflon pellet with 5% of TLD embedded, diameter 12 mm, thickness 0.4 mm; and (3) ^{6}LiF, LiF, ^{7}LiF each in a sintered disc, diameter 4.5 mm, thickness 0.85 mm. Neutron energies range from 2.1 to 14.7 MeV[18].

Higher enhancement of relative response in the case of Al_2O_3 shows that the yield of luminescence after energy absorption from neutrons is greater than that from gamma radiation, and greater for this material than for LiF. The increment of relative response also increases with neutron energy when other conditions are constant. If the thickness of the detector is reduced, the increment of relative response is increased. All these three types of TLD have roughly the same LET dependence of TL yields. However, luminescent yield of LiF and $CaSO_4$:Dy starts to decrease with LET above about 5 keV.μm^{-1}.

In addition to the external proton radiator, an internal proton radiator (i.e. a mixed inorganic phosphor–polyethylene detector) can also be used. The combination of external and internal proton radiators is another approach[19]. The TL materials can be mixed with cross-linked polyethylene. The cross linking is achieved through a special peroxidic procedure and can be successively heated to temperatures of about 553 K. The heating will cause some temporary softening. However, after cooling the polyethylene disc regains its original optical properties and, in the case of mixed discs, its original TL properties.

The experimental results show that: (1) it is possible to prepare mixed TLD + PE detectors capable of being used at least ten times in succession without significant changes in their TL properties; (2) after mixing with PE, the fast neutron relative response of the TLDs increases more for the reactor beam than for 14.7 MeV neutrons; (3) this increase could still be enhanced by adding an external PE radiator; (4) the increase in response is clearly higher for the Al_2O_3 than for $CaSO_4$ (it indicates that the relative TL yield for Al_2O_3 depends on LET more favorably than for $CaSO_4$); (5) some further studies such as the influence of grain size for the same TLD, the origin of Al_2O_3, etc. are needed to understand all properties of such detectors in neutron dosimetry.

Heating rates in excess of 10^4 K.s^{-1} have been achieved for thin layers of TLDs by laser heating. The high heating rate improves the S/N ratio. This technique is quite promising for personnel dosimetry of fast neutrons since: (1) it allows simple badge construction and easy dose determination; (2) it is more convenient than the technique employing liquid radiators; (3) the hydrogenous radiator is not heated during the dose readout; (4) the figure of merit is comparable to that obtained with other methods; and (5) the readout time is of the order of milliseconds[20].

9.7.   **Thermoluminescence dosimetry of heavy charged particles**

The relative TL response, the dose–TL response, and the relative glow peak intensities (in LiF) appear to be significantly dependent on the heavy charged particle (HCP) charge, mass and energy and not only on the $\overline{LET}_\infty$ as is widely believed. The HCP-TL properties can be described in the framework of a modified track structure theory (TST) via their relationship to low energy electron TL response.

Many phyical, chemical and biological effects are related to the energy dissipation mechanisms of ionisation and excitation. In HCP interaction with matter, these two mechanisms are the main channels of energy loss for HCP specific energies greater than ~ 0.1 MeV.amu^{-1}. The direct energy transfer from the HCP (first generation of energy transport agents) takes place within less than a nanometer of the HCP path. Subsequently the ejected electrons mainly dissipate their kinetic energy via excitations and ionisations (second and higher order generations of energy transport agents). The physical track of a HCP is thus determined by the spatial distribution of the primary localised events (e.g. capture of charge carriers at particular defect centres) following the physical stage of the radiation action and the subsequent localisation of the energy transport agents. The spatial distribution of the primary localised events can be expected to be closely related to the radial distribution of dose around the HCP path.

The first stage in the TL process following the physical stage of the radiation action is the localisation of the energy transport agents at particular locations called TL traps. The multi-stage nature of the TL mechanism leading to the final effect (i.e. TL photon emission) does not allow the possibility of measuring or calculating the probability of creation of the final effect in the same volume. The final effect produced by the radiation action such as emission of TL photons, radical production, mutation induction, etc., is believed to be strongly influenced by the spatial distribution of the localised primary events around the HCP path.

The major premise of TST is that the concentration of liberated charge carriers around the path of the HCP is the only parameter that governs the dependence of the relative TL properties on the type of HCP radiation. Though TST does provide a convincing and convenient framework to understand HCP and the neutron induced TL, it is premature to conclude that TST is the whole truth underlying HCP induced TL[21].

Hoffmann and Prediger[22] have studied the LET dependence of the low and high temperature peaks of CaF_2:Tm (140°C and 230°C) and ^{7}LiF phosphors irradiated along the depth dose distribution of a Ne^{+10} and He^{+2} beam. The sensitivity of the 140°C peak decreases rapidly with increasing LET while that of the 230°C peak, remains constant up to 50 keV.μm^{-1} and then decreases. Similar observations are made for the main peak (sensitivity decreases) and the 260°C peak (sensitivity increases with increasing LET) in case of ^{7}LiF. It has been shown that CaF_2:Tm can be used in HCP dosimetry as a reliable two component dosemeter allowing separation of the low and high LET contents of the radiation field. ^{7}LiF could also be used in the same way but the dose determination is less accurate in this case.

9.8.   **Thermoluminescence materials**

The sodium content in the reagent $CaSO_4.H_2O$ has been found to have a profound effect on its response to ionising radiation. The higher the sodium content in the reagent, the more serious the environmental effects are, particularly the rapid fading observed in the high-sodium sample of TLD $CaSO_4$:Dy. One of the methods for reducing the sodium content is to use water washing and recrystallisation in the process of crystal growth.

The combination of washing and recrystallisation can remove 90% of sodium from the sample. This is attributed to the redistribution of sodium in the polycrystal during recrystallisation. The washing process can dissolve more sodium than the recrystallisation[23].

High signal-to-noise ratio LiF:Mg:Cu:P solid chips have been developed recently. The lower detection threshold is 60 nGy (6 μrad) which is about 100 times less than TLD-100 chips. The dosimetric characteristics are quite different from LiF:Mg:Ti in many respects, and their anomalous photon energy dependence needs further experimental and theoretical studies[24,25].

References

1. Stoebe, T. (ed) *Solid State Dosimetry*, Proceedings of VII Int. Conf. on Solid State Dosimetry, Ottawa, 27-30 September 1983. Radiat. Prot. Dosim. **6** (1-4) (1984).

2. Marshall, T. O. and Dennis, J. A. (eds) *Solid State Dosimetry*, Proceedings of VIII Int. Conf. on Solid State Dosimetry, Oxford, 26-29 August 1986. Radiat. Prot. Dosim. **17** (1-4) (1986).

3. Oberhofer, M. *Dosimetry of Non-ionising Radiation*. Radiat. Prot. Dosim. **6**, 109 (1984).

4. Busuoli, G. *UV Dosimetry with TLD materials,* Proc. ISPRA course on Applied Thermolominescence Dosimetry, Commission of the European Community (1982).

5. Diffey, B. IN *Radiation Dosimetry: Physical and Biological Aspects,* Ed. C. G. Orton. (New York: Plenum Press) (1986).

6. Richmond, R. G., Ogunleye, O. T., Cash, B. L. and Jones, K. L. *On the Phototransferred Thermoluminescence in MgB_4O_7:Dy.* Int. J. Appl. Radiat. Isot. **38**, 313 (1987).

7. Nehate, A. K., Joshi, T. R., Kathuria, S. P. and Joshi, R. V., *Effect of Variable Doses of Ultraviolet Radiation (253.7 nm) on Thermoluminescence NaCl:Ca(T) Material.* Health Phys. **50**, 485 (1986).

8. Iacconi, P., Lapraz, D., Keller, P., Portal, G. and Barthe, J. *Photo-induced Thermoluminescence of X-irradiated α-Al_2O_3 Dosimetric Properties.* Radiat. Prot. Dosim. **17**, 475 (1986).

9. Horowitz, Y. S., Moscovitch, M. and Wilt, M., *Computerized Glow Curve Deconvolution Applied to Ultralow Dose LiF Thermoluminescence Dosimetry.* Nucl. Instrum. Methods **244**, 556 (1986).

10. Horowitz, Y. S. and Moscovitch, M. *LiF TLD in the Microgray Dose Range via Computerized Glow Curve Deconvolution and Background Smoothing.* Radiat. Prot. Dosim. **17**, 337 (1986).

11. Moscovitch, M. *Automatic Method for Evaluating Elapsed Time Between Irradiation and Readout in LiF TLD*. Radiat. Prot. Dosim. **17**, 165 (1986).

12. Prokic, M. S., *Beta Dosimetry with Newly Developed Graphite Mixed TL Detectors*. Phys. Med. Biol. **30**, 323 (1985).

13. Sherbini, S. Sykes, III, J. Porter, S. W. and Lodde, G., *Experimental Evaluation of a Method for Performing Personnel β Dosimetry Using Multi-element Thermoluminescent Dosimeters*. Health Phys. **49**, 55 (1985).

14. Shen, L., Catchen, G. L. and Levine, S. H., *Experimental and Computational Techniques for β-Particle Dosimetry*. Health Phys. **53**, 37 (1987).

15. Shen, L., Levine, S. H. Catchen, G. L., *An Optimized Computational Method for Determining the β Dose Distribution Using a Multi-element Thermoluminescent Dosimeter System*. Health Phys. **53**, 49 (1987).

16. Uchrin, G., *Beta Dosimetry, Different Solutions*. Radiat. Prot. Dosim. **17**, 99 (1986).

17. Wang, T. K., Hsu, P. C. and Weng, P. S. *Application of TLD-200 Dosimeters to the Discrimination of α, β and γ Radiation*. Radiat. Prot. Dosim. **16**, 225 (1986).

18. Spurny, F. *Enhancement of Fast Neutron Response of Some Luminescent Detectors due to External Proton Radiators*. Radiat. Prot. Dosim. **9**, 257 (1984).

19. Spurny, F., Votočková, I., Gelev, M., Mishev, I. T., Apostolova, M. and Shopov, J. *Fast Neutron Detection using Thermoluminescent Phosphors with Internal and/ or External Proton Radiator*. Radiat. Prot. Dosim. **16**, 219 (1986).

20. Mathur, V.K., Brown, M. D. and Bräunlich, P. *Thermoluminescence Fast Neutron Dosimetry by Laser Heating*. Radiat. Prot. Dosim. **6**, 163 (1984).

21. Kalef-Ezra, J. and Horowitz, Y. S. *Heavy Charged Particle Thermoluminescence Dosimetry: Track Structure Theory and Experiments*. Int. J. Appl. Radiat. Isot. **33**, 1085 (1982).

22. Hoffmann, W. and Prediger, B. *Heavy Particle Dosimetry with High Temperature Peaks of CaF_2:Tm and ^7LiF Phosphors*. Radiat. Prot. Dosim. **6**, 149 (1984).

23. Weng, P. S., Hsu, P. C., Lin, Y. M. and Huang, J. T. *Removal of Sodium in TLD $CaSO_4$:Dy Through Washing and Crystallization*. Paper No. TAM-G7, 32nd Annual Meeting of the Health Physics Society, Salt Lake City (U.S.A.), 5-9 July (1987).

24. Nakajima, T., Murayama, Y., Matsuzawa, T. and Koyana, A. *Development of a New Highly Sensitive LiF Thermoluminescent Dosimeter and its Applications*. Nucl. Instrum. Methods **157**, 155 (1978).

25. Wang, S., Chen, G., Wu, F., Li, Y., Zha, Z. and Zhu, J. *Newly Developed Highly Sensitive LiF(Mg,Cu,P) TL Chips with High Signal-to-Noise Ratio*. Radiat. Prot. Dosim. **14** 223 (1986).

Phosphor Terminology

K. Mahesh

Activators and fluxes

French chemist A. Verneuli showed that while a pure calcium sulphide was non-luminescent, after heating with a trace of a compound of bismuth, it gave a violet emission. Thus the necessity was established for what is now known as an activator (bismuth in this case) which must be added in small proportions in the host crystal (calcium sulphide). Zinc sulphide phosphor with an added activator of a few thousandths of a per cent of copper emitted green light. Besides this role as activators, it was also observed that the addition of small quantities of certain salts in the preparation of phosphors helped improve their efficiency. For instance, sodium chloride was particularly effective as an additive in making an efficient $ZnS:Cu$ phosphor. Thus these salts acted as catalytic agents to facilitate efficient phosphor making. These additives are called fluxes.

Luminescence killers/poisons

In the course of investigations, the laboratory-prepared phosphors, in which purity could be achieved with precision, were found more efficient than natural mineral phosphors in which there was no control on impurities and other spurious ingredients. In synthetic phosphors it is possible to eliminate spurious ingredients, purify thoroughly, mix the required ingredients properly and crystalise the final mixture at an optimum high temperature (about 1000°C). It was also found that there are certain other undesirable impurities, particularly iron, that decreased the efficiency of phosphors. These efficiency-reducing impurities are known as killers or poisons. Individual phosphors are found to be hypersensitive to certain poisons but relatively insensitive to others. Another feature concerns fluxes. While these are necessary in making sulphide type phosphors, they are unnecessary, or even act as poisons, in the making of most of oxide type phosphors.

Luminescent centres and classes of activators

1. Centres

Luminescence action takes place at positions where activators are located. P. Lenard called these atoms or group of atoms positioned within the host lattice, luminescent centres. Thus, the host crystal acts as a medium in which the suspended activator atoms, in interaction with the neighbour atoms of the host crystal, serve as discrete centres for localised absorption of excitation energy and for the subsequent emission of light. There is an optimum proportion of the activator which is around $1 \times 10^{-3}\%$ to $300 \times 10^{-3}\%$ of the weight of the host crystal for the most efficient luminescence yield. Activator added in excess of the optimum value decreases the light yield. This later role of the activator is termed *concentration quenching*.

2. Intensifier activator

Pure crystals by themselves can act as phosphors. Calcium tungstate ($CaWO_4$) and pure zinc silicate (Zn_2SiO_4) are examples in which the octahedral WO_4 group and tetrahedral silicate (SiO_4) group act as luminescent centres. However, if 0.2% of titania (TiO_2) is incorporated in pure Zn_2SiO_4 or in some other pure silicates, the violet emission of the pure SiO_4 centre is substantially enhanced. Thus we find that addition of TiO_2 actuates rather more of the silicate groups to luminesce. An impurity, when it itensifies a latent host crystal luminescence, is termed an intensifier activator.

3. Co-activators

In many phosphors (LiF:Mg,Ti and LiF:Cu,Ag) incorporation of some other impurity in addition to an activator also helps increase the luminescent yield. Such an impurity is called a co-activator. One of the activators acts as trapping centres while the other, located close to the crystal defects, is related to the luminescence centres.

4. Self activators

The ability of a material to luminescence is associated with the presence of activator atoms. Some specific impurity atoms naturally occuring in relatively small concentrations in the host material or a small stoichiometric excess of one of the constituents of the material may act as activators. This is an example of self activators.

Radiation Quantities and Units

K. Mahesh

This appendix provides selected definitions used in radiation protection which may be useful to those concerned with the use of thermoluminescence processes for radiation dosimetry and dating.

1. Radioactivity

The *activity*, A, of an amount of radioactive nuclide in a particular energy state at a given time is the quotient of dN by dt, where dN is the expectation value of the number of spontaneous nuclear transitions from the energy state in the time interval dt.

$$A = \frac{dN}{dt}$$

The SI unit of radioactivity is s^{-1} and the special name for the unit of activity is bequerel (Bq).

$$1\,Bq = 1s^{-1}$$

The special unit of activity, curie (Ci) may be used temporarily.

$$1\,Ci = 3.7 \times 10^{10}\,s^{-1}\ (exactly)$$

2. Kerma

The *kerma*, K, is the quotient of dE_{tr} by dm, where dE_{tr} is the sum of the initial kinetic energies of all the charged ionising particles liberated by uncharged ionising particles in a material of mass dm.

$$K = \frac{dE_{tr}}{dm}$$

The SI unit of kerma is $J.kg^{-1}$ and the special name for the unit of kerma is gray (Gy)

$$1\,Gy = 1\,J.kg^{-1}$$

The special unit of kerma, rad, may be used temporarily

$$1\,rad = 10^{-2}\,J.kg^{-1}$$

Revised and updated by E. P. Goldfinch.

3. Absorbed Dose

The *absorbed dose, D,* is the quotient of d$\bar{E}$ by dm, where d$\bar{E}$ is the mean energy imparted by ionising radiation to matter of mass dm.

$$D = \frac{\mathrm{d}\bar{E}}{\mathrm{d}m}$$

The SI unit of absorbed dose is J.kg^{-1} and the special name for the unit of absorbed dose is gray (Gy)

$$1\ \mathrm{Gy} = 1\ \mathrm{J.kg^{-1}}$$

The special unit of absorbed dose, rad, may be used temporarily.

$$1\ \mathrm{rad} = 10^{-2}\ \mathrm{J.kg^{-1}}$$

Note: If charged particle equilibrium obtains and bremsstrahlung is negligible absorbed dose and kerma are numerically equal.

4. Exposure

The *exposure, X,* is the quotient of dQ by dm where the value of dQ is the absolute value of the total charge of ions of one sign produced in air when all the electrons (negatrons and protons) liberated by photons in air of mass dm are completely stopped in air

$$X = \frac{\mathrm{d}Q}{\mathrm{d}m}$$

The SI unit of exposure is C.kg^{-1}. There is no special name for the unit of exposure.

The special unit, roentgen (R), may be used temporarily.

$$1\ \mathrm{R} = 2.58 \times 10^{-4}\ \mathrm{C.kg^{-1}}\ \text{(exactly)}.$$

5. Dose Equivalent

The *dose equivalent, H,* is the product of D, Q and N at the point of interest in tissue, where D is the absorbed dose, Q is the quality factor, and N is the product of all other modifying factors.

$$H = DQN$$

The SI unit of dose equivalent is J.kg^{-1}. The special name for the unit of dose equivalent is sievert (Sv).

$$1\ \mathrm{Sv} = 1\ \mathrm{J.kg^{-1}}$$

The special unit of dose equivalent, rem, may be used temporarily.

$$1 \text{ rem} = 10^{-2} \text{ J.kg}^{-1}$$

6. Quality Factor

The radiation quality factor, Q, is intended to allow for the effect on radiation detriment of the microscopic distribution of absorbed energy. It is defined as a function of the collision stopping power, L, in water, at the point of interest in accordance with the following table:

L in water keV/μm	Q
3.5	1
7	2
23	5
53	10
175	20

In practice, for a spectrum of radiation, an effective value, $\bar{Q}$, can be assigned. The values currently assigned are as follows:

Radiation	$\bar{Q}$
Photons and electrons	1
Neutrons, alpha particles and heavy ions	20

Where the energy is known, Q may be used in the place of $\bar{Q}$. Thus for thermal neutrons, where L is uniquely defined, a value for Q of 4.6 may be used and for tritium beta radiation a value for Q of 2 should be used.

Note: The above values of Q and $\bar{Q}$ take into account the ICRP interim recommendation made at its Paris meeting during March 1985 to increase Q for neutrons by a factor of two.

7. **Other Dose Equivalent Quantities**

The forgoing definition of dose equivalent relates to that quantity at a specified point in tissue. To be meaningful in practical application it must be further qualified in relation to that application. ICRP and ICRU recommendations cover practical radiation protection applications in the use of such quantities as effective dose equivalent, organ dose equivalent, ambient dose equivalent, directional dose equivalent, individual dose equivalent etc. It is not the purpose of this appendix to examine the radiation protection aspects in any depth. The reader is referred to the Bibliography.

Bibliography

1. ICRP. Publication 26 - *Recommendations of the International Commission on Radiological Protection* - Annals of the ICRP 1 (3) 1977.
2. ICRP. *Statement of the 1978 Stockholm Meeting of the ICRP.* Annals of the ICRP 2, (1) 1978.
3. ICRP. *A Compilation of the Major Concepts and Quantities in Use by ICRP.* Annals of the ICRP 14, (4) 1984.
4. ICRP. *Statement from the 1985 Paris Meeting of the ICRP.* Annals of the ICRP 15, (3) 1985.
5. ICRP. *Data for Use in Protection Against External Radiation.* Annals of the ICRP 17, (2/3) 1987.
6. ICRU Report 33. *Radiation Quantities and Units.* ICRU April 1980.
7. ICRU Report 39. *Determination of Dose Equivalents Resulting from External Radiation Sources.* ICRU February 1985.
8. ICRU Report 40. *Quality Factor in Radiation Protection.* ICRU April 1986.

Scalar and Vector Potentials

K. Mahesh

The charge and current distributions varying with time produce an electromagnetic field. The intensity E at a point due to an electrostatic field is related to the space derivative of the electrostatic potential ϕ (scalar potential), that is

$$E = - \operatorname{grad} \phi \tag{1}$$

where

$$\phi = (1/4\pi\epsilon_o) \sum_{i=1}^{n} (q_i/R_i) \tag{2}$$

If the charge distribution in a volume is considered uniform and ρ is the charge density,

$$\phi = (1/4\pi\epsilon_o) \int (1/R)\, \rho \, d\upsilon \tag{3}$$

ϵ_o $(1/4\pi)$ $(1/9\times10^9)$ F.m^{-1} is the electric constant (permittivity of free space).

Direct current in a conductor will produce a steady magnetic field. The intensity H of the field at a point is obtained by taking the space derivative of the potential A (vector potential) at that point.

$$H = \operatorname{curl} A \tag{4}$$

The magnetic field generated by direct current J in vacuum is given by Maxwell's equation

$$\operatorname{curl} H = \mu_o j \tag{5}$$

using Equation 4,

$$\operatorname{curl} \operatorname{curl} A = \mu_o j$$

that is,

$$\operatorname{grad} \operatorname{div} A - \nabla^2 A = \mu_o j$$

using the condition that div $A = 0$

$$\nabla^2 A = - \mu_o j$$

Its solution is

$$A = (\mu_o/4\pi) \int (1/R) \, j \, d\upsilon \tag{6}$$

$\mu_o = 4\pi \times 10^{-7}$ H.m^{-1} is the magnetic constant (permeability of free space). The term $(\epsilon_o \mu_o)^{-\frac{1}{2}}$ is equal to the velocity c of light. Note that the dimensions of the farad are m^{-2}.kg^{-1}.s^4.A^2 and of the henry (inductance) m^2.kg.s^{-2}.A^{-2} and thus c $= 3 \times 10^8$ m.s^{-1}.

For alternating current, the current density can be written as

$$j = j_o \cos \omega t$$

Hence

$$A = (\mu_o/4\pi) \int (1/R) \, j_o \cos \omega t \, d\upsilon \tag{7}$$

R is the distance of the observation point from the volume element $d\upsilon$ in the conductor. Further, in Equation 7 it is assumed that the vector potential at the observation point will be in phase with $j_o \cos \omega t$ regardless of the value of R. This assumption is valid only if the velocity of propagation of the electromagnetic wave is infinite. This being not so, there will be a time delay between any change in the current density and the effect of this change to be seen at the observation point. This time delay $t' = R/c$ and the corresponding phase delay is $\omega t'$. Hence

$$A = (\mu_o/4\pi) \int (1/R) \, j_o \cos \omega(t-t') d\upsilon \tag{8}$$

and

$$\phi = (1/4\pi\epsilon_o) \int (1/R) \rho_o \cos \omega(t-t') d\upsilon \tag{9}$$

Potentials given by Equations 8 and 9 are called the retarded potentials.

The energy flux density of electromagnetic radiation (rate of flow of energy density across 1 m^2 normal to the direction of propagation) is given by the Poynting vector S which is the vector product of E and H of the radiation field at a given point.

$$S = E \times H \tag{10}$$

Dipole and quadrupole transitions

For the allowed radiative transitions, the matrix element of the dipole moment (electric or magnetic, as the case may be) or the electric quadrupole moment of the system is non-zero. The transition is, otherwise, optically forbidden. The Einstein coefficient A_{21} which is the probability of the transition rate is given as:

Electric dipole transition: $A_{D21} = (4\omega^3/3\,\hbar c^3)\,|\,(\boldsymbol{d})_{21}|^2$ $\qquad \boldsymbol{d} = -e\boldsymbol{r}$

Magnetic dipole transition: $A_{M21} = (4\omega^3/3\,\hbar c^3)\,|\,(\boldsymbol{n}\times\boldsymbol{\mu})_{21}|^2$

where

$\boldsymbol{r}$ = radius vector of the orbital electron with respect to the nucleus.

$\boldsymbol{\mu} = (-e/2m)\,\boldsymbol{L}$ is the electron magnetic moment vector

and $\boldsymbol{L}$ is the electron orbital angular momentum vector.

Electric quadrupole transition: $A_{Q21} = (4\omega^3/3\hbar c^3)\,|\,(\boldsymbol{k}\,Q)_{21}|^2$

$$where\ Q = (-e/2)\begin{pmatrix} x^2 & xy & xz \\ yx & y^2 & yz \\ zx & zy & z^2 \end{pmatrix}$$

is the electric quadrupole moment tensor.

$(\quad)_{21}$ denotes the matrix element for the $2 \rightarrow 1$ transition. $\boldsymbol{d}$ = electric dipole moment vector of the electron. $\boldsymbol{k}$ = wave vector of radiation and $\boldsymbol{n} = -(\boldsymbol{k}/k)$.

For a radiation of 1 μm wavelength, $(A_{M21}/A_{D21}) \approx 10^{-4}$ and $(A_{Q21}/A_{D21}) \approx 10^{-6}$.

Laser Heating of TL Phosphors

K. Mahesh

Heating of TL phosphors constitutes the central feature of the TL glow curve recording. As such, the heating unit is an important component of a TLD reader. The glow curve profile and its subsequent analysis can be significantly influenced by the heating rate, the presence of any thermal gradient between the heater and the phosphor sample, and by any lack of reproducibility of the heating cycle (temperature–time profile). Figure 1[1] illustrates these temperature–time profiles obtainable with different heating methods and the corresponding glow curve shapes of a typical TL phosphor.

Heating of TL phosphors by a CO_2 laser beam of 10.6 μm wavelength (having strong absorption in Al_2O_3, $CaSO_4$:Dy, $CaSO_4$:Mn, quartz) has been reported by Braunlich *et al*[2]. Details of various features of laser heating have been published in a series of papers[3–6] by Braunlich and co-workers.

Although basically simple, the technique of laser heating requires sophisticated technology such as microprocessor controlled beam power stabilisation[3] and optics to shape the laser beam[6]. This mode of heating provides, however, a very fast readout with an efficient heat transfer and eliminates thermal gradients which are otherwise present in conventional heating. Laser heating is particularly suitable for thin layers of TLD material and is, therefore, especially useful for beta dosimetry. Several advantages of laser heating of TLD can be enumerated as follows:

(i) Fast heating rates allow a perceptible increase in TL sensitivity of phosphors owing to an improved signal to noise ratio due to the short time interval, enough for the production of glow curve. This feature extends the area of application of TL to true microdosimetry. Braunlich *et al*[2] have shown that useful TL signals can be obtained from a $CaSO_4$:Mn powder sample of mass less than 20 μg with an exposure of only 15 mR of 662 keV γ rays from a ^{137}Cs source.

(ii) A two dimensional dose distribution profile of ionising radiation

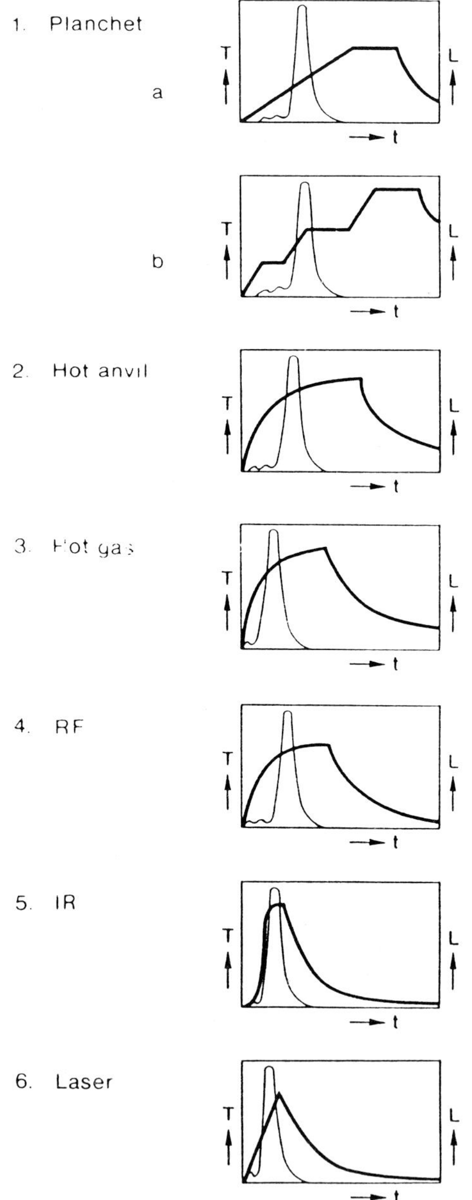

Figure 1. Schematic diagram of temperature–time profiles (thick lines) obtained with different heating methods and the corresponding glow curves (thin lines) for a typical TLD[1].

in a thin TL phosphor is possible.

(iii) Monitoring of two dimensional temperature spread on a thin TL phosphor surface (about 100 μm) can be done by heating it with a localised laser beam. This is possible by recording the TL glow curves after each successive short interval (of a few hundred ms or so). The glow curve profile build-up across the sample width (of course larger than the beam width) indicates the temperature spread $T(r,t)$, where r represents distance from the centre of the beam incident on the optically flat plate onto which a thin layer of phosphor powder is settled and t denotes the time intervals.

(iv) There is extremely rapid evaluation of Tl glow curves.

(v) On account of fast heating rates available by laser heating of TLD, it was possible to find[5] that the thermoluminescence efficiency (area under the glow curve) remains nearly independent of the heating rate q over a large range ($4 \leq q \leq 5500\ °C.s^{-1}$).

(vi) Laser heating allows improved black-body radiation discrimination as the only heated surface is the TL phosphor surface itself.

(vii) Post-readout heating eliminates residual signals more effectively and quickly than the conventional heating.

At low doses the inherent limitation in the light collection efficiency of monochromators makes elevated heating rates an attractive option for increasing the precision of the spectrum measurement. Heating rates between 10 and 40 $K.s^{-1}$ were found[7] to be optimum for LiF giving an adequate signal to noise ratio while maintaining sufficient glow peak separation (the uniform laser beam profile as against the Gaussian beam profile is especially helpful for the latter aspect). Such heating rates are difficult to obtain reproducibly using conventional heating.

The ability to obtain measurable TL signals at microdosimetry levels through laser heating has made possible feasibility studies for X ray imaging using $CaSO_4$:Mn TL film[8,9]. TLDs are about 100 times as sensitive as a film badge consisting of silver halide emulsion film. Microcrystalline $CaSO_4$:Mn grains of about 30 μm size in polyimide resin binder formed a one layer thick film of 100×100 mm^2 area. The objects in between the film and X ray tube (60 kV, without filter) consisted of TLDs and metals. The exposed TL film was wrapped around a drum which could rotate at 30 rpm. The 0.4 W CO_2 laser beam spot of 0.2 mm size scanned the film surface at a rate of 0.3 mm per line. The emitted TL having a wavelength of 496 nm was collected by an optical guide of 2 mm diameter which was then detected by a photomultiplier. The photocurrent signals after DC amplification were fed to a data recorder and displayed on an oscilloscope as X ray images of the objects. The image resolution had not been quite satisfactory but seems to have a chance of improving with a better collection efficiency. Likewise, the sensitivity of the TLD films could be increased from the present exposure level of 100 mR to 1 mR by developing a high efficiency TL material.

Kelly *et al*[10] have studied the problems associated with thermoluminescence kinetics, such as carrier pile-up in bands, under the conditions of very fast heating rates obtainable with the laser beam. Despite the experimental compromises that were inevitable on account of unusually high heating rate (e.g. non-linear heating rates and the difficulty of measuring the temperature rise during rapid heating), they observed a good resemblance of the experimental glow curves to the ones calculated on the basis of classical Randall-Wilkins kinetics (formulated for low heating rates). Complete absence of background noise and incandescent emission is also observed in the glow curves.

The fact that laser heating allows the use of micro-amounts of TL phosphor (a few tens of μg) and the achievement of high signal to noise ratio from only very small levels of absorbed dose (a few hundred μGy), enabled Braunlich *et al*[11] to construct a microdosemeter probe for use in work involving radiation therapy of cancer in which remote *in vivo* measurement of the absorbed dose will be required. The heating pulse of a 6 W CO_2 laser is delivered to the $CaSO_4$:Mn phosphor held at the tip of a 15 cm long hypodermic needle by a metre-long flexible KRS-5 (thallium bromo-iodide) fibre passing through the needle. The emitted TL is collected by the thin-walled quartz tube inserted in the needle and is focussed onto the photomultiplier with the help of a 25 mm focal length spherical mirror positioned appropriately in the PM mount enclosure.

Mathur and co-workers[12] have used laser heating of $CaSO_4$:Dy phosphor film in order to determine its dose response to fast neutrons (in the mixed n-γ field) from a Pu–Be source and 14.6 MeV neutron generator. Half of the film is covered with a polyethylene radiator of 1 mm thickness in which the neutrons produce recoil protons which characterise the neutron dose received by the dosemeter[13] while the remaining half of the dosemeter film registers only the γ dose. The difference between the integrated TL signals of the two halves of the phosphor film will give the neutron dose. It is shown that the neutron dose of 300 mrem can be measured with an accuracy of 50%, and 60 mrem at an accuracy of ≤ 70%.

Braunlich and Tetzlaff[14] have constructed a fully automated and computer controlled laser heated TLD reader (see also Abhold[15]) employing a uniform laser beam profile; a schematic diagram of the reader is shown in Figure 2.

The laser power delivered to the dosemeter is controlled by a microprocessor in order to provide a 'time-shaped' reproducible heating cycle. The uniform intensity profile of the laser beam, in contrast to a Gaussian profile, is especially attractive in that the glow curves obtained are, except for a slowly decreasing tail at the high temperature side of the peak, nearly identical to those obtained with conventional heating. Furthermore, the limitation of observing complex glow curves exhibiting limited peak separation involved with Gaussian beam heating is also largely absent. It is possible to select different heating programmes depending upon the layer thickness of the phosphor spots on the substrate and the pre- and post-anneal treatment requirements. The dosimetric response of $CaSO_4$:Dy to low energy β rays has been determined and its application in environmental, fast neutron, and γ dosimetry has been demonstrated.

In concluding, it will be pertinent here to list some limitations in

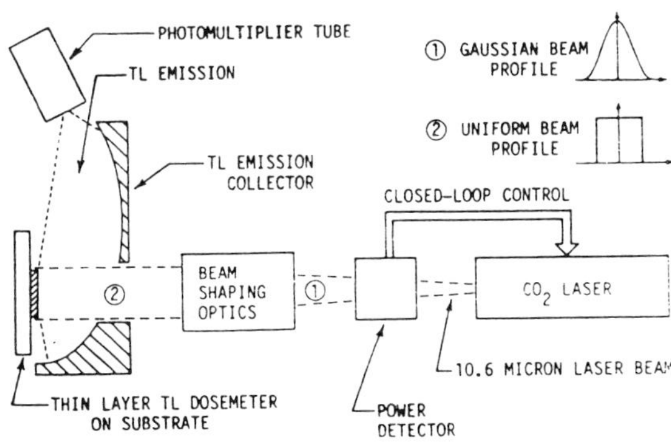

Figure 2. Schematic diagram showing technical features of an automated and microprocessor controlled laser heated TLD reader[14].

laser heating of TL phosphors (see also Jones[16], who has compared features of laser heating with those of the hot anvil method: in the latter case the TL phosphor is brought into contact with a hot metallic block kept at a fixed temperature). These are:

(1) high cost and complexity of the readout system;

(2) difficulty in obtaining the temperature profile of the laser heated TL phosphor on account of its very rapid heating;

(3) need for precise adjustment of the beam and positioning of TLDs;

(4) especially suitable only for thin TLDs for which thermal capacity is negligible (this factor ensures instant compatibility between the laser power delivered to the TLD and its resulting final temperature).

References

1. Julius, H. W. Radiat. Prot. Dosim. **17**, 267 (1986).
2. Braunlich, P., Gasiot, J., Fillard, J. P. and Castagne, M. Appl. Phys. Lett. **39**, 769 (1981).
3. Abtahi, A., Braunlich, P., Kelly, P. and Gasiot, J. J. Appl. Phys. **58**, 1626 (1985).
4. Abtahi, A., Braunlich, P. and Kelly, P. J. Appl. Phys. **60**, 3417 (1986).
5. Abtahi, A., Braunlich, P., Hagan, T. and Kelly, P. Radiat. Prot. Dosim. **17**, 313 (1986).
6. Kelly, P., Abtahi, A. and Braunlich, P. J. Appl. Phys. **61**, 738 (1987).
7. Abhold, M. E. Radiat. Prot. Dosim. **17**, 325 (1986).
8. Yasuno, Y., Tsutsui, H., Yamamoto, O. and Yamashita, T. Jap. J. Appl. Phys. **21**, 967 (1982).
9. Yasuno, Y., Tsutsui, H., Yamamoto, O. and Yamashita, T. Radiat. Prot. Dosim. **6**, 341 (1984).
10. Kelly, P., Braunlich, P., Abtahi, A., Jones, S. C. and de Murcia, M. Radiat. Prot. Dosim. **6**, 25 (1984).
11. Braunlich, P., Jones, S. C. and Tetzlaff, W. Radiat. Prot. Dosim. **6**, 103 (1984).
12. Mathur, V. K., Brown, M. D. and Braunlich, P. Radiat. Prot. Dosim. **6**, 163 (1984).
13. Chassede-Baroz, Ph., Braunlich, P., Tetzlaff, W. and Gasiot, J. Radiat. Prot. Dosim. **17**, 119 (1986).
14. Braunlich, P. and Tetzlaff, W. Radiat. Prot. Dosim. **17**, 321 (1986).
15. Abhold, M. E. Radiat. Prot. Dosim. **17**, 325 (1986).
16. Jones, A. R. Radiat. Prot. Dosim. **17**, 307 (1986).

Subject Index

A

B

C

U

V

W

X

Y

Z